HINWEISE ZUR HANDHABUNG DES BUCHES

Dieses Buch besteht aus drei Teilen vei
Länderspezifische Angaben und *Teil Drei:* zig
in diesem Buch beschriebenen Länder sir der
Länder - USA und Japan - liegen zwar ausse it in
die Betrachtung einbezogen.

Teil Eins: Bauen in Europa

In Teil Eins werden drei Aufsätze vorgestellt: *Die Bauindustrie in Europa* wird der Wandel der Bauindustrie beschrieben der sich derzeit sowohl in Westeuropa als auch in Osteuropa bemerkbar macht. In *Die Gesetzgebung der Europäischen Gemeomschaft* wird ein kurzer Einblick in die gesetzgeberischen Maßnahmen der EG gegeben die das Bauwesen betreffen. In *Die Förderungsmaßnahmen der EG* wird ein Überblick über die wichtigsten Förderprogramme der EG im Bereich des Bauwesens gegeben.

TEIL Zwei: Länderspezifische Angaben

In Teil Zwei sind siebenundzwanzig Länder in alphabetischer Reihenfolge aufgeführt und nach einem einheitlichen Schema unter folgenden Überschriften beschrieben:

- *Schlüsseldaten*: Hier sind wesentlichen Daten zur Ökonomie und zur Bauwirtschaft der Länder aufgeführt
- *Bauinudustrie*: Hier werden die Struktur der Bauindustrie sowie die in den Ländern üblichen Ausschreibungs - und Vergabeverfahren und die geltenden Vorschriften und Normen beschrieben
- *Baukostendaten*: Hier werden Lohn - und Materialkosten genannt, Einheitspreise von wesentlichen Bauleistungen (Positionen) aufgeführt und Kennwerte für frühe Phasen der Kostenermittlung für unterschiedliche Gebäudenutzungen ausgewiesen
- *Wechselkurse und Inflation*: Hier werden die Wechselkurse der jeweiligen Währung zum Pfund Sterling und zum US-Dollar aufgeführt und Angaben zur Preisentwicklung der Lebenshaltungskosten und der Baupreise gemacht
- *Nützliche Adressen*: Hier werden Adressen von Stellen der öffentlichen Hand, von Berufsverbänden und Handelsorganisationen genannt.

In der Einführung (4) werden allgemeine Hinweise, Erläuterungen und Begriffsbestimmungen gegeben.

Die ausführlichen Positionstexte zu den im Abschnitt Baukostendaten aufgeführten Positionen finden sich in Abschnitt 6. Sie sind in fünf Sprachen aufgeführt (Englisch, Französisch, Deutsch, Italienisch und Spanisch).

Teil Drei: Vergleichende Daten

Um den schnelle Vergleich der Daten zwischen den einzelnen in diesem Buch behandelten Ländern zu ermöglichen, werden in Teil Drei die Daten aus Teil Zwei gegenübergestellt. Dies geschieht unter drei Gesichtspunkten:

- *Schlüsseldaten der Macroökonomie* mit Wirtschfatsdaten und demographischen Daten
- *Bauwirtschaftsdaten* mit Daten zu Umsätzen und Produktivität
- *Baukostendaten* mit Basisdaten zu Lohn und Material, Preisen des Baustoffhandels und Preisen von Einzelleistungen in Angeboten der Bauunternehmen und Handwerksbetriebe.

Spon's European Construction Costs Handbook

Edited by

DAVIS LANGDON & EVEREST

Chartered Quantity Surveyors

First edition

E & FN SPON

An Imprint of Chapman & Hall

London · New York · Tokyo · Melbourne · Madras

Published by E & FN Spon, an imprint of Chapman & Hall, 2–6 Boundary Row, London SE1 8HN

Chapman & Hall, 2–6 Boundary Row, London SE1 8HN, UK

Blackie Academic & Professional, Wester Cleddens Road, Bishopbriggs, Glasgow G64 2NZ, UK

Van Nostrand Reinhold Inc., 115 5th Avenue, New York NY10003, USA

Chapman & Hall Japan, Thomson Publishing Japan, Hirakawacho Nemoto Building, 6F, 1-7-11 Hirakawa-cho, Chiyoda-ku, Tokyo 102, Japan

Chapman & Hall Australia, Thomas Nelson Australia, 102 Dodds Street, South Melbourne, Victoria 3205, Australia

Chapman & Hall India, R. Seshadri, 32 Second Main Road, CIT East, Madras 600 035, India

First edition 1992

Printed in Great Britain by
Richard Clay Ltd, Bungay, Suffolk

ISBN 0 419 17480 X

A catalogue record for this book is available from the British Library

Contents

page

PART TWO: INDIVIDUAL COUNTRIES

page

PART THREE: COMPARATIVE DATA

Preface

In 1988 the Editors compiled the first *Spon's International Construction Costs Handbook*, covering thirty two countries worldwide. That was intended to be the first in a series of such handbooks which would become progressively more detailed and extensive. It has now been decided to produce two volumes; this one, covering Europe and including the USA and Japan, and a companion volume, available later this year, covering Asia and the Pacific Basin.

The format has been changed to include considerably more information than was provided in the first book. In particular, detailed cost data is provided for individual items of construction work. Additionally, the section on unit costs for different building types, Approximate Estimating, has been greatly enlarged.

This book has been produced against a background of momentous change in Europe, particularly in Eastern Europe and the former Soviet Union. It is always difficult to collect reliable statistical or cost data on construction industries; not only are the figures often hard to come by, but there are significant differences in specification standards, terminology and definitions. Recent economic, social and political changes have compounded such difficulties. It is not yet possible to present reliable construction cost information for many former communist countries where a free market in construction is being established or re-established after years of central planning. In particular, it is too early to present data on the constituent republics of the former Soviet Union, which was formally disbanded at the end of 1991. Similarly, reliable construction data for the re-unified Germany is not yet available. Data on Germany is therefore presented in the old East/West context.

The publication of this book is particularly timely; it marks the end of one era and the start of a new one. The integration of the European Community economies after 1992 will provide a theoretically single construction market currently valued at some £400bn (580bn ECU) annually. In the longer term, further expansion of the European Community may be expected with the addition of some of the European Free Trade Association (EFTA) states and, possibly a number of former Comecon countries.

The book is designed to be a convenient reference. Its purpose is to present coherent snapshots of the economies and construction industries of Europe; it also puts this information in an international context with the inclusion of the USA and Japan. It is not a substitute for local knowledge and professional advice. It will, however, be extremely useful as an introduction to a country and its construction industry for clients, consultants, contractors, manufacturers of construction materials and equipment and others concerned with development, property and construction in Europe.

Davis Langdon & Everest
Consultancy Group
London
1992

Acknowledgements

The contents of this book have been gathered together from a variety of sources - individuals, organizations and publications. Construction cost data and general background information on local construction industries is based on contributions from a network of professional colleagues and associates worldwide. These include:

CFE Contractors B and L, Belgium
JMV QS Service, Nicosia, Cyprus
Ustav Ekonomiky a Organizacie Stavebnczva, Bratislava, Czechoslovakia
Enemaerke and Petersen A/S. Denmark
VTT Technical Research Centre, Laboratory of Building Production Technology, Finland
Davis Langdon & Weiss, Frankfurt, Germany
Toeothe AE Construction Company, Athens, Greece
Patterson Kempster and Shortall, Dublin, Ireland
Davis Langdon & Copper, Rome, Italy
Futaba Quantity Surveyors, Tokyo, Japan
Ballini & Pitt, Luxembourg
Zammit and Associates, Malta
Berenschot Osborne bv, Netherlands
ORGBUD, Instytut Organizacji Zarzadzania I Ekonomiki Przemyslu Budowlanego, Poland
Davis Langdon Edetco, Gerona, Spain
ECAS AG., Zurich, Switzerland
Schweizerische Zentralstelle Fur Baurationalisierung, Zurich, Switzerland
Mimarlik Fakultesi Teknik Universite, Istanbul, Turkey
RS Means Company Inc, USA.

Much of the statistical data is from World Bank Development Reports, The *Economist World in Figures* and UN Housing and Construction Statistics. The background on individual countries has come from local sources, Lloyds Bank Economic Reviews, supplements in the *London Times* and the *Financial Times* newspapers, and national yearbooks and annual reports.

Important sources of general and construction industry data have been the British Overseas Trade Board Library, the Library of the Royal Institution of Chartered Surveyors, the Westminster Reference Library and various Embassies, High Commissions, Trade Missions, statistical offices, and Government Departments in the UK and overseas. Information on international contracting is largely based on surveys undertaken by *Engineering News Record* and *Building* magazines. Data on exchange rates and consumer price indices comes mainly from the *Financial Times* or International Monetary Fund publications.

The published sources of data for the general statistics are:

The Economist, World in Figures, London; Hutchinson, 1991
The World Bank, World development report 1990: Poverty Oxford: Oxford University Press, 1990
The Statesman Yearbook 1990/91
OECD in Figures
European Community, Eurostat: General Statistics 1989, Luxembourg: European Community, 1990.

The principal published sources on the construction industry are as follows:

1. Euroconstruct, Baukonjunktur 1990/91 in Westeuropa, IFO Institut für Wirtschaftsforschug München, Dezember 1990: for 1989 output figures for countries covered.
2. The series of reports by CIRIA on the construction industry eg The French Construction Industry - A Guide for UK Professionals. London: CIRIA, 1987.
3. The series of reports by the RIBA on architectural practice eg J L Meikle and P M Hillebrandt, Architectural Practice in Europe: France. London: RIBA, 1990.
4. The series of reports on engineering produced by the Association of Consulting Engineers (ACE), Foreningen af Raadgivende Ingenioret (FRI) and Orde van Nederlandse Raadgevende Ingenieurs (ONRI) eg Engineering Consultancy in the European Community. Republic of Ireland London: Telford, 1990.
5. Davis Langdon & Everest, Spons International Construction Costs Handbook, London: Spons, 1985.
6. UNECE Annual Bulletin of Housing and Building Statistics for Europe New York: UN, 1989.
7. Department of the Environment, Planning Control in Western Europe London: HMSO, 1989.
8. ISOCARP International Manual of Planning Practice, Vol 1 1989.
9. United Nations ECE, Building Regulations in ECE Countries, Second Report, New York: UN, 1985.
10. "The Top 500 European Companies: 300 Contractors, 200 Materials Producers ", Building, 16 November 1990.
11. "The Top 250 International Contractors" Engineering News Record Special Report, 5, July 1990.

Specific acknowledgements and sources as well as further reading are given where appropriate in each country. The research and compilation of this book were undertaken by the Davis Langdon & Everest Consultancy Group. Special acknowledgements are due to Dr Patricia Hillebrandt.

How to use this book

This book is in three parts - *Part One: Construction in Europe; Part Two: Individual Countries* and *Part Three: Comparative Data*. The twenty seven countries covered in the book are listed in the *Contents*. Two of these - the USA and Japan - are outside of Europe but are included for comparative purposes.

Part One: Construction in Europe

Three essays are presented in Part One. *The Construction Industry in Europe* describes the changes occurring in construction industries across Europe, East and West. *European Community Legislation* presents a brief guide to legislative changes in the EC which affect construction. *European Community Financial Support Programmes* outlines the main funds for development aid available from the EC.

Part Two: Individual countries

In Part Two twenty seven countries are arranged in alphabetical order, and each is presented in a similar format under the following main headings:

- *Key data* provides main economic and construction indicators
- *The construction industry* outlines the structure of the industry, tendering and contract procedures, and regulations and standards
- *Construction cost data* includes data on labour and materials costs, measured rates for items of construction work and unit costs (approximate estimating) for different building types
- *Exchange rates and inflation* presents data on exchange rates with the £ sterling and $US, and includes data on the main indices of price movements for retail prices and construction
- *Useful addresses* gives the names and addresses of authorities, professional institutions, trade associations, etc.

Introductory notes (4) provide general notes, explanations and definitions. All of the construction items appearing in *Construction cost data* are described fully in section 6 in five languages (English, French, German, Italian and Spanish).

Part Three: Comparative data

To allow rapid comparison between countries covered in the book, Part Three brings together data from Part Two and presents them under three main headings:

- *Key macroeconomic indicators* including financial and demographic data
- *Construction output indicators* including output per capita
- *Construction cost data* including input costs, unit costs and measured rates.

Abbreviations

LENGTH

kilometre	km
metre	m
decimetre	dm
millimetre	mm
yard	yd
foot	ft
inch	in

AREA

hectare	ha

VOLUME

kilolitre	kl
hectolitre	hl
litre	l
millilitre	ml

WEIGHT (MASS)

tonne	t
kilogram	kg
gram	g
hundredweight	cwt
pound	lb
ounce	oz

FORCE

kilonewton	kN
newton	N

not available	n.a.

Conversion factors

LENGTH

Metric		**Imperial equivalent**
1 kilometre	1000 metres	0.6214 miles
		1093.6 yards
1 metre	100 centimetres	1.0936 yards
	1000 millimetres	3.2808 feet
		39.370 inches
1 centimetre	10 millimetres	0.3937 inches
1 millimetre		0.0394 inches
Imperial		**Metric equivalent**
1 mile	1760 yards	1.6093 kilometres
	5280 feet	1609.3 metres
1 yard	3 feet	0.9144 metres
	36 inches	914.40 millimetres
1 foot	12 inches	0.3048 metres
		304.80 millimetres
1 inch		25.400 millimetres

AREA

Metric		**Imperial equivalent**
1 square kilometre	100 hectares	0.3861 square miles
	10^6 square metres	247.11 acres
1 hectare	10 000 square metres	2.4711 acres
		11960 square yards
1 square metre	10 000 square centimetres	1.1960 square yards
		10.764 square feet
1 square centimetre	100 square millimetres	0.1550 square inches
1 square millimetre		0.0016 square inches
Imperial		**Metric equivalent**
1 square mile	640 acres	2.5900 square kilometres
		259.00 hectares
1 acre	4840 square yards	0.4047 hectares
		4046.9 square metres
1 square yard	9 square feet	0.8361 square metres
1 square inch		6.4516 square centimetres
		645.16 square millimetres

VOLUME

Metric		Imperial equivalent
1 cubic metre or 1 kilolitre	10 hectolitres 1000 cubic decimetres 1000 litres	1.3080 cubic yards 35.315 cubic feet
1 hectolitre	100 litres	3.5315 cubic feet 21.997 gallons
1 cubic decimetre or 1 litre	1000 cubic centimetres 1000 millilitres	61.023 cubic inches 0.2200 gallons 1.7598 pints 0.2642 US gallons 2.1134 US pints
1 cubic centimetre or 1 millilitre	1000 cubic millimetres	0.0610 cubic inches
Imperial		**Metric equivalent**
1 cubic yard	9 cubic feet	0.7646 cubic metres
1 cubic foot	1728 cubic inches 6.2288 gallons 7.4805 US gallons	28.317 litres
1 cubic inch		16.387 cubic centimetres
1 gallon	8 pints	4.5461 litres
1 pint		0.5683 litres
US		
1 barrel	42 gallons	158.99 litres
1 gallon	8 pints	3.7854 litres
1 pint		0.4732 litres

WEIGHT (MASS)

Metric		Imperial equivalent
1 tonne	1000 kilograms	0.9842 tons 1.1023 US tons 2204.6 pounds
1 kilogram	1000 grams	2.2046 pounds 35.274 ounces
1 gram		0.0353 ounces

Imperial		Metric equivalent
1 ton	20 hundredweights	1.0160 tonnes
	2240 pounds	1016.0 kilograms
1 hundredweight	112 pounds	50.802 kilograms
1 pound	16 ounces	0.4536 kilograms
		453.59 grams
1 ounce		28.350 grams
US		
1 ton	20 hundredweights	0.9072 tonnes
	2000 pounds	907.18 kilograms
1 hundredweight	100 pounds	45.359 kilograms

FORCE

Metric		Imperial equivalent
1 kilonewton	1000 newtons	0.1004 tons force
		0.1124 US tons force
1 newton		0.2248 pounds force
Imperial		**Metric equivalent**
1 ton force	2240 pounds force	9.9640 kilonewtons
1 pound force		4.4482 newtons
US		
1 ton force	2000 pounds force	8.8964 kilonewtons

PRESSURE

Metric	Imperial equivalent
1 newton per square millimetre	145.04 pounds force per square inch
1 kilonewton per square metre	20.885 pounds force per square foot
Imperial	**Metric equivalent**
1 pound force per square inch	6.8948 kilonewtons per square metre
	0.0069 newtons per square millimetre
1 ton force per square inch	107.25 kilonewtons per square metre
	0.1073 newtons per square millimetre
US	
1 ton force per square foot	95.761 kilonewtons per square metre
	0.9576 newtons per square millimetre

Davis Langdon & Everest

Profile

Davis Langdon & Everest is an independent firm of Construction Cost Consultants and is a member of the Davis Langdon & Seah International Group which operates throughout the UK, Europe, the Middle East, South East Asia and Australia, and has associations in the USA and Japan. There are over 1600 staff in 50 offices around the world.

In Europe these offices are:

- In the UK: Davis Langdon & Everest in 21 offices covering England, Scotland and Wales
- In Germany: Davis Langdon & Weiss in Stuttgart, Frankfurt and Berlin
- In Spain: Davis Langdon Edetco in Gerona, Barcelona, Banyoles, Seville and Bilbao
- In Italy: Davis Langdon & Copper in Rome and Milan.

In addition the firm is active in France, Holland, Belgium, Luxembourg and Czechoslovakia and our Consultancy division undertakes detailed research projects throughout Europe.

Services

Our core service in the financial management of construction projects includes:

- Initial cost estimates
- Strategic advice on design economics
- Detailed cost planning
- Advice on cost of alternative design solutions
- Advice on Pan-European construction procurement and the selection of contractors
- Handling the procurement process
- Financial administration of construction contracts

In order to satisfy the changing needs of our clients we provide additional specialist services including:

- Project management
- Construction management
- Provision of employer's agent services
- Project programming
- Advice on development economics and appraisals
- Value analysis and management studies
- Quality management reviews
- Construction litigation services

- Advice on tax treatment of construction expenditure
- Planned maintenance and facilities management
- Fire insurance valuations
- Acting for receivers on liquidations
- Computer software development
- Research and consultancy in all aspects of construction economics.

Projects

The range of work undertaken covers new building and refurbishment projects of all sizes in all fields including:

- Arts and cultural buildings
- Business park developments
- Civic buildings
- Civil engineering
- Educational buildings
- Health and welfare buildings
- Hotels
- Industrial/warehouse developments
- Manufacturing/research buildings
- Office developments and fit-outs
- Petro-chemical installations
- Power stations
- Residential developments
- Retail developments
- Sports and leisure developments
- Transportation.

Davis Langdon & Everest with its associated firms offers a wide range of services to its clients around the world. Our work in preparing this Price Book is evidence of one aspect of the expertise which we make available to our clients.

DAVIS LANGDON & SEAH INTERNATIONAL

SOUTH EAST ASIA

BRUNEI

DAVIS LANGDON & SEAH
Suite 506
Badi'ah Complex
Jalan Tutong
PO Box 313
Bandar Seri Begawan 1903
Negara Brunei Darussalam
Tel : (010 6732) 222982
Fax : (010 6732) 222973

PETROKON UTAMA SDN. BHD
Suites T1-T8
Hongkong Bank Chambers
Jalan McKerron
PO Box 811
Kuala Belait 6008
Negara Brunei Darussalam
Tel : (010 6733) 335718
Fax : (010 6733) 335717

CHINA

DAVIS LANGDON & SEAH CHINA LTD
c/o Davis Langdon & Seah
Hong Kong Ltd

SHANGHAI GENIUS QUANTITY SURVEYING LTD
1111 Yan An Road (M)
Shanghai
China
Tel : (010 8621) 2477149 ext 329
Fax : (010 8621) 3291790

HONG KONG

DAVIS LANGDON & SEAH HONG KONG LTD
21st Floor
Leighton Centre
77 Leighton Road
Hong Kong
Tel : (010 852) 576 3231
Fax : (010 852) 576 0416

INDONESIA

DAVIS LANGDON & SEAH INDONESIA PT
Wisma Metropolitan 1
Level 13
Jalan Jendral Sudirman Kav. 29
PO Box 3139/Jkt
Jakarta 10001
Indonesia
Tel : (010 6221) 514745
Fax : (010 6221) 514764

MALAYSIA

DAVIS LANGDON & SEAH MALAYSIA
124 Jalan Kasah
Damansara Heights
50490 Kuala Lumpur
Tel : (010 603) 254 3411
Fax : (010 603) 255 9660

DAVIS LANGDON & SEAH MALAYSIA
Suite 8A
8th Floor
Wisma Pendidikan
Jalan Padang
PO Box 11598
88817 Kota Kinabalu
Tel : (010 6088) 223369
Fax : (010 6088) 216537

DAVIS LANGDON & SEAH MALAYSIA
2nd Floor
Lot 250
Jalan Tunku Abdul Rahman
93100 Kuching
Sarawak
Tel : (010 6082) 429861
Fax : (010 6082) 426416

DAVIS LANGDON & SEAH MALAYSIA
2nd Floor
10 Bishop Street
10200 Penang
Tel : (010 604) 619051
Fax : (010 604) 619028

DAVIS LANGDON & SEAH MALAYSIA
3rd Floor Wisma Zainah
95 Jalan Tun Abdul Razak
8000 Johor Bahru
Tel : (010 607) 236229
Fax : (010 607) 235975

PHILIPPINES

DAVIS LANGDON & SEAH PHILIPPINES INC
7th Floor, Electra House
Esteban Street
Legaspi Village
Makati, Metro Manila
Philippines
Tel : (010 632) 812 2679
Fax : (010 632) 815 6460

SINGAPORE

DAVIS LANGDON & SEAH
135 Cecil Street # 12 - 00
LKN Building
Singapore 0106
Tel : (010 65) 2223888
Fax : (010 65) 2247089

THAILAND

DAVIS LANGDON & SEAH (THAILAND) LTD
8th Floor
Kian Gwan Building
140 Wireless Road
Bangkok 10330
Thailand
Tel : (010 662) 253 7390
Fax : (010 662) 253 4977

I'm Sorry but the Contract has been Awarded!

PART ONE

CONSTRUCTION IN EUROPE

1. The construction industry in Europe

INTRODUCTION

The late 1980s and early 1990s are witnessing what is probably the greatest economic transformation in the European scene ever to take place in so short a time. In the West there is the development of the European Community and, particularly, of the Single European Market on 1st January 1993. In the East there is the breakup of the old economic and political order with the rapid decline in influence of the USSR and its disintegration into more or less autonomous republics, and the winding up of the Warsaw Pact (the military alliance) and Comecon (the common political and economic organization) in their old form. At the same time there are moves away from centrally planned economies towards some form of market economy, and away from authoritarian government towards various political reforms and more democracy.

This section looks at changes both in Eastern and Western Europe. It describes the way the construction industry operated, how it is changing and makes an assessment of the market. The market for construction in western Europe is then described and some changes in the way the industry operates are noted. Finally, some comparisons are made between the East and the West as well as with the USA and Japan.

EASTERN EUROPE

Political and economic transformation

The changes which have taken place over the last few years both on the political and economic fronts have been dramatic in their impact on the countries of Eastern Europe and on the rest of the world.

Politically,the introduction by President Gorbachev of *perestroika* (restructuring) and *glasnost* (openness) paved the way in the Warsaw Pact countries for a number of popular uprisings which overthrew the old regimes. In their place various degrees of democratization have been introduced. The old German Democratic Republic has become part of the new Germany, for example, and its old institutions are being replaced by those in the style of the Federal Republic. In most countries, there are moves towards democracy with elections which may not be really free. In others, the democracy established in the first flush of enthusiasm has been sorely tried by conflicts within the country and there has been back-tracking towards a more authoritarian rule. In a third group of countries, the struggle still goes on to overthrow the old regime. However, no European old Communist bloc country has been unaffected by these mass movements.

Most notable among the political problems in Eastern Europe and in the USSR and Yugoslavia in particular, are various degrees of separatism of the component parts of the old countries - the republics in the USSR and the regions of Yugoslavia. There are incipient problems in a number of other countries particularly where political boundaries do not necessarily coincide with ethnic boundaries.

Economically, there is movement towards some form of market economy. Some countries, such as Poland, have tried to make the transformation very quickly and are having great difficulties in adjustment, evidenced, for example, by inflation and by high unemployment. Other countries have talked of introducing a market economy but have not really done so. The USSR, for example, has disbanded much of the economic planning machinery but has not put another system in its place and the results are serious economic imbalances and drastic falls in the standard of living.

A further problem in Eastern Europe as a whole is that the economic problems of the Soviet Union have closed down much trade between it and other Eastern European countries so that the viability of their industries are seriously threatened.

It has been a matter for dispute whether it is practical or wise to try to introduce a market economy until some stabilization of the broader economy is achieved in terms of prices, balance of payments, the monetary system and the fiscal system. Even then, the question is whether it is better to do it quickly over all sectors of the economy simultaneously, or by a more gradual approach. In the countries which have disbanded the old system before proper planned introduction of a new one, the prospects for the next few years are not good and they are left with few options. Even where the changeover is better implemented according to some logical plan, it will probably be a decade or more before the changed economies are operating smoothly. Popular demands for improvement will be well ahead of what can realistically be produced for many years - perhaps a whole generation.

Some western support has already been provided to reduce hardship and to help with the restructuring of the economies and more will be in place very soon. The European Bank for Reconstruction and Development will provide funds for studies of various sectors and for investment projects. There are a number of bilateral schemes, for example, the UK know-how fund for selected Eastern European countries and a similar fund for the USSR. There is talk of a "Marshall Aid" type plan for the USSR and Eastern Europe similar to that given to Western Europe after the Second World War.

The construction industries: Common characteristics

The planned economies of USSR and Eastern Europe had many characteristics in common in their construction industries. Some of them are listed below:

- the design, contracting, and material producing organizations were of mammoth size but the scope for small private firms differed between countries
- the whole process for construction, from the desire to have a building or works, to commissioning was normally very long - several times that in the market economies
- the techniques of construction emphasized the use of heavy concrete slab construction and consequent high cement consumption
- with a few exceptions quality was not high
- there was very little choice in the system, for example, the range of components was restricted, and thus design options were limited; the choice of designers and contractors was small or nil
- procurement policies were mainly a matter of allocation rather than competition
- pricing of materials and other inputs and of the construction products did not reflect the supply and demand situation nor the costs of production.

Changes taking place

At the time of writing (mid-1991), it is too early to assess the implications of the changes that are taking place across the whole of Eastern Europe because the countries are in very different stages of transition, from the very fundamental changes in Poland, for example, to relatively little change of the former economic system in Albania. The most advanced countries seem to be Poland, Czechoslovakia and Hungary. However, trends may be identified which are likely to be followed in most countries as follows:

- a move to a more market orientated construction process
- fragmentation and privatization in some way or other of the construction organizations, ie into design, contracting and material producing organizations
- encouragement for new small firms, especially for small building and maintenance work
- attempts to get western capital investment to participate in development. In some countries this includes taking over unfinished buildings
- concern at construction quality and technology and the duration of construction projects.

For these changes in many countries, the role of the market system presents problems. In the USSR, there is no memory of a market system in which success was dependent on knowledge or judgement of the level of costs, the price which could be charged and the profit to be made. There is no understanding that failure to take the right decisions could lead to closure of a business. Middle management are unused to taking their own decisions rather than consulting the book of norms. There is no culture of private initiative or enterprise. In other countries, after over 40 years of centrally planned economies, the middle aged and elderly have the memory but not necessarily the inclination to change.

The past methods for estimation of costs have borne little relationship to real costs. Thus it is necessary, not only to build up a new data system of actual costs, but to do so in a way which keeps pace with the rapidly changing prices which will operate in the transition to a market economy. It is necessary to learn about the balancing of price against the chances of getting a job in a competitive contracting system, or the amount which will be bought in a competitive market. It is necessary, in running enterprises, to cost properly the capital equipment and overheads, to assess working capital requirements and cash flow, and ultimately to cope with the inevitable higher cost of money which makes both these so important; it is necessary to be prepared to appraise and accept risk as part of everyday life.

Much will need to be imported - materials, skills management and capital. There will be problems of exchange rates and balance of payments. All this will have implications for costs over and above the problems generated internally.

The construction market

The individual sections on each East European country in the main part of the book conform as far as possible to the format adopted generally, except that for the USSR, the old system has been more fully discussed. The past system has not been described in detail elsewhere because it has so much in common with the USSR and is changing so rapidly now that it is not very relevant.

To compare construction output in Eastern Europe with that in the west, the monetary value of output must be estimated in comparable currencies (ECUs). However, statistical data of output suffer from two major problems. First, the exchange rate is, for all Eastern European countries, an artificial one and is unlikely to give a real comparison with other countries of physical volume of output. Second, inflation sometimes means that if data are not available for the same years, the indications for a previous year may not be even close for the later years. Third, prices at which output is calculated are not the real prices and do not include the same components. Fourth, output, especially small works, is often under-recorded.

An alternative approach is to use some real measure or indicator of output. Two methods have been adopted here:

1. The total number of dwellings constructed, cement consumed and persons employed in both Eastern and Western Europe is adjusted for the estimated lower quality and size of dwellings, the higher cement consumption per volume of work and the lower productivity in Eastern Europe. The resultant output for Eastern Europe is about 35% of total European output equivalent to some 335 billion ECUs.

2. The USSR output of 220 billion ECUs (converted at the official exchange rate) equates to about 60% of the total output for the whole of Eastern Europe; this gives a total of some 310 billion ECUs.

Both methods involve bold assumptions and can give only approximations to the real figures. For comparative purposes a figure of 300 billion ECUs for the whole of Eastern Europe has been assumed. This is probably a maximum.

It is not possible to derive accurate estimates of the output for individual countries in Eastern Europe from this data. The following ranking can, however, be made:

1. The USSR is by far the largest market, up to ten times the size of that of any other East European country.
2. Poland and Rumania, as well as thc old GDR, all havc substantial output.
3. Czechoslovakia and Yugoslavia have a lesser output than the countries above, although the various indices for Yugoslavia are not consistent with each other.
4. Bulgaria and Hungary have the lowest output of this group.
5. There are no figures available for Albania.

WESTERN EUROPE

The construction market

The total market for construction is about 550 to 600 billion ECUs. On average, Western European countries spend about 11% of their GDP on construction, with Norway, Finland, Switzerland and Cyprus being highest; the UK, Luxembourg and Italy lowest. The distribution of work by types varies substantially between countries but on average is as follows:

New residential	26%	New civil engineering	17%
New non-residential	24%	Repair and maintenance	33%

Countries which have relatively higher or lower than average proportions of the main types of output are shown below.

Breakdown of construction output: relative proportions by country

	New residential		*New non -residential*		*New civil engineering*		*Repair and maintenance*	
High	Turkey	50%	Belgium	37%	Portugal	30%	Denmark	46%
	Portugal	40%	UK	32%	Spain	30%	Netherlands	44%
	Belgium	38%	Ireland	30%	Norway	25%	FRG	43%
			France	24%	UK	41%		
			Italy	40%				
Low	Denmark	15%	Spain	16%	Switzerland	9%	Portugal	10%
	UK	17%	France	17%	Belgium	9%	Belgium	16%
			Italy	18%	UK	10%	Norway	18%
					Ireland	10%	Spain	24%
					Netherlands	11%		

The proportion of repair and maintenance may be affected by differences of definition and of collection of the data. Repair and maintenance output for the UK is actually higher than the high figure recorded and other countries may be similarly variable.

It is tempting to find relationships with other indicators. For example, many of the countries with the highest repair and maintenance expenditure also have high per capita income. However, factors which often play a major part in determining the total amount of construction undertaken and its distribution by work types may vary considerably from country to country.

In Building magazine's "Top 300 European Contractors" the numbers in the top 100 and 300 respectively are shown in the table below, which also shows the numbers of European contractors in Engineering News Record's "Top 250 International Contractors".

Number of major contractors by country

	Top 100 European	*Top 300 European*	*Europeans in Top 250 International*
UK	30	79	13
Italy	7	51	38
France	16	43	10
Germany	12	27	16
Netherlands	8	22	7
Switzerland	3	19	3
Sweden	9	15	1
Denmark	2	14	3
Finland	4	8	4
Spain	6	7	2
Belgium	2	5	2
Norway	1	5	1
Portugal	-	4	-
Greece	-	1	1
Yugoslavia	-	-	8
Turkey	-	-	6
Austria	-	-	3
Poland	-	-	1
Hungary	-	-	1
Cyprus	-	-	1

Source: Building; Engineering News Record

In 1989, according to the Engineering News Record Survey there were 117 European firms in the top 250 international contractors who did 32% of the work compared with 42 American contractors doing 23% and 34 Japanese doing 8% of the work.

The structure of the industry

Most countries have a large number of small firms and relatively few large ones. Large firms employing more than 500 employees normally account for no more than a quarter of the work done. However, exceptions in Western Europe are Finland and Sweden, where large firms account for 44% of the work. By comparison, in the UK large firms account for 25% of the work.

The construction process

Changes are taking place across Western Europe in the construction process. Traditionally, in some countries, the separate trades system operated under which the architect arranged contracts directly with separate trades contractors and had therefore to co-ordinate the work on site. This system is gradually being replaced by the main contractor system, at least for large contracts. For many years now there has been an increasing diversity in procurement methods, especially in the private sector. This is noticeable not only in the UK, where management contracting and design-and-build contracts are common, but also in other countries where for various reasons, such as duration of the process and the need for flexibility during construction, alternative methods are being deployed.

COMPARISONS

The estimated construction outputs in 1989 of Eastern and Western Europe compared with the USA and Japan are shown below.

Construction output 1989: Eastern and Western Europe

	All work (bn ECUs)	*New work (bn ECUs)*	*Repair and maintenance (%)*
Western Europe	575	390	32
Eastern Europe	300	225 (estimate)	25 (estimate)
USA	665 (estimate)	560	16
Japan	440	395 (estimate)	10 (estimate)

Breakdown of new construction output in 1989

	*Western Europe** (%)	*USA* (%)	*Japan* (%)
Residential	39	48	33
Non-residential building	36	32	28
Civil engineering	25	20	39
Total	100	100	100

* *country averages*

It is not possible to quote figures for Eastern Europe with any degree of confidence, but it seems that residential construction in most countries is lower than in the West. A feature of Eastern European countries is the importance of construction for industry, but this might be civil works as well as industrial building.

The level of repair and maintenance is generally regarded as very low in Eastern European countries. Indeed some of the statistics purporting to represent the repair and maintenance output of Western European countries do not reflect the real situation. In some countries for 1989, figures of around 40% have been quoted, showing larger increases on those of the early 1980s. There is, however, a major problem of definition of repair and maintenance between, and even within countries. The problems in comparing Eastern with Western Europe are even greater and no numerical comparisons can be made.

Efficiency in eastern Europe is undoubtedly low compared with the West, for although the East has about a half of the output:

- the cement consumption in Eastern Europe is about the same as in Western Europe. This is due to choice of technology but also partly due to overdesign and waste

- there are some 16.6 million persons employed in the industry in Eastern Europe, compared with about 10.6 million in Western Europe. (Data on Eastern European may include some or all building material production)

- it is also thought that the investment in construction plant and equipment is very high in Eastern Europe, but not very effective. In 1987, 14% of the expenditure on building construction work was plant and equipment. The proportion in the UK in the early 1980s was about 5%. In part, the difference is due to the relatively low pay of workers which emphasizes the contribution of other inputs.

The relative levels of construction inputs between the UK and the USSR are shown below.

Construction inputs: UK and USSR

	UK (%)	*USSR (%)*
Manpower	35	13
Materials	37	55
Plant	5	14
Other	23	18
Total	100	100

Construction times in the USSR and, it is believed, in other Eastern European countries, are very long. Large projects may take upwards of four times as long to complete as in the USA, and the USSR's own norms for construction times are exceeded two to three times. There is a large stock of uncompleted buildings in the USSR. Norms in the USSR for construction costs are similarly exceeded two to three times (for further details see the country section on USSR).

The structure of the construction industry is quite different between East and West. In the East, the dominant organizational type equates to a giant conglomerate - undertaking design, construction and frequently responsible for production of materials. There are, in most countries, no medium sized firms and only a very few small firms. In Western Europe, the USA and Japan, the situation is virtually reversed with a predominance of small firms, some medium sized and a few large ones, but not many equivalent in size to the mammoth East European organizations.

However, in Japan, there are six very large companies with turnovers as large or larger than the large Western European and US companies. They are very highly integrated companies involved in the whole process of construction from design to completion and handover. Because there is no separation of the design and construction process, they are also able to undertake large research programmes.

The eleven largest international contractors, in terms of total contracts won in 1989, are listed in the table below.

Major international contractors

	Country of origin	*Total contracts won in 1989 (billion $)*
Fluor Daniels Inc	USA	16.6
Shimizu Corporation	Japan	13.2
Bechtel Group	USA	12.0
Kajima Corporation	Japan	11.9
Taisei Corporation	Japan	11.5
Takenaka Corporation	Japan	11.3
Brown and Post Inc	USA	11.0
Ohbayashi Corporation	Japan	10.0
Parsons Corporation	USA	9.7
The M W Kellogg Co	USA	9.4
Bouygues	France	8.5

Source: Engineering News Record: "Top 250 International Contractors"

Major differences exist in the way in which the process operates, from the client's decision to commission a building, to the delivery of the completed project. In Eastern Europe there is a regulated process at all stages and a situation where delays can be imposed or arise at any stage for a great number of reasons, notably lack of funds and lack of real resources. Contract conditions have little meaning - there are always external reasons for non-completion in the time and cost allocated. In the Western world there is competition at all stages of the process between the potential participants. The process, even in the public sector, is quicker, more flexible and with allocated responsibilities for performance. Failure to perform brings penalties, if not on the particular project, then on the ability to compete successfully for further work. So great are the differences between the construction products, the construction processes and the construction industries in the two groups of European countries that it seems the only really common factor is that both produce buildings and works with men, materials and plant and equipment. Who does it, how they do it, how long it takes, how much it costs and what it is like when it is completed are all very different.

2. European Community legislation

INTRODUCTION

European Community (EC) legislation takes different forms:

- regulations
- decisions
- directives.

Regulations have general application and are directly applicable in each Member State. Decisions (eg antitrust decisions) are also directly applicable, but only to the specific individuals or enterprises to whom they are addressed.

A directive is binding on Member States, not on individuals. Moreover, a directive is binding solely for the end result, leaving each Member State free to choose the most appropriate legislation to achieve it. In principle, a directive becomes applicable to individuals only after the appropriate national legislation is in place (usually 18 months to 2 years later). Thus, new directives or proposals for directives are an interesting source of information on what future national law may be. In certain circumstances, directives have direct effect, even if a Member State has failed to implement them in due time; in such cases, directives have the same status as national legislation. The most important (draft) EC directives with impact on the construction industry are described below.

PUBLIC PROCUREMENT

The EC's effort to open up the public procurement markets of its Member States received new impetus after the publication of the Commission's 1985 White Paper on the completion of the Internal Market.

There are five principal directives that cover or will cover public sector procurement described below. It should be noted that several of them are in fact revisions of earlier directives.

Public works directive

This directive (89/440/EEC) applies to works contracts and concessions of 5m ECUs net VAT (1 million ECU until 1.3.1992). It provides that:

- the tender notices must be prepared in accordance with a prescribed layout and must be published in the Official Journal

- the minimum time limits for the tender procedure must be respected
- technical specifications for the works must be specified in the general documentation on the tender and in the contract documents and must not discriminate against contractors from other Member States
- disqualification of applicants to bid or bidders is allowed only on the basis of specific qualitative criteria
- evidence of technical, financial and economic capacity is subject to detailed rules.

After 1 March 1992, the date on which the 1989 amendments to the works directive must be implemented in each Member State, the scope of application of the directive will be broader; it will also apply to works contracts in which the consideration is a franchise to operate the works, and to contracts which are awarded by private clients but financed more than 50% by public authorities.

In addition, the amended directive will require early publication of annual public works programmes. It will clarify tendering and award procedures by requiring contracting authorities to explain to an eliminated contractor why his or her tender was rejected. It will require that the Commission be supplied with a report on each award decision and a notice published on the outcome of each award decision.

Excluded services directive

This directive covers utilities previously excluded. It is sometimes known as the Utilities Directive or WETT (Water, Energy, Transport and Telecommunications) Directive. It is due for implementation at the beginning of December 1991. The directive is similar to that of the Public Works directive. It will apply to utilities even if they are not in the public sector.

Services directive

The Commission is preparing a directive on the co-ordination of procedures for the award of public service contracts. In general, the draft attempts to create rules for the award of public service contracts which are as close as possible to those for public supply and public works contracts. The rules will apply to both public service contracts and public service concessions.

The services which the directive would cover include:

- architectural services; engineering services; urban planning and landscape architectural services; related scientific and technical consulting services; technical testing and analysis services
- computer and related services
- management consulting and related services.

Contracts for the construction-related design services above would be subject to the directive's rules when they represent a value of at least 400 000 ECU. Contracts for other services would be covered by the directive when their estimated value is 200 000 ECU or more.

The preliminary draft for the directive lists the cases in which a public service contract may be awarded by negotiated procedure, with, or in some cases without, prior publication of a tender notice. In all other cases, public service contracts would have to be awarded after an open or restricted award procedure. Detailed rules for design competitions are also laid down in the draft.

The draft requires contracting authorities to define technical specifications by reference to national standards implementing European standards, by reference to European technical approvals, or to common technical specifications. Contracting authorities will be required to publish annually their intended total procurement for all the above services and, for construction-related and computer services, the estimated total need for services by the contracting authority itself or by another contracting authority.

The intention to award a public service contract must be published, as must the results of the award procedure. The draft lays down models for the notices, time limits for the award procedure, rules for participation and selection, and criteria for the award of contracts. It is expected that the draft will be adopted and that the directive will take effect in the course of 1992.

Compliance directive

A directive of 21 December 1989 on application of review procedures creates more effective measures for preventing and sanctioning infringements of the public procurement rules (Directive 89/665/EEC, OJ L395/89). The directive requires Member States to provide interim measures such as suspension of the award procedure or any decision taken by the contracting authority, the setting aside of unlawful decisions and the award of damages. Furthermore, the Commission may intervene directly when it considers that a clear and manifest infringement of Community procurement rules has taken place.

Public supplies directive

This directive (88/295/EEC) came into force on 1 January 1989 and covers supplies and materials purchased by central government departments and other listed bodies. The directive covers all contracts with 0.2m ECUs or more.

HARMONIZATION AND MUTUAL RECOGNITION

Different national standards, rules and codes are considered to constitute a serious barrier to trade not only of products, but also of services related to construction or other technical industries. The Community has the authority to harmonize these rules through directives (and has done so, for example, for construction plant and equipment), but has found impractical the process of adopting long, detailed and complex technical directives which needed continuous updating. It has, therefore, adopted a new harmonization policy, based on the recognition that all technical regulations have similar objectives such as protection of health and public security or consumer protection. The Community has decided to regulate the objectives only, and to do this for whole sectors at once or for a wide range of products with similar protection requirements. The Construction Products Directive to be implemented by national law by 27 June 1991 lays down the six essential requirements (mechanical resistance and stability; fire safety; hygiene, health and environment; safety in use; protection against noise; energy economy, and heat retention) which must ultimately be complied with by the projects in which the construction materials will be incorporated.

Detailed technical specifications for specific products will eventually be worked out by the European Committee for Standardization (CEN) and the European Committee for Electro-technical Standardization (CENELEC). These are the European standardization bodies to which the EC has given a formal mandate to prepare harmonized European standards. The standards will be voluntary, not mandatory, but goods complying with them will be guaranteed free movement throughout the EC. Similarly, harmonized technical rules for the design of building and civil engineering works, called Eurocodes, have been drafted under the auspices of the EC Commission and will be prepared in the future by CEN. They are intended initially as an alternative to the different rules in force in the various Member States (so-called optional harmonization) and, ultimately, to replace them.

SERVICES OF CONSTRUCTION PROFESSIONALS

Freedom to provide services and even to establish a practice in other EC Member States in the field of architecture is guaranteed by a Council Directive of 10 June 1985. It requires all Member States to recognize diplomas awarded in other Member States which fulfil specific requirements of education and training listed in the directive, and to grant the holders of the diplomas the right to act as an architect in their territory. The directive also contains provisions which oblige the host Member State to accept certificates from the other Member State certifying that the migrant professional fulfils any other requirements for practice of the profession in the host country (e.g. good reputation, professional discipline, authorization from a professional body, absence of bankruptcy, insurance etc).

It is uncertain whether a special directive for engineers' diplomas will be adopted, since a general framework directive has been drafted providing for the mutual recognition of higher educational diplomas which came into force in January 1991.

There are no special rules for contractors, except those laid down in the 1971 directive on the co-ordination of procedures for the award of public works contracts (as recently amended). Articles 23 to 28 provide criteria for qualitative selection of contractors i.e. how a contractor is to prove his or her financial, economic and technical capacity to perform public works in other EC Member States.

For construction technicians, technical barriers to their practising in other Member States can be set aside, since a list of comparable vocational qualifications throughout the EC has been published (OJ C292 of 20 November 1989).

LIABILITY

The Community has either adopted or is currently preparing directives which, when implemented in the Member States, will change the rules on professional liability.

The product liability directive of 1985 has no immediate effect on consultants, since it applies only to the producers or suppliers of products. Only when consultants enter into contracts where they take responsibility for the actual construction activities, would they come under the scope of this directive, which introduces the principle of strict liability (i.e. the injured person does not have to prove negligence).

A draft for a parallel directive is being prepared on liability for services with a safety defect. It provides for strict liability of the supplier of a service which causes damage within a period of one year after it was supplied, as soon as the injured party can show a plausible link between his or her damage and a safety defect. The injured party has three years following the discovery or "discoverability" of the defect, the damage and the identity of the supplier during which to bring an action, to which a "long stop" prescription period of five years applies.

A draft provides for even longer limitation periods for services for the design or construction of immovable property (for damage occurring within ten years, and for a "long stop" of twenty years). Construction is to be included (in February 1991). However the Directorate General (DG) III of the EC Commission has launched an extensive study on the need for, and feasibility of, harmonizing the professional liability rules in construction. There is yet no agreement, but if agreement is reached then construction may be excluded from the services directive.

The Community is expected simultaneously to look at preventive measures which may improve the safety of consumers and the quality of construction, e.g.:

- state or private control of all constructional activities
- quality assurance
- certification of professionals
- new insurance practices.

Many of the measures can be initiated by professionals themselves. In several countries, consultants have started work on them already, in the knowledge that only self-regulation may prevent over-regulation by the Brussels authorities.

CORPORATE ACTIVITIES IN MEMBER STATES

The Community has adopted twelve directives so far, aimed at harmonizing national company legislation. Although the main aim is to protect shareholders and creditors of corporations, for industry at large, and for consultants wanting to operate abroad in the form of a local company, the harmonized rules render the foreign corporate environment more familiar. The eleventh and twelfth directives, adopted on 21 December 1989, relate to disclosure requirements for branches and for single-member private limited-liability companies, respectively. The Community also attempts to create special legal corporate instruments for trans-national co-operation.

The European Economic Interest Grouping (EEIG) attempts to foster a truly European structure. It aims to facilitate the economic activities of its members by enabling them to co-operate effectively across frontiers and to overcome legal, fiscal and psychological difficulties. Its members must come from at least two Member States. The EEIG is flexibile and has few administrative requirements for its foundation: it is ancillary in nature and must not replace its members' activities, and the members remain entirely independent. The Commission envisages the EEIG as a useful means of forming multidisciplinary consortia in the construction industry, to tender for public or private contracts.

A proposal for a European Company (*Societas Europeana*) statute has been in existence for almost twenty years. It would allow for the setting up of a limited liability company which would not have to embody the nationality of any of the Member States. It could be established by a merger or by creation of a holding company or a joint subsidiary. The proposed *Societas Europeana* closely resemble existing public limited companies (although certain specific tax rules are applicable to them). They would be governed by a general meeting of shareholders and either a management board and a supervisory board (two tier system) or an administrative board (one tier). Accompanying this proposal is a draft directive on employee participation which leaves a broad range of options open to both employers and Member States.

ENVIRONMENT

The Community has established complex programmes on the environment and has adopted an impressive number of directives for preventing air, water, noise and soil pollution and for fighting pollution through control of waste, specific dangerous substances, dangerous industrial activities, etc. An important general measure was Directive 85/337/EEC (OJ L175/85), requiring environmental impact assessments for all public and private projects which are likely to have significant effects on the environment.

ANTITRUST (MONOPOLIES)

All companies are subject to the EC's antitrust rules, which prohibit all agreements and concerted practices between enterprises which have as their object or result an appreciable effect on competition between Member States. The EC Commission can exempt certain restrictive agreements or practices provided they have been notified and that certain conditions are fulfilled.

Several agreements concluded in connection with the Channel Tunnel have been notified to, and closely examined by the EC Commission. Although there were several restrictive features in the agreements, the Commission did not object because their sole aim was the establishment of consortia for the performance of certain obligations which the participating enterprises could not deal with on their own. The agreements were not found to imply any restriction on the freedom of the parties for activities other than those related to the tunnel project.

Anti-competitive agreements on construction supply products (e.g. cement) have come under much closer scrutiny, and consultants facing trade restrictions for certain construction products they want to specify, may want to suggest to their clients that they file a complaint with the Commission's anti-trust division.

EUROPEAN CURRENCY UNITS (ECUs)

The European Monetary System (EMS) came into force on 13 March 1979 and with it the ECU (European Currency Unit) as the unit of account for the Community. The ECU is based on a "basket" of currencies of the Member States of the Community, each currency contributing a percentage of the ECU's value. These percentages, which were last reviewed on 20 September 1989, are shown below together with the value of the ECU against each national currency as at January 1991.

Composition and value of the European currency unit

Country	*Currency*	*Percentage contribution to ECU's value**	*Value of 1 ECU in members currency as at April 1990*
Belgium	Belgian franc	7.6	42.24
Denmark	Danish krone	2.45	7.89
France	franc	19.0	9.94
Greece	drachma	0.8	216.5
Italy	lira	10.15	154.1
Ireland	punt	1.1	0.768
Luxembourg	Luxembourg franc	0.3	42.24
Netherlands	guilder	9.4	2.31
Portugal	escudo	0.8	182.7
Spain	peseta	5.3	128.9
UK	pound sterling	13.0	0.701
West Germany	mark	30.1	2.05

**Since 20 September 1989.*
Source: Lloyds Bank and Financial Times.

MISCELLANEOUS

Consultants' work or work opportunities may be influenced, sooner or later, by other EC measures, including:

- a Council recommendation of 22 December 1986 on fire safety in hotels
- a proposal for a Council directive on information on the energy efficiency of buildings, providing that every prospective buyer or tenant of a building must be informed of the energy efficiency of that property; for energy audits an annex to the proposal contains a requirement for experts carrying out audits to be approved
- a proposal for a Council decision on the establishment of a network of high-speed trains (OJ C34/90)
- Council decisions on financing major Community infrastructure projects.

ACKNOWLEDGEMENTS

The authors are grateful to Foreningen af Raadgivende Ingenioret, Crown and Association of Consulting Engineers, Orde van Nederlandse Raadgevende Ingenieurs and Ministerei von Economische Zakan for permission to use material for this section which previously appeared in the Engineering Consultancy in the European Community series of 11 country profiles, published by Thomas Telford Ltd, 1990.

3. European Community financial support programmes

INTRODUCTION

Financial support from the Community comprises a number of programmes all of which may impact on the construction sector. These include:

- European Regional Development Fund (ERDF)
- European Social Fund (ESF)
- European Agricultural Guidance and Guarantee Fund (EAGGF)
- Integrated Mediterranean Programmes (IMPs)
- European Coal and Steel Community (ECSC)
- European Investment Bank (EIB).

EC structural funds comprise three funds: the European Regional Development Fund (ERDF), the European Social Fund (ESF) and the European Agricultural Guidance and Guarantee Fund (EAGGF). The aim of structural funds is to focus action on instruments of economic development. Structural policies should enable all the regions of the EC to share increasingly in the major economic benefits derived from the frontier-free market.

The Commission of the European Communities, stated in 1989 that assistance was to be concentrated on the following five priority objectives (the main instruments that may be used are indicated in parenthesis):

1. Promoting the development and adjustment of the regions whose development is lagging behind (i.e. where per capita GDP is less than, or close to, 75% of the Community average) (ERDF, ESF, EAGGF).
2. Converting the regions, frontier regions or part of regions... seriously affected by industrial decline (ERDF, ESF).
3. Combating long-term unemployment (ESF).
4. Facilitating the occupational integration of young people (ESF).

5a. Adapting production, processing and marketing structures in agriculture and forestry (EAGGF, Guidance section only).

5b. Promoting the development of rural areas (EAGGF, ERDF, ESF).

The financial resources available for structural funds have been increased substantially. The regions covered by Objective 1 should benefit significantly from this increase in resources. The financial commitment is to be a total of 60 billion ECU (about 420 billion fr.) from 1989 to 1993. The distribution of fund resources for objectives 1-5 is indicated in the table below.

Distribution of EC structural funds for priority objectives 1-5 for 1989-93

	Distribution (billion ECU)	*Percentage distribution*
Objective 1	38 300	63.5
Objective 2	7 205	12.0
Objective 3	7 450	12.4
Objective 4	3 415	5.7
Objective 5(a)	2 795	4.6
Objective 5(b)	1 150	1.9
Total	60 315	100.0

Source: Commission of the European Communities.

Approximately 48 billion ECU has been allocated by fund and objective.

EUROPEAN REGIONAL DEVELOPMENT FUND (ERDF)

In mid-1988 a major review of the Community's structural funds identified the ERDF as the main instrument for ensuring the development and adjustment of regions which are lagging behind; it also pointed to the central role in the conversion of regions (including employment areas and urban communities) seriously affected by industrial decline. The reform of the structural funds will result in decreased support for individual projects, the emphasis in future being on strategic long-term programmes.

For the ERDF a total of 24 billion ECU has been assigned to cover the priority objectives 1 and 2. The ERDF is to devote approximately 80% of its appropriations to objective 1, i.e. the less developed regions which require aid for industry, training or national infrastructure, in order to bring them to a level which more closely approximates the more highly developed regions of Europe. These regions are Greece, Portugal, Ireland, Northern Ireland, parts of Spain and Italy, Corsica and the French Departments d'Outre Mer.

All the recipient countries were required to submit their regional development plans by March 1989 which, after approval from the Commission, now form the framework for allocation and approval of project funding. These funds represent an enormous opportunity for construction projects and for the design and planning services which go with them.

Projects which provide basic services or communications to existing and future industry are most likely to receive financial assistance. Eligible projects include those connected with industrial estate provision; most forms of transport investment; power supplies; and water, drainage, sewage and flood protection works. Projects that will benefit only single users or one type of industry may also be eligible, but normally at a reduced grant rate.

Any proposed projects must lead to an improvement of facilities and must cost more than 50 000 ECUs in order to be eligible. For projects costing less than 15 million ECUs, the rate of grant is 50%; for those costing more than this, the rate of grant is between 30% and 50%. In exceptional cases, where projects are of significant importance to the economic development of a region, grant aid of up to 55% is available, irrespective of the cost of the project.

Direct access to the ERDF on the part of industry is limited, most of the finance being tapped by local and regional authorities and public utilities. It is expected that, under the regulations under review, ERDF funds will have to be additional to national governments' own spending in the regions.

EUROPEAN SOCIAL FUND (ESF)

The ESF is an employment fund originally established under the Treaty of Rome. Article 123 of the Treaty described its purpose as "to improve employment opportunities for workers in the Common Market and to contribute thereby to raising the standard of living ... it shall have the task of rendering the employment of workers easier and of increasing their geographic and occupational mobility within the Community".

Assistance may be granted from the fund only for expenditure intended to cover:

- incomes of persons undergoing vocational training
- the costs of:
  - the preparation, operation and administration of vocational training measures including the vocational guidance of the recipients and the cost of training teaching staff and of depreciation
  - board, lodging and travelling expenses for the recipients of vocational training
  - the adaption of workplaces in the case of vocational integration of the handicapped
- the granting for a period not exceeding twelve months per person, of aid for recruitment to additional jobs or for employment in projects for the creation of additional jobs which fulfil a public need, for young people under 25 who are seeking employment and long term unemployed; the jobs referred to must be stable or likely to provide additional training or experience with a vocational content giving access to the labour market and facilitating employment in a stable job
- benefits designed to facilitate the transfer and integration of migrant workers and members of their families
- carrying out preparatory or evaluation operations or studies.

EUROPEAN AGRICULTURAL GUIDANCE AND GUARANTEE FUND (EAGGF)

The EAGGF was set up in 1962, and has two divisions - the Guidance and Guarantee Sections. The Guidance Section is one of the structural funds.

The Guarantee Section deals with the expenditure required for the operation of markets and the guaranteeing of prices.

The Guidance Section finances "common measures" decided upon with a view to achieving the basic aims of the Treaty of Rome with regard to increased agricultural productivity. The majority of measures are financed through the reimbursement of part of the eligible expenditure by Member States (indirect measures); other measures are financed by direct subsidy for projects (direct action). In the first case the expenditure involved is incurred by the Member State concerned and the Guidance Section refunds a proportion. In the case of direct action, applications for aid must still be approved by the Member State concerned, and aid is granted through specially designated agencies.

Direct action aid may be given for:

- rationalization or development of storage, market preparation, preservation, treatment or processing of agriculture and of fish products
- improvement of marketing channels
- a better knowledge of the facts relating to prices and to their formation on the markets for agricultural products
- Annex II of the Treaty of Rome lists those products to which the projects must relate. Products which are eligible include fish and processed fish, horticultural products and the majority of agricultural products grown or produced within the Community.

Projects must benefit the primary producer and guarantee him an "adequate and lasting share in the benefits arising from projects". Projects must also offer adequate guarantees of profitability. In addition only those projects which are within the framework of an investment programme drawn up by Member States and approved by the Commission will be eligible for aid.

INTEGRATED MEDITERRANEAN PROGRAMMES (IMPs)

The objectives of IMPs are the speeding up of the development of the Mediterranean regions, particularly Greece, but also France and Italy and assistance for these countries in facing the impact of the accession of Spain and Portugal. The programmes are concerned with investment in productive industry, infrastructure and the better use of human resources. Funds come from various sources - all the structural funds plus loans from the European Investment Bank. There will also be a special item in the IMP's budget to supplement other funds or to help geographical areas not covered by these. The programmes started in 1987 and will last for seven years. The first IMP to be adopted was for Crete, followed by three for France in April 1987. The regions which will receive aid are primarily agricultural except for Greece where the entire country is included. In the manufacturing and services sector the aim is to help small and medium sized companies, to promote tourism, to create the relevant infrastructure and to provide vocational training. Financial assistance takes the form of a contribution of 2500 million ECUs from the ERDF, European Social Fund and the Guidance Section of EAGGF, plus an additional 1600 million ECUs. EIB loans could amount to a further 2500 million ECUs. The rate of assistance will be up to 70% of the total cost of a project.

EUROPEAN COAL AND STEEL COMMUNITY (ECSC)

The ECSC finances investment in the coal and steel industries by raising the necessary funds and granting loans. It may also help finance investment upstream and downstream of the actual plants; for example, port facilities for unloading raw materials for use in the iron and steel industry. It may also provide funds for the construction of low-cost housing and help finance investment by firms for the creation of new activities which would productively re-employ redundant coal and steel workers.

Loans or guarantees for investment programmes are available both for those directly involved in coal and steel production in the Community, and for projects which contribute directly towards production, growth or reduced production costs, or which ensure re-employment for those made redundant in the industries.

Loans fall into four categories: industrial loans, reconversion loans, loans for converting from oil/gas to coal, and loans for workers' housing.

Loans can be made up to a maximum of 50% of a project's gross capital; interest rates are fixed with flexible repayment terms. Interest rebates are available to support the community dimension of certain loans.

EUROPEAN INVESTMENT BANK (EIB)

The EIB lends to both public and private sectors. Loans are for capital investment in industry, infrastructure and energy which help to further a European Community objective. The EIB was established under the Treaty of Rome to assist in:

- the economic development of less favoured regions
- improving communications between Member States (transport and telecommunications)
- the energy objectives of the Community
- protection of the environment
- the modernization of industry
- the introduction of advanced technology.

Following its decision to double the structural funds allocation by 1992, the EIB will continue to expand its role as a source of loan finance for regional development and European infrastructure projects to accompany grants from the EC. Loans are available for up to 50% of the fixed capital cost of the project. The loan duration depends on the project life.

Nuclear projects - nuclear power stations and industrial installations using nuclear fuel - are eligible for loans of up to 20% of the total project cost. The interest rate and the repayment period will depend on market conditions. Administration of the loan is carried out by the European Investment Bank.

OTHER EC FUNDS

Apart from the more general funds described above, various Directorate Generals of the EC Commission have funds for special purposes e.g. for research and development projects such as:

- RACE, ESPRIT, BRITE
- bio-technology
- alternative energy sources
- environment, natural resources and materials technology
- clean technologies.

The total annual investments in these projects is several million ECU.

Further information is available through Directorate General XII.

ACKNOWLEDGEMENTS

The authors are grateful to Foreningen af Raadgivende Ingenioret, Crown and Association of Consulting Engineers, Orde van Nederlandse Raadgevende Ingenieurs and Ministerei von Economische Zakan for permission to use material for this section which previously appeared in the Engineering Consultancy in the European Community series of 11 country profiles, published by Thomas Telford Ltd, 1990.

PART TWO
INDIVIDUAL COUNTRIES

4. Introductory notes to country sections

INTRODUCTION

In this part of the book, twenty seven countries are arranged alphabetically, and each country is presented as far as possible in a similar format, under five main headings - Key data, The construction industry, Construction cost data, Exchange rates and inflation, and Useful addresses.

These notes introduce the five main sections of information presented on the individual countries and provide, in one place, general notes, definitions and explanations, in order to keep the individual country sections as succinct as possible.

KEY DATA

The key data sheet at the start of each country lists main population, geographic, economic and construction indicators and thus provides a brief statistical overview of that country. Most of these indicators are taken from official World Bank and United Nations publications and are therefore authoritative, but it takes time for international agencies to collate and publish such data so that it can on occasions be a little dated. Official figures are used here to be consistent and to ensure that the summaries of key data in Part Three are presented on a comparable basis.

THE CONSTRUCTION INDUSTRY

The main topics covered in this section are the contribution of the industry to the economy; the structure of the industry; the availability of and constraints on, construction labour and materials; tendering and contract procedures and standards.

Although construction is often fragmented and tends to be labour intensive with low capital investment, it is invariably the single largest industry in a country. In most countries it contributes between 5% and 15% to Gross Domestic Product (GDP) and a similar percentage to direct construction employment (indirect employment - in the construction materials industries and other related activities - can more than double the contribution).

CONSTRUCTION COST DATA

This section includes both construction costs incurred by contractors and the costs they charge their clients. The costs of labour and materials are input costs of construction, i.e. the costs incurred by contractors. Unit costs, measured rates for construction work and approximate estimating costs per square metre are output costs, i.e. the costs contractors charge their clients. Problems of definition make meaningful and consistent presentation of unit rates extremely difficult. For unit rates to be useful it is essential to be clear what is included and what excluded, for example, how preliminary and general items are dealt with.

Cost of labour and materials

Typical costs for construction labour and materials are given in most country sections. Two figures are generally given for each grade of labour. The wage rate is the basis of an employee's income - his basic weekly wage will be the number of hours worked multiplied by his wage rate. The cost of labour, on the other hand, is the cost to the employer of employing that employee; it is also based on the wage rate but includes (where applicable) allowances for:

- incentive payments
- travelling time and fares
- lodging and subsistence
- public and annual holidays with pay
- training levies
- employer's liability and third party insurances
- health insurance
- payroll taxes
- other mandatory and voluntary payments.

The costs of main construction materials are given as delivered to site in quantities appropriate to a reasonably substantial building project. It is presumed that there are no particular difficulties of access etc., which would significantly affect costs. Generally tax, and particularly any value added tax, is excluded from material costs mainly because the rate of tax to be levied may depend on the type of work the material is to be incorporated in.

Unit rates

Rates for a variety of commonly occurring construction items are provided for most countries. They are usually based on a major, if not the capital, city and the relevant date is always January 1991. Rates include all necessary labour, materials, plant and equipment and, where appropriate, allowances for contractors' overheads and profit, preliminary and general items associated with site set-up, etc., and contractors' profit and attendance on specialist trades. Value added tax and other taxes are excluded. The rates are appropriate to a reasonably substantial building project.

A full description of each work item is presented in section 6 in five languages (English, French, German, Italian and Spanish).

Approximate estimating

Approximate estimating costs per unit area (square metres and square feet) are given for most countries for a variety of building types. Notes on the method of measurement and what is or is not included in unit rates are provided in each country section. Areas are measured on all floors inside external walls and with no deduction for internal walls, columns, etc.. Generally tax, and particularly value added tax, is excluded in approximate estimating costs.

When making comparisons of construction costs between countries it is important to be clear about what is being compared. In terms of approximate estimating costs there are two main methods of comparison: first the comparison of identical buildings in each country and, second, the comparison of functionally similar buildings in each country.

To some extent, this book permits comparisons to be made of both identical and functionally similar buildings in European countries. The EEC Statistical Office (Eurostat) exercise, described and presented in Part Three, is an example of "identical" comparisons - exactly the same construction projects have been priced in each of the Community Countries and the results converted to a common currency (European Currency Units - ECUs). In the individual country sections in this book, the approximate estimating rates given are for the standard of building of each type normally built in that country. Rates are therefore closer to the "functionally similar" approach described above. The rate per square metre given for an office building, for example, or a warehouse in any particular country refers to the normal type of office building or warehouse built in that country. In country sections they are presented in national currencies. A selection of approximate estimating costs are also presented in pound sterling, US dollar and ECU equivalents in Part Three thus enabling comparisons on a common currency basis to be made.

EXCHANGE RATES AND INFLATION

Exchange rates

Currency exchange rates are important when comparing costs between one country and another. While it is most useful to consider costs within a country in that country's currency, it is necessary, from time to time, to use a common currency in order to compare one country's costs with another. But exchange rates fluctuate dramatically and few currencies (even those considered strong) can be considered really stable. It can be risky to think in terms, for example, of one country being consistently a set percentage more or less expensive than another.

Different rates of internal inflation affect the relative values of currencies and, therefore, the rates of exchange between them. However, the reasons behind exchange rate fluctuations are complex and often political as much as economic; they include such factors as interest rates, balance of payments, trade figures and, of course, government intervention in the foreign exchange markets, and, for that matter, other government actions.

Graphs of exchange rates since 1980 against the US dollar and the pound sterling are included for most countries. They have been calculated by averaging the published weekly values in each quarter. The values given are therefore smoothed - the most dramatic peaks and troughs have been ironed out. They are, however, useful for indicating long term trends.

Exchange rates for each country are plotted against the pound sterling and the US dollar. As far as possible, the form of the graph is kept the same, hence the vertical scale is adjusted to accommodate different currencies. It should always be checked whether marked movement in a graph is a result of erratic exchange rates or merely the selected vertical scale.

If a line moves up from left to right (for example, the Danish Kroner against the US dollar up to 1985 - see below) it indicates that the subject currency (the Danish Kroner) is declining in value against the currency of the line (the American dollar). The higher the line is, the more subject currency is required to purchase the line currency. If, on the other hand, a line moves down from left to right (for example, the Danish Kroner against the US dollar after 1985 - see below) it indicates the subject currency strengthening against the line currency. Where there is virtually no movement at all, that is the line is horizontal, this usually indicates a currency effectively "tied" or "pegged" to the line currency. A number of currencies within the Exchange Rate Mechanism (ERM) of the European Monetary System (EMS) have this relationship.

Exchange rate graph

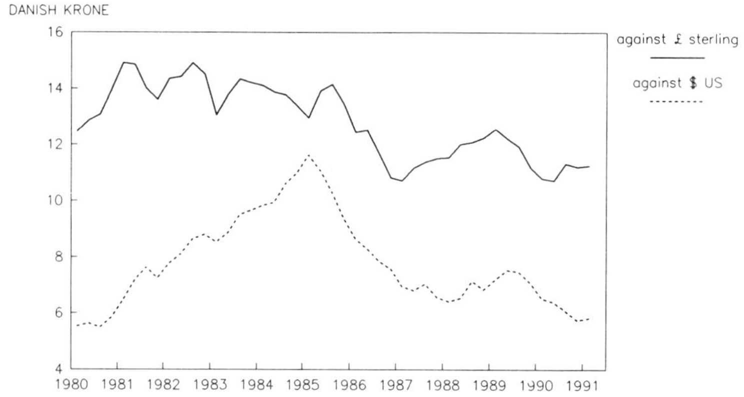

Inflation

General inflation has been measured using consumer or retail price indices. These reflect price changes in a basket of goods and services weighted according to the spending patterns of a typical family. Weights are changed periodically, and new items inserted. General inflation indices usually rise and, in so doing erode the purchasing power of a given currency unit. Other measures of inflation tend to be related to specific items. The two most commonly prepared for the construction industry are discussed below.

Building cost and tender price indices

Cost and tender price indices measure different types of inflation which occur within the construction industry. Building costs are the costs actually incurred by a contractor in the course of his business, the major ones being labour and materials; tender prices are the prices for which a contractor offers to erect a building. Tender prices include building costs but also take into account the prevailing market situation. When there is plenty of construction work tender prices may increase at a greater rate than building costs while, when work is scarce, tender prices may actually fall even if building costs are rising.

Most countries have building cost indices - the method of compilation is generally relatively simple, basically comprising a weighted basket of the main inputs to construction. Rather fewer countries have tender price indices - their method of compilation is more complex usually involving a detailed analysis of accepted tenders for construction work and comparing these with a common base schedule of prices.

USEFUL ADDRESSES

At the end of each country section, a list of addresses is given. This usually comprises main government, contracting, professional, standards and research organizations involved in the construction industry.

5. Individual countries

Austria

KEY DATA

Population

Population (1989)	7.61m
Urban population	57%
Population under 15	17%
Population over 65	15%
Average annual growth rate (1983 to 1988)	0.1%

Geography

Land area	84 000 km²
Agricultural area	32 374 km²
Capital city	Vienna (population 1 531 346)

Economy

Monetary unit	Schilling (ASch)
Exchange rate (January 1991) to	
the ECU	ASch 14.4
the pound sterling	ASch 20.6
the US dollar	ASch 10.6
Average annual inflation (1984 to 1989)	2.2%
Gross Domestic Product (GDP)	ASch 1 567bn
GDP per capita (1989)	ASch 205 900
GDP growth (real) (1980 to 1988)	1.8%
Private consumption (% of total)	56.4%
Public consumption (% of total)	17.4%
Investment (% of total)	28.2%
Central government expenditure as percentage of GDP	33.2%

Construction

Total value of construction output (1989)	Asch 264bn
Construction output per capita	Asch 34 690
Construction output as a percentage of GDP	16.8%

THE CONSTRUCTION INDUSTRY

Construction output

In 1989 the value of the gross output of the Austrian construction industry was about ASch264 billion or 18.1 billion ECUs. This represents 16.8% of GDP. The breakdown by type of work is shown below.

Construction output 1989

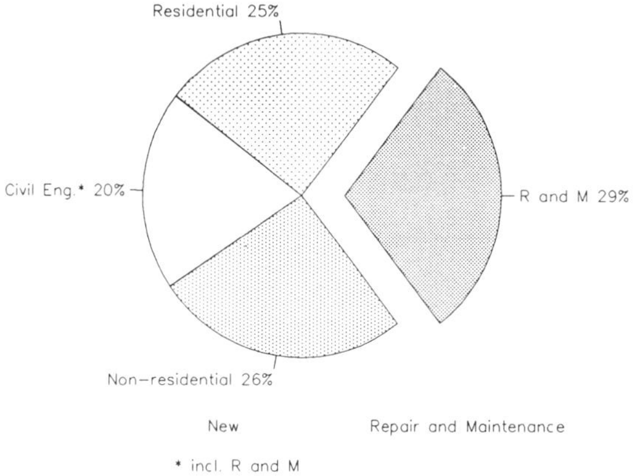

The construction sector is expected to continue growing in the next year or so.

The geographical distribution of construction work in 1989 compared with that of population is shown below.

Population and construction output 1989: regional distribution

	Population (%)	*Building and civil engineering work (%)*
Burgenland	3.5	1.8
Kärnten	7.1	7.2
Niederösterreich	18.8	18.3
Oberösterreich	17.1	14.6
Salzburg	6.1	8.0
Steiermark	15.5	13.3
Tirol	8.1	8.1
Vorarlberg	4.2	3.8
Wien	19.5	24.9
	100.0	100.0

Source: *Beiträge zur Östereichischen Statistic,* *Baustatisik 1989 and Demographisches Jahrbuch 1989*

Characteristics and structure of the industry

In 1989 there were about 3000 contracting firms including 10 with 1000 employees and over who did 15% of the work. The 27 firms with 500 employees and over did 24% of the work. Most contracting is carried out on a trade-by-trade basis with overall management responsibility being the province of the *Bauleiter* (building leader) who is usually an architect. General contracting has, however, grown to take up approximately 9% of total building volume, relieving the architect of certain of his management and co-ordination responsibilities.

There are no Austrian companies in Building magazine's "Top 300 European Contractors" and there are only three in the Engineering News Record's "Top 250 International Contractors".

Major Austrian contractors

Major contractors	*Head office*	*Place in ENR's "Top 250 International Contractors"*	*Exports (% of turnover)*	*Main work types*
Universale-Bau AG	Vienna	165	15	Building and Civil engineering
Voest-Alpine Industrieanlagenbau GmbH	Linz	184	n.a.	Industrial and Petroleum
Wanmpex Handels AG	Vienna	213	n.a.	Building

Source: Engineering News Record

Austrian contractors have strong ties with Eastern European countries and are operating in Hungary and Czechoslovakia.

Clients and finance

At one time the amount of work financed directly or indirectly by government was about 60% of construction output but this is now changing and the emphasis is switching to private financing. Major public projects are increasingly managed by private organizations with private funding.

However, government offers incentives for investments at federal, provincial and local level. These are aimed mainly at employment generation and especially in areas in need of special assistance. In general, the central government assistance is in the form of loans to supplement the available commercial credit. An important example is that the *Östereichische Kommunal Kredit* AG grants loans for up to 35 years at a low rate of interest (8.75% in 1991) for the development of industrial sites. Provincial authorities have similar schemes. The local authorities provide help often in the form of accommodation or sites at low rents or even free of charge.

Contractual arrangements

In the case of large contracts to be awarded by central or local government agencies, public advertisement and open tendering is generally the method of obtaining competitive tenders. Private sector work is usually awarded as a result of selective tendering, tenders typically being invited from between three and five firms. In the public sector, contracts are invariably awarded to the lowest tenderer. In the private sector, particularly where design is at a less advanced stage, the capability and suitability of the tenderer may, on occasion, be considered more important than his tender bid alone.

Standard conditions of contract exist but are revised as appropriate for public works contracts thus, in effect, every government department has its own conditions of contract. Similarly, where an architect is appointed to handle the project management of private sector work, contract conditions, though derived from published conditions, are produced on a "one-off" basis. Contract documents consist of general and special conditions of contract, drawings and bills of quantities. Specifications and quantities of work are often approximate and subject to remeasurement by the architect. No standard methods of measurement are published though a degree of uniformity does exist among offices.

The successful contractor is selected by the architect and the building owner, but it is the architect who places the order for work and therefore enters into a contractual relationship with the contractor.

During construction, interim payments are made as detailed in the contract and, although not standardized, a retention of 5% to 10% of the monies due is usually held. No advance payments are made. On completion, usually 5% is retained by the architect during a defects liability period. This period is typically two years long though this retention may be released earlier if a letter of warranty is received from a bank. The final account is prepared by the contractor and checked by the architect, taking into account remeasured quantities. Variations are valued using the tender information as a basis.

In the event of a contractual dispute arising, a settlement will be reached as laid down in the contract. Typically this is reached in the first instance by a decision from a prominent architect. Failing this, an arbitrator may be nominated by both parties, or ultimately, by the courts.

Development control and standards

Building regulations covering the satisfactory construction and maintenance of buildings - including requirements for energy conservation - are laid down and administered by the individual federal provinces, though they tend to follow a common pattern. The conditions laid down are usually in the form of functional requirements. The regulations for civil engineering are different from those for building but the two are complementary. For many types of civil works, e.g. road building, power stations, military works mining, etc. the regulations are not under the control of local authorities.

The Austrian Standards Institute attributes the classification "ONORM" to the standards that it establishes. The standards are, in principle, recommendations but they may be declared legally binding or as "deemed to satisfy" either in whole or in part through reference or inclusion in laws or statutory regulations. The Standards Institute may also recommend foreign or international standards.

CONSTRUCTION COST DATA

Cost of labour

The figures below are typical of labour costs in Vienna as at January 1991. The "wage rate" is the basis of an employee's income, while the "cost of labour" indicates the cost to a contractor of employing that employee. The difference between the two covers a variety of mandatory and voluntary contributions - a list of items which could be included is given in section 4.

	Wage rate (per hour) ASch	*Cost of labour (per hour) ASch*
Unskilled labour	70.2	220
Semi-skilled labour	76.4	240
Skilled labour	86.1	270

Cost of materials

The figures that follow are the costs of main construction materials, delivered to site in the Vienna area, as incurred by contractors in January 1991. These assume that the materials would be in quantities as required for a medium sized construction project and that the location of the works would be neither constrained nor remote.

All the costs in this section exclude value added tax (VAT - see below).

	Units	*Cost ASch*
Cement and aggregate		
Ordinary portland cement in 50kg bags	tonne	1250
Coarse aggregates for concrete	m^3	200
Ready mixed concrete (20N/mm^2)	m^3	1000
Ready mixed concrete (30N/mm^2)	m^3	1300
Steel		
Mild steel reinforcement	tonne	10 000
Bricks and blocks		
Medium quality facing bricks	1000	8360
Hollow concrete blocks	1000	13 100
Timber		
Softwood for carpentry	m^3	3160
Softwood internal door complete with frames and ironmongery	nr	960
Glass and ceramics		
Float glass (4mm)	m^2	1080
Sealed double glazing units	m^2	3300

Approximate estimating

The building costs per unit area given below are averages incurred by building clients for typical buildings in the Vienna area as at January 1991. They are based upon the total floor area of all storeys, measured between external walls and without deduction for internal walls.

Approximate estimating costs generally include for mechanical and electrical installations but exclude furniture, loose or special equipment, and external works; they also exclude fees for professional services. The costs shown are for specifications and standards appropriate to Germany and this should be borne in mind when attempting comparisons with similarly described building types in other countries. A discussion of this issue is included in section 4. Comparative data for countries covered in this publication, including construction cost data are presented in section Part Three.

Approximate estimating costs must be treated with reserve; they cannot provide more than a rough guide to the probable cost of building.

All the rates in this section exclude value added tax (VAT - see below).

	Cost m^2 *ASch*	*Cost* ft^2 *ASch*
Industrial buildings		
Factories for letting	8000	740
Warehouses, low bay (6 to 8m high) for letting	6500	600
Administrative and commercial buildings		
Offices for letting, 5 to 10 storeys	18 000	1680
Offices for owner occupation, 5 to 10 storeys	27 000	2515
Residential buidlings		
Private sector apartment building (standard specification)	13 200	1220

Value added tax (VAT)

The standard rate of value added tax (VAT) is currently 20%. VAT is chargeable on general building work.

EXCHANGE RATES AND INFLATION

The combined effect of exchange rates and inflation on prices within a country and price comparisons between countries is discussed in section 4.

The graph below plots the movement of the Austrian Schilling against sterling and the US dollar since 1980. The figures used for the graph are quarterly and the method of calculating these, and other related issues are discussed in section 4. The exchange rate at January 1991 was ASch 20.6 to the pound sterling and ASch 10.6 to the US dollar.

The Austrian Schilling against Sterling and the US Dollar

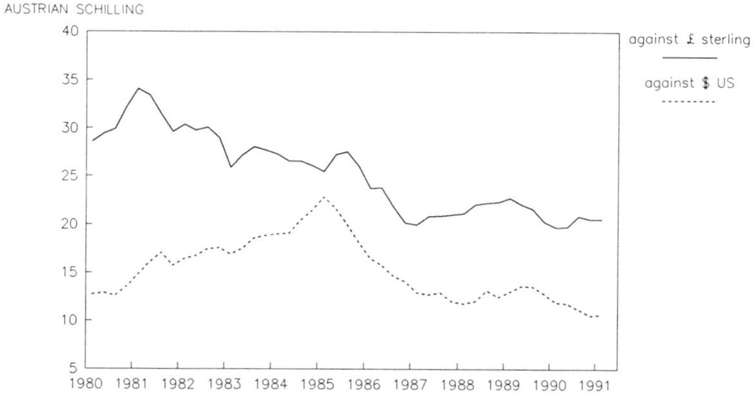

Inflation

The graph and the table below present the consumer price index showing the trend in inflation since 1980. The index has been rebased to 1982=100. Notes on inflation and the distinction between costs and prices are included in section 4.

Consumer price inflation

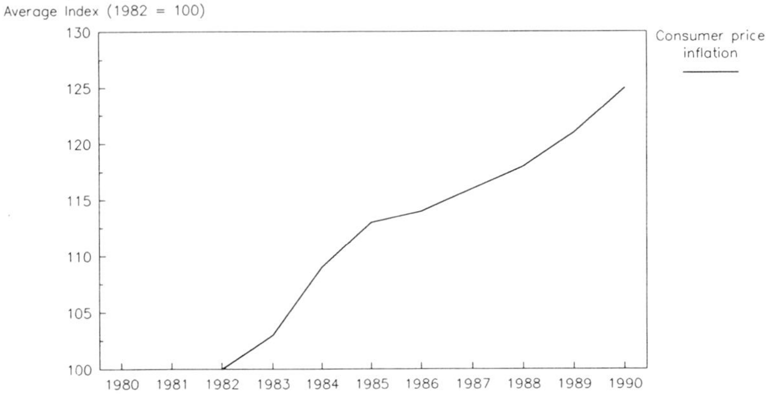

The basis of the figures is the official consumer price index converted to quarterly averages. Between 1984 and 1987 the annual rate of inflation fell steadily, but rose in 1989 and again in 1990.

Notes on inflation are included in section 4.

Consumer price inflation 1982 = 100

	1983	1984	1985	1986	1987	1988	1989	1990
average index	103	109	113	115	116	118	122	126
annual change (%)	3.3	5.6	3.2	1.7	1.4	2.0	2.6	3.3

USEFUL ADDRESSES

BM fuer wirtschaftliche Angelegenheiten
(Federal Ministry of Economic Affairs)
Abt. Bauten und Technik, Stubenring 1
A-1010 Wien
Tel: 1-71100
Fax: 1-71393/11

BM fuer oeffenliche Wirtschaft und Verkehr
(Federal Ministry of Public Economy and Traffic)
Radetzkystrasse 2
A-1030 Wien
Tel: 1-71162
Fax: 1-713 7876

Technische Universitaet Wien
Karlsplatz 13
A-1040 Wien
Tel: 1-58801
Fax: 1-587 8905

Technische Universitaet Graz
Rechbauerstrasse 12
A-8020 Graz
Tel: 316-873 6111

Fachverband der Bauindustrie
Engelsbergaasse 4/III
A-1030 Wien
Tel: 1-713 6504
Fax: 713 6504/19

Ingenieurkammer, Karlsplatz 13
A-1040 Wien
Tel: 1-505 1781
Fax: 1-505 1005

Magistratsabteilung 37 - Baupolizei
Dresdner Strasse 25
A-1200 Wien
Tel: 1-356611

Oesterreichisches Normungsinstitut
Heinestrasse 38
A-1021 Wien
Tel: 1-267535
Fax: 1-267552#

Statistisches Zentralamt, Hintere Zollamtsstrasse 2b,
A-1030 Wien
Tel: 1-71128
Fax: 1-71128/7728

Verband Beratender Ingenieure Osterreichs, VBIO
Karlsgasse 9
1040 Wien
Tel: (222) 651 781

Verband Osterreichischer Beton und Fertigteilwerke
1 Hasnerstrasse 36/1/4, 4020 Linz/Donau
Bundesingenieurkammer, Bundesfachgruppe - Architekten
(International Union of Architects)
Karlsgasse 9
A-1040 Wien

Belgium

KEY DATA

Population

Population (1989)	9.9m
Urban population	97%
Population under 15	18.1%
Population over 65	14.7%
Average annual growth rate (1983 to 1988)	0.1%

Geography

Land area	33 000 km^2
Agricultural area	15 000 km^2
Capital city	Brussels
	(population 976 536)

Economy

Monetary unit	Belgian Francs (Bfr)
Exchange rate (January 1991) to	
the ECU	Bfr 42.2
the pound sterling	Bfr 60.3
the US dollar	Bfr 31.2
Average annual inflation (1984 to 1989)	2.4%
Gross Domestic Product (GDP)	Bfr 5604bn
GDP per capita (1989)	Bfr 566 000
GDP growth (real) (1980 to 1988)	1.5%
Private consumption (% of total)	64.0%
Public consumption (% of total)	16.0%
Investment (% of total)	18.0%
Central government expenditure as percentage of GDP	50.6%

Construction

Total value of construction output (1989)	Bfr 772bn
Construction output per capita	Bfr 77 979
Construction output as a percentage of GDP	13.8%

THE CONSTRUCTION INDUSTRY

Construction output

The value of gross construction output in Belgium in 1989 was FB 772 billion or 17.8 billion ECUs. This represents 13.8% of GDP. Output in the early 1980s fell substantially but since 1986 has greatly improved. The level of output is now expected to stabilize. The breakdown by type of work is shown below.

Construction output 1989

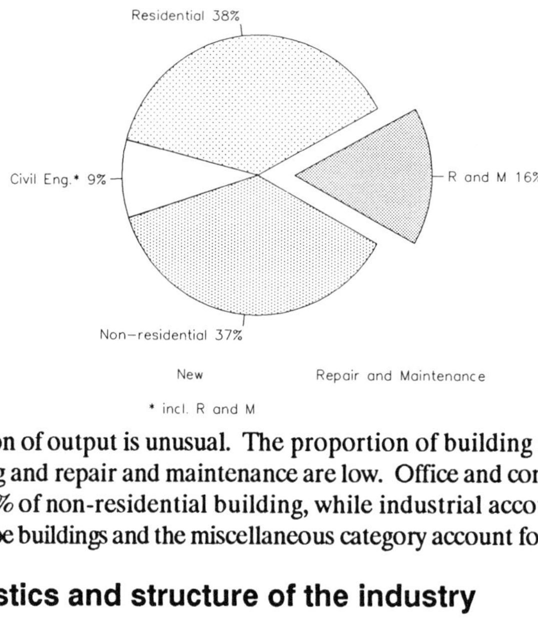

This distribution of output is unusual. The proportion of building is very high while civil engineering and repair and maintenance are low. Office and commercial buildings account for 55% of non-residential building, while industrial accounts for a further 30%. Social-type buildings and the miscellaneous category account for 15% of this total.

Characteristics and structure of the industry

The title of engineer is protected by law in relation to the holding of the appropriate higher educational diploma. Some engineers are employed in state organizations; others are employed in engineering *bureaux* or operate as freelance consultants. The choice of a *bureau* for a particular project is influenced by political factors, and by their regional location (partly because of language).

The title of architect is also protected. It is obligatory, according to laws of February 1939 and 26 June 1963, to employ an architect for design and execution of works for which planning permission is necessary. The *Ordre des Architectes* is responsible for maintaining a register of qualified architects. The size of practices is generally small: e.g. a practice of 9 to 12 architects would be considered fairly large.

Contractors are required to be registered by law. In 1985 there were about 1,350 firms in the industry. Only 9 firms had more than 500 employees and these did 13% of the work. Over the past few years there has been a fall in the number of construction firms. In 1989 Belgium had two firms in the top 100 European firms and five in the top 300.

Major Belgian contractors

Major contractors	*Place in Building's "Top 300 European Contractors"*	*Exports (% of turnover)*
SBB	96	20-49
CFE	99	Over 50
Franki	135	20-49
De Nul	261	n.a.
Dredging International	262	n.a.

Source: Building

There are two Belgian companies in Engineering News Record's "Top 250 International Contractors" and both are in specialized fields outside normal building and civil engineering.

Clients and finance

Housing in Belgium is often produced by co-operatives. Low cost housing, designed and operated by communal low cost housing companies, is financed partly by the regions. There are private single family houses built by individuals and housing units built by real estate developers.

Since regionalization, the communities have assumed responsibility for education, culture, health care and social services. The services of the government controlled company which manages and finances new projects - the *Régie des Bâtiments* - have been dispersed between the various regions.

Selection of design consultants

Public sector design work is often undertaken by in-house design teams. Alternatively, if an external designer is employed, he is selected on his reputation via a wide ranging interview process. It is important for architects to be well known. Competitions are rare.

In the private sector the process for selection of designers is varied but is normally based on reputation and past association with the client.

Contractual arrangements

Open tendering is the method of contractor selection used for Government work and the Government is bound to pay damages if the lowest tenderer's offer is not accepted. Contractors cover the risks involved in open tendering by a well established system of price adjustment before tenders are submitted; tenders are adjusted to the arithmetical average and allowance is also made for the cost of tendering. There can also be a considerable amount of post-tender negotiation. It is normal to employ a general contractor who may sublet work to specialist trades; however, this does not relieve him of his obligations to the client and he is directly responsible for the trades work contracts. There is also normally a fluctuations clause that involves adjustments by a pre-agreed formula based upon monthly price indices published by the Government Statistical Office.

Selective tendering is most common in private work but negotiated tendering is on the increase. Another method, mainly for private work is where contractors submit unit prices for labour, materials and plant with a percentage addition for on-costs, and valuations are based on an assessment of resources used.

There is no generally accepted standard form of contract in use, although a standard form has been issued by the *Federation Royale des Societes d'Architectes* and agreed by the *Confederation Nationale de la Construction.* The three most frequently used contracts, defined according to the method of establishing the cost of the works, are:

- *Marché à Forfait Absolu* (the fixed lump sum contract) with no provision for variations

- *Marché à Forfait Relatif* which includes a variation and fluctuation clause and a remeasurement contract

- *Marché à Borderaux de Prix* based upon a schedule of rates.

There are also government standard forms of contract for public works and dwellings built with government subsidies.

Liability and insurance

Both the contractor and the architect are jointly responsible for the condition of a building for a period of ten years. In addition, it is necessary to be insured for injury and accidents.

Insurance is obligatory, although the amount of the insurance is not stipulated. Moreover, it is advisable that the client or his representative (e.g. the architect) ensures that, at least for larger projects, all the parties are insured. The client will often take out a blanket insurance to cover all parties. Insurance is considerably cheaper than in France and probably adds about 1%, or a little more, to the cost of a project.

Development control and standards

The planning system is largely controlled by the three regional governments. The basic rules and regulations for obtaining planning consent are well formulated and if they are adhered to planning permission will normally be given.

A planning application must be signed by a registered architect. Moreover, the *Ordre des Architectes* receives a copy of every building application to check that the architect is in fact registered. The decision on granting a planning application is taken by the mayor. There are, however, frequently specific local peculiarities and early consultation with the mayor is advisable.

Building control in Belgium is also exercised through local authorities, each of which is authorized to issue by-laws on planning, and on safety and hygiene which may affect design and equipment. There are no national laws covering technical requirements for building construction, though the local authority may request submission of technical calculations and drawings. There are technical controllers, (*bureau de contrôle*) who are employed by the client alone. The quality of their advice is high.

The *Institut de Normalisation* centralizes regulations produced by several institutions and produces a large number of standards for construction. Standards produced by the Institute are not, in general, mandatory unless referred to by a body with legal authority.

The Inspectorate for the Safety of Building in Belgium (SECO) carries out the technical inspection of plans, calculations, construction, and of products used in construction. It operates its own (SECO) approval scheme for certain products; it also operates two schemes in conjunction with other organizations. One of these schemes is the SECO-CSTC (*Le Centre Scientifique et Technique de la Construction* : the Scientific and Technical Centre for Building) scheme which is administered jointly by SECO and CSTC; the other is the BENOR (*l'Institut Belge de Normalisation*) approval scheme which is sponsored by the Belgian Standards Institution. SECO tests products in its own laboratories; SECO inspectors will then make periodic visits to manufacturers' works to check the production of approved products, which must be subjected to quality control by the manufacturer.

For new products or products for which there are no Belgian standards, *Agrément* certificates issued by the National Housing Institute are widely accepted.

Research and development

The CSTC is a non-profit making organization financed by government grants and a levy on the building industry. It conducts research into building construction, components and materials. It provides an information service and publishes research reports and information notes : the notes contain recommendations and rules for good practice.

Further reading

Association of Consulting Engineers et al, Engineering Consultancy in the European Community: Belgium and Luxembourg, London: Telford 1990.

CONSTRUCTION COST DATA

Cost of materials

The figures that follow are the costs of main construction materials, delivered to site in the Bruxelles area, as incurred by contractors in January 1991. These assume that the materials would be in quantities as required for a medium sized construction project and that the location of the works would be neither constrained nor remote.

	Units	*Cost Bfr*
Cement and aggregate		
Ordinary portland cement in 50kg bags	tonne	4000
Coarse aggregates for concrete	m^3	650
Fine aggregates for concrete	m^3	750
Ready mixed concrete (250kg Cement/m^3)	m^3	2520
Steel		
Mild steel reinforcement	tonne	33 000
High tensile steel reinforcement	tonne	34 800
Structural steel sections	tonne	25 500
Bricks and blocks		
Common bricks (215 x 100 x 55mm)	1000	4500
Good quality facing bricks (210 x 100 x 50mm)	1000	6000
Hollow concrete blocks (290 x 140 x 190mm)	1000	18 500
Solid concrete blocks (210 x 150 x 85mm)	1000	11 200
Precast concrete cladding units with exposed aggregate finish	m^2	7500

	Units	*Cost Bfr*
Timber and insulation		
Softwood sections for carpentry	m^3	15 000
Softwood for joinery	m^3	21 000
Hardwood for joinery	m^3	24 500
Exterior quality plywood (18mm)	m^2	1400
Plywood for interior joinery (18mm)	m^2	850
Softwood strip flooring	m^2	500
Chipboard sheet flooring	m^2	160
100 mm thick quilt insulation (55kg/m^3 density)	m^2	110
100 mm thick rigid slab insulation	m^2	150
Softwood internal door complete with frames and ironmongery	nr	9500
Glass and ceramics		
Sheet/float glass (5 mm)	m^2	800
Good quality ceramic wall tiles (150 x 150)	m^2	850
Plaster and paint		
Plasterboard (12.5mm thick)	m^2	130
Emulsion paint in 5 litre tins	litre	110
Gloss oil paint in 5 litre tins	litre	200
Tiles and paviors		
Vinyl floor tiles (300 x 300 x 1.5mm)	m^2	500
Precast concrete roof tiles	1000	26 000
Drainage		
WC suite complete	nr	14 000
Lavatory basin complete	nr	8000
100mm diameter clay drain pipes	m	250
150mm diameter cast iron drain pipes	m	1100

Unit rates

The descriptions below are generally shortened versions of standard descriptions listed in five languages (English, French, Italian, German and Spanish) in section 6. Where an item has a two digit reference number (e.g. 05 or 33), this relates to the full description against that number in section 6. Where an item has an alphabetic suffix (e.g. 12A or 34B) this indicates that the standard description has been modified. Where a modification is major the complete modified description is included here and the standard description should be ignored; where a modification is minor (e.g. the insertion of a named hardwood) the shortened description has been modified here but, in general, the full description in section 6 prevails.

The unit rates below are for main work items in a typical construction project in the Bruxelles area in January 1991. The rates include all necessary labour, materials, equipment and, where appropriate, allowances for contractors overheads and profit, preliminary and general items and contractors profit and attendance on specialist trades. All the rates in this section exclude value added tax (VAT - see below).

		Unit	Rate Bfr
Excavation			
01	Mechanical excavation of foundation trenches	m^3	170
02	Hardcore filling making up levels	m^2	135
03	Earthwork support	m^2	250
Concrete work			
04	Plain insitu concrete in strip foundations in trenches	m^3	4100
05	Reinforced insitu concrete in beds	m^3	3900
06	Reinforced insitu concrete in walls	m^3	4800
07	Reinforced insitu concrete in suspended floor or roof slabs	m^3	3900
08	Reinforced insitu concrete in columns	m^3	5200
09	Reinforced insitu concrete in isolated beams	m^3	4600
10	Precast concrete slab	m^2	3650
Formwork			
11A	Metal formwork to concrete walls	m^2	3200
12A	Softwood formwork to concrete columns	m^2	3600
13A	Metal formwork to horizontal soffits of slabs	m^2	3000
Reinforcement			
14	Reinforcement in concrete walls	tonne	42 000
15	Reinforcement in suspended concrete slabs	tonne	55 000
16	Fabric reinforcement in concrete beds	m^2	260
Steelwork			
17	Fabricate, supply and erect steel framed structure	tonne	80 000
Brickwork and blockwork			
19	Solid (perforated) concrete blocks	m^2	2500
21	Facing bricks	m^2	3000
Roofing			
22	Concrete interlocking roof tiles 430 x 380mm	m^2	900
26	Particle board roof coverings	m^2	1100
27	3 layers glass-fibre based bitumen felt roof covering	m^2	1200
28	Bitumen based mastic asphalt roof covering	m^2	1000
29	Glass-fibre mat roof insulation 160mm thick	m^2	750
31	Troughed galvanized steel roof cladding	m^2	1850
Woodwork and metalwork			
34	Single glazed casement window in hardwood, size 650 x 900mm	m^2	16 000
35	Two panel glazed door in hardwood, size 850 x 2000mm	each	18 500

		Unit	*Rate Bfr*
36	Solid core half hour fire resisting hardwood internal flush doors, size 800 x 2000mm	each	10 000
37	Aluminium double glazed window, size 1200 x 1200mm	m^2	18 000
38	Aluminium double glazed door, size 850 x 2100mm	each	14 000
40	Framed structural steelwork in universal joist sections	tonne	130 000
41	Structural steelwork lattice roof trusses	tonne	180 000

Electrical work

53	13 amp unswitched socket outlet	each	1800
54	Flush mounted 20 amp, 1 way light switch	each	2500

Finishings

55	2 coats gypsum based plaster on brick walls	m^2	500
56	White glazed tiles on plaster walls	m^2	1500
58	Cement and sand screed to concrete floors	m^2	450
59	Thermoplastic floor tiles on screed	m^2	800
60A	Mineral fibre tiles on concealed suspension system	m^2	1100

Glazing

61	Glazing to wood	m^2	2000

Painting

62	Emulsion on plaster walls	m^2	300
63	Oil paint on timber	m^2	450

Painting

62	Emulsion on plaster walls	m^2	300
63	Oil paint on timber	m^2	450

Approximate estimating

The building costs per unit area given below are averages incurred by building clients for typical buildings in the Bruxelles area as at January 1991. They are based upon the total floor area of all storeys, measured between external walls and without deduction for internal walls.

Approximate estimating costs generally include for mechanical and electrical installations but exclude furniture, loose or special equipment, and external works; they also exclude fees for professional services. The costs shown are for specifications and standards appropriate to Belgium and this should be borne in mind when attempting comparisons with similarly described building types in other countries. A discussion of this issue is included in section 4. Comparative data for countries covered in this publication, including construction cost data, are presented in Part Three.

Approximate estimating costs must be treated with reserve; they cannot provide more than a rough guide to the probable cost of building.

	Cost m^2 *Bfr*	*Cost* ft^2 *Bfr*
Industrial buildings		
Factories for letting	19 000	1750
Factories for owner occupation (light industrial use)	24 000	2250
Factories for owner occupation (heavy industrial use)	32 000	2950
Factory/office (high-tech) for letting (shell and core only)	42 000	3900
Factory/office (high-tech) for letting (ground floor shell, first floor offices)	50 000	4650
Factory/office (high tech) for owner occupation (controlled environment, fully finished)	64 000	5950
Warehouses, low bay (6-8m high) for letting (no heating)	16 000	1500
Warehouses, low bay for owner occupation (including heating)	20 000	1850
Warehouses, high bay for owner occupation (including heating)	25 000	2300
Administrative and commercial buildings		
Civic offices, non air conditioned	40 000	3700
Civic offices, fully air conditioned	47 000	4350
Offices for letting, 5-10 storeys, non air conditioned	40 000	3700
Offices for letting, 5-10 storeys, air conditioned	52 000	4850
Offices for letting, high rise, air conditioned	59 000	5500
Offices for owner occupation 5-10 storeys, non air conditioned	44 000	4100
Offices for owner occupation 5-10 storeys, air conditioned	59 000	5500
Offices for owner occupation, high rise, air conditioned	66 000	6150
Prestige/headquarters office, 5-10 storeys, air conditioned	70 000	6500
Prestige/headquarters office, high rise, air conditioned	75 000	6950
Health and education buildings		
General hospitals (400 beds)	80 000	7450
Teaching hospitals (600 beds)	90 000	8350
Private hospitals (200 beds)	85 000	7900
Health centres	46 000	4250
Secondary/middle schools	42 000	3900
University (science) buildings	45 000	4200
Recreation and arts buildings		
Theatres (over 500 seats) including seating and stage equipment	72 000	6700
Concert halls including seating and stage equipment	70 000	6500
Sports halls including changing and social facilities	69 000	6400

	Cost m^2 Bfr	Cost ft^2 Bfr
Swimming pools (international standard) including changing and social facilities	38 000	3550
Swimming pools (schools standard) including changing facilities	35 000	3250
City centre/central libraries	42 000	3900
Residential buildings		
Social/economic single family housing (multiple units)	27 000	2500
Private/mass market single family housing 2 storey detached/semidetached (multiple units)	33 000	3050
Purpose designed single family housing 2 storey detached (single unit)	48 000	4450
Social/economic apartment housing, low rise (no lifts)	22 000	2050
Social/economic apartment housing, high rise (with lifts)	28 000	2600
Private sector apartment building (standard specification)	32 000	2950
Private sector apartment buildings (luxury)	38 000	3550
Hotel, 5 star, city centre	78 000	7250
Hotel, 3 star,city/provincial	66 000	6150

Regional variations

The approximate estimating costs are based on projects in Bruxelles. Adjust these cost by the following factors for regional variations:

Antwerp :+ 3%
Liege : - 5%

Value added tax (VAT)

Standard rates of value added tax (VAT) are currently 17% and 19%. VAT is chargeable at these standard rates on new construction work and consultants fees: 19% on supply items and 17% on supply and fix items.

EXCHANGE RATES AND INFLATION

The combined effect of exchange rates and inflation on prices within a country and price comparisons between countries is discussed in section 4.

Exchange rates

The graph below plots the movement of the Belgian Franc against sterling and the US dollar since 1980. The figures used for the graph are quarterly and the method of calculating these, and other related issues are discussed in section 4. The exchange rate at January 1991 was Bfr 60.3 to the pound sterling and Bfr 31.2 to the US dollar.

Exchange rates

The Belgian Franc against Sterling and the US Dollar

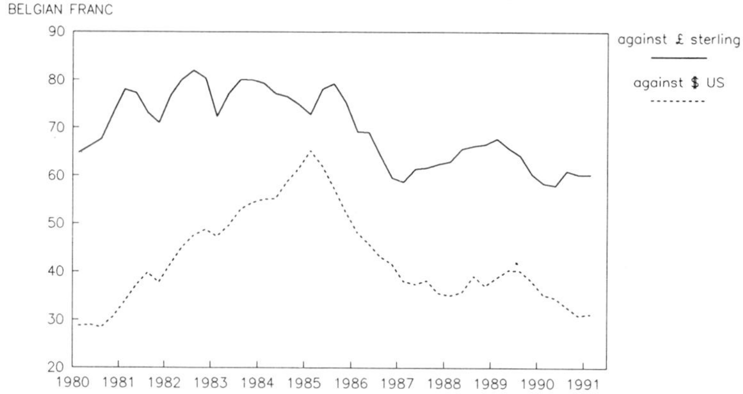

Inflation

The graph and the table below present the consumer price index showing the trend in inflation since 1981. The index has been rebased to 1981=100. Notes on inflation and the distinction between costs and prices are included in section 4.

Consumer price inflation

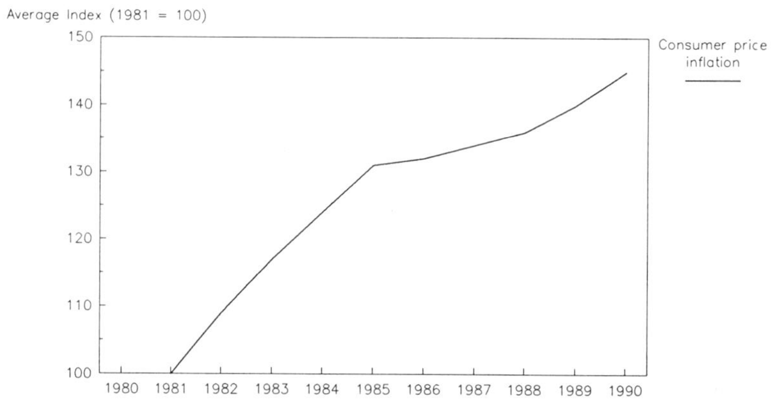

The basis of the figures is the official consumer price index converted to quarterly averages. Inflation peaked in 1982. Efforts to reduce inflation have succeeded in holding the annual rate to below 5% since 1985.

Notes on inflation are included on section 5.

Consumer price inflation 1981 = 100

	1982	1983	1984	1985	1986	1987	1988	1989	1990
average index	109	117	125	131	132	134	136	140	145
annual change (%)	8.7	7.7	6.3	4.9	1.3	1.6	1.2	3.1	3.5

USEFUL ADDRESSES

Le Ministère des Travaux Publics
Rue de la Roi 207
1040 Bruxelles

Centre Scientifique et Technique de la Construction (CSTC)
Rue du Lombard 41
1000 Bruxelles
Tel: (2) 511 0633

Confédération Nationale de la Construction (CNC)
Rue du Lombard 34 - 42
B-1000 Bruxelles
Tel: (2) 513 6532

Fédération Royale des Societies d'Architectes de Belgique
21 Rue Ernest Allard
Conseils de Belgique CICB
Hotel Ravenstein, 3 Rue Ravenstein
1000 Bruxelles
Tel: (2) 511 5294

Confédération Nationale de la Construction
Rue du Lombard 34-bte 42
1000 Bruxelles

Fédération des Entreprises de Belgique
Rue Ravenstein 4
B-1000 Bruxelles
Tel: (2) 511 5880

Institut Belge de Normalisation
Avenue de la Brabanconne 29
B-1040 Bruxelles

Belgische Bouwcentrum VZW
Centre Belge du Bâtiment ASBL
Rue de la Fourche 17-19
1000 Bruxelles

National Bouwcentrum
Centre National du Bâtiment
Jan Van Rijswijcllaan 191
2020 Antwerpen

Cyprus

KEY DATA

Population

Population (1989)	0.70m
Urban population	n.a.
Population under 15	n.a.
Population over 65	10.3%
Average annual growth rate (1983 to 1988)	1.2%

Geography

Land area	9000 km^2
Agricultural area	n.a.
Capital city	Nicosia (population 167 000)

Economy

Monetary unit	Cyprus Pound (C£)
Exchange rate (January 1991) to	
the ECU	C£0.583
the pound sterling	C£0.833
the US dollar	C£0.431
Average annual inflation (1984 to 1989)	3.7%
Gross Domestic Product (GDP)	C£ 1.89bn
GDP per capita (1989)	C£ 2700
GDP growth (real) (1980 to 1988)	n.a.
Private consumption (% of total)	n.a.
Public consumption (% of total)	n.a.
Investment (% of total)	n.a.
Central government expenditure as percentage of GDP	n.a.

Construction

Total value of construction output (1989)	C£ 390 m
Construction output per capita	C£ 557
Construction output as a percentage of GNP	20.6%

THE CONSTRUCTION INDUSTRY

Construction output

The value of gross construction output in Cyprus in 1989 was C£390 million. In 1990 it was C£432 million or 18% of GNP. The breakdown by type of work is shown below.

Construction output in 1989

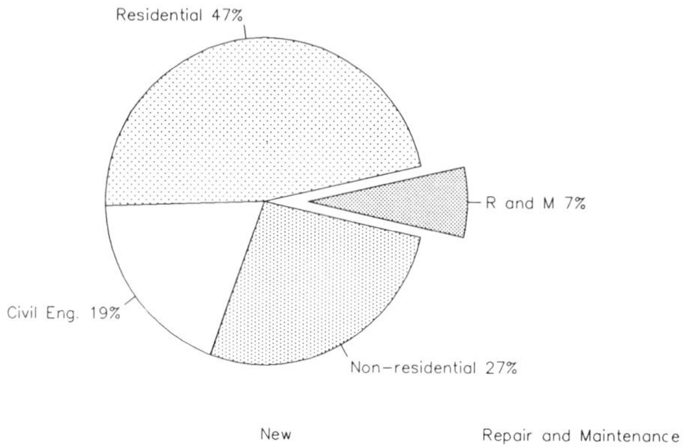

Tourism is a major industry in Cyprus and much construction has been tourist related.

The contribution of the public sector to construction activity is mainly in civil engineering. In 1988 only 8% of housing was in the public sector. The public sector will increase its share in the years to come as considerable infrastructure developmment is planned.

Structure of the industry

In 1990 there were 19 construction firms in Cyprus with over 50 employees. The top seven firms operate internationally, undertaking contracts mainly in Arab countries. One of these - J P Zachariades Ltd - is number 186 in the Engineering News Record's Top 250 International Contractors. It undertakes only building work.

The Cyprus Civil Engineers and Architects Regulation Board is charged, by the Architects and Civil Engineers Law of 1962, with the duty of maintaining suitable professional standards.

Contractual arrangements

Tenders for Government contracts are usually invited by public advertisement. Contractors must hold the appropriate licence for the class of work involved to be eligible to participate. In the private sector, most contracts are awarded as a result of competitive tenders received from a short list of firms selected by the client's consultants.

The following forms of contract are used in Cyprus:

- the standard form of contract issued by the Association of Architects and Civil Engineers and the Association of Building Contractors. This is the form most widely used on all types of construction projects and is based on a 1947 RIBA form. An updated version was introduced in May 1991
- the Government standard form. This is a form based on a 1930s edition of JCT and has survived from colonial times with almost no revision
- FIDIC (3rd Edition)
- the ICI form.

The tender documentation under the standard forms of contract (public and private) comprise:

- drawings
- specifications or schedules of works
- bills of quantities.

Contracts under the standard forms are fluctuating for increases in the cost of labour and involve the use of adjustment formulae and quarterly published indices.

The contractor and nominated subcontractors receive a 12% advance payment based on the contract value and proportionate deductions are made from subsequent payments. The client retains 5% of the contract sum until practical completion. The contractor and subcontractors provide the client with Performance and Advance Payment guarantees. The general contractor carries out the work with his own labour and subcontractors and with specific subcontractors appointed by the client's consultants.

Development control and standards

The only existing law is the Streets and Building Regulations Law which lays down certain standards, many of which are necessary to the proper functioning of buildings. Any building work is subject to a permit from the local district office or municipality. The application must be signed by an architect or engineer registered with the Cyprus Civil Engineers and Architects Regulation Board. Examination of the submission is undertaken by trained building inspectors who also inspect the building work at stages throughout the operations.

The only control on quality of materials is that imposed by the manufacturers.

CONSTRUCTION COST DATA

Cost of labour

The figures below are typical of labour costs in Nicosia as at January 1991. The "wage rate" is the basis of an employee's income, while the "cost of labour" indicates the cost to a contractor of employing that employee. The difference between the two covers a variety of mandatory and voluntary contributions - a list of items which could be included is given in section 4.

	Wage rate (per hour) C£	*Cost of labour (per hour) C£*	*Number of hours worked per year*
Site operatives			
Mason/bricklayer	3.20	4.48	1960
Carpenter	3.20	4.48	1960
Plumber	3.20	4.48	1960
Electrician	3.20	4.48	1960
Structural steel erector	4.60	6.44	1960
HVAC installer	3.20	4.48	1960
Semi-skilled worker	2.90	4.06	1960
Unskilled labourer	2.60	3.64	1960
Equipment operator	3.30	4.62	1960
Watchman/security	2.60	3.64	1960
Site supervision			
General foreman	3.80	5.32	1960
Trades foreman	3.50	4.90	1960
Clerk of works	3.80	5.32	1960
Contractors' personnel			
Site manager	6.00	7.92	1960
Resident engineer	5.00	6.60	1960
Resident surveyor	4.50	5.94	1960
Junior engineer	3.00	3.96	1960
Junior surveyor	3.00	3.96	1960
Planner	4.00	5.28	1960
Consultants' personnel			
Senior architect	5.00	6.60	1960
Senior engineer	5.00	6.60	1960
Senior surveyor	5.00	6.60	1960
Qualified architect	3.00	3.96	1960
Qualified engineer	3.00	3.96	1960
Qualified surveyor	3.00	3.96	1960

Cost of materials

The figures that follow are the costs of main construction materials, delivered to site in the Nicosia area, as incurred by contractors in January 1991. These assume that the materials would be in quantities as required for a medium sized construction project and that the location of the works would be neither constrained nor remote.

	Units	*Cost C£*
Cement and aggregate		
Ordinary portland cement in 50kg bags	tonne	27.00
Coarse aggregates for concrete	m^3	4.75
Fine aggregates for concrete	m^3	7.00
Ready mixed concrete ($25N/mm^2$)	m^3	22.00
Ready mixed concrete ($20N/mm^2$)	m^3	20.00
Steel		
Mild steel reinforcement	tonne	185.00
High tensile steel reinforcement	tonne	185.00
Structural steel sections	tonne	340.00
Bricks and blocks		
Common bricks (300 x 200 x 100mm)	1000	105.00
Good quality facing bricks (200 x 100 x 75mm)	1000	300.00
Hollow concrete blocks (400 x 200 x 150mm)	1000	300.00
Precast concrete cladding units with exposed aggregate finish	m^2	24.00
Timber and insulation		
Softwood sections for carpentry	m^3	180.00
Softwood for joinery	m^3	230.00
Hardwood for joinery	m^3	650.00
Exterior quality plywood (20 mm)	m^2	3.40
Plywood for interior joinery (20 mm)	m^2	3.40
Softwood boarding (12 mm)	m^2	8.00
Chipboard sheet flooring (12 mm)	m^2	1.60
100 mm thick quilt insulation	m^2	3.50
100 mm thick rigid slab insulation	m^2	4.50
Softwood internal door complete with frames and ironmongery	nr	110.00
Glass and ceramics		
Float glass (6 mm)	m^2	10.50
Sealed double glazing units	m^2	18.00
Good quality ceramic wall tiles (100 x 100mm)	m^2	7.00

	Units	Cost C£
Plaster and paint		
Plaster in 50 kg bags	tonne	50.00
Plasterboard (12.5mm thick)	m^2	2.20
Emulsion paint in 5 litre tins	litre	1.35
Gloss oil paint in 5 litre tins	litre	2.65
Tiles and paviors		
Clay floor tiles (200 x 100 x 20mm)	m^2	9.00
Vinyl floor tiles (400 x 400 x 2mm)	m^2	4.00
Precast concrete paving slabs (400 x 400 x 40mm)	m^2	2.50
Clay roof tiles	1000	420.00
Drainage		
WC suite complete	nr	85.00
Lavatory basin complete	nr	75.00
150mm diameter cast iron drain pipes	m	10.50
Terrazzo floor tiles (400 x 400 x 40)	m^2	10.00
Stone cladding (600 x 300 x 30)	m^2	10.00

Unit rates

The descriptions below are generally shortened versions of standard descriptions listed in five languages (English, French, Italian, German and Spanish) in section 6. Where an item has a two digit reference number (e.g. 05 or 33), this relates to the full description against that number in section 6. Where an item has an alphabetic suffix (e.g. 12A or 34B) this indicates that the standard description has been modified. Where a modification is major the complete modified description is included here and the standard description should be ignored; where a modification is minor (e.g. the insertion of a named hardwood) the shortened description has been modified here but, in general, the full description in section 6 prevails.

The unit rates below are for main work items on a typical construction project in the Nicosia area in January 1991. The rates include all necessary labour, materials, equipment and, where appropriate, allowances for contractors overheads and profit, preliminary and general items and contractors profit and attendance on specialist trades.

		Unit	Rate C£
Excavation			
01	Mechanical excavation of foundation trenches	m^3	3.10
02	Hardcore filling making up levels	m^2	0.60
03	Earthwork support	m^2	0.75

		Unit	Rate C£
Concrete work			
04	Plain insitu concrete in strip foundations in trenches	m^3	25.00
05	Reinforced insitu concrete in beds	m^3	32.00
06	Reinforced insitu concrete in walls	m^3	27.00
07	Reinforced insitu concrete in suspended floor or roof slabs	m^3	27.00
08	Reinforced insitu concrete in columns	m^3	35.00
09	Reinforced insitu concrete in isolated beams	m^3	29.00
10	Precast concrete slab	m^2	102.00
Formwork			
11A	Softwood formwork to concrete walls	m^2	3.10
12A	Softwood formwork to concrete columns	m^2	8.00
13	Softwood or metal formwork to horizontal soffits of slabs	m^2	3.30
Reinforcement			
14	Reinforcement in concrete walls	tonne	320.00
15	Reinforcement in suspended concrete slabs	tonne	320.00
16	Fabric reinforcement in concrete beds	m^2	2.20
Steelwork			
17	Fabricate, supply and erect steel framed structure	tonne	1050.00
Brickwork and blockwork			
18	Precast lightweight aggregate hollow concrete block walls	m^2	8.50
19	Solid (perforated) concrete blocks	m^2	6.70
21	Facing bricks	m^2	32.00
Roofing			
23	Plain clay roof tiles 260 x 160mm	m^2	16.00
25	Sawn softwood roof boarding	m^2	9.80
26	Particle board roof coverings	m^2	16.50
27	3 layers glass-fibre based bitumen felt roof covering	m^2	6.80
28	Bitumen based mastic asphalt roof covering	m^2	8.00
29	Glass-fibre mat roof insulation 160mm thick	m^2	3.80
30	Rigid sheet loadbearing roof insulation 75mm thick	m^2	7.20
31	Troughed galvanized steel roof cladding	m^2	15.00
Woodwork and metalwork			
32	Preservative treated sawn softwood 50mm x 100mm	m	2.25
33	Preservative treated sawn softwood 50 x 150mm	m	3.25
34	Single glazed casement window in hardwood, size 650 x 900mm	each	65.00
35	Two panel glazed door in hardwood, size 850 x 2000mm	each	220.00
36	Solid core half hour fire resisting hardwood internal flush doors, size 800 x 2000mm	each	130.00

		Unit	Rate C£
37	Aluminium double glazed window, size 1200 x 1200mm	each	110.00
38	Aluminium double glazed door, size 850 x 2100mm	each	165.00
39	Hardwood skirtings	m	2.20
40	Framed structural steelwork in universal joist sections	tonne	1000.00
41	Structural steelwork lattice roof trusses	tonne	1100.00
Plumbing			
42	UPVC half round eaves gutter	m	7.50
43	UPVC rainwater pipes	m	4.25
44	Light gauge copper cold water tubing	m	4.00
45	High pressure plastic for cold water supply	m	3.50
46	Low pressure plastic pipes for cold water distribution	m	3.50
47	UPVC soil and vent pipes	m	5.50
48	White vitreous china WC suite	each	85.00
49	White vitreous china lavatory basin	each	44.00
50	Glazed fireclay shower tray	each	46.00
51	Stainless steel single bowl sink and double drainer	each	47.00
Electrical work			
52	PVC insulated and copper sheathed cable	m	0.90
53	13 amp unswitched socket outlet	each	7.50
54	Flush mounted 20 amp, 1 way light switch	each	24.50
Finishings			
55	2 coats gypsum based plaster on brick walls	m^2	5.50
56	White glazed tiles on plaster walls	m^2	10.00
57	Red clay quarry tiles on concrete floor	m^2	14.00
58	Cement and sand screed to concrete floors	m^2	3.50
59	Thermoplastic floor tiles on screed	m^2	12.00
60	Mineral fibre tiles on concealed suspension system	m^2	9.50
X1	400 x 400 x 40mm terrazzo floor tiles bedded and jointed in mortar	m^2	8.50
X2	Stone cladding 30mm thick	m^2	17.50
Glazing			
61	Glazing to wood	m^2	6.00
Painting			
62	Emulsion on plaster walls	m^2	1.30
63	Oil paint on timber	m^2	3.50

Approximate estimating

The building costs per unit area given below are averages incurred by building clients for typical buildings in the Nicosia area as at January 1991. They are based upon the total floor area of all storeys, measured between external walls and without deduction for internal walls.

Approximate estimating costs generally include for mechanical and electrical installations but exclude furniture, loose or special equipment, and external works; they also exclude fees for professional services. The costs shown are for specifications and standards appropriate to Cyprus and this should be borne in mind when attempting comparisons with similarly described building types in other countries. A discussion of this issue is included in section 4. Comparative data for countries covered in this publication, including construction cost data, are presented in Part Three.

Approximate estimating costs must be treated with reserve; they cannot provide more than a rough guide to the probable cost of building.

All rates in this section exclude value added tax (VAT - see below).

	Cost m^2 *C£*	*Cost* ft^2 *C£*
Industrial buildings		
Factories for letting	85	8
Factories for owner occupation (light industrial use)	105	10
Factories for owner occupation (heavy industrial use)	125	12
Factory/office (high-tech) for letting (shell and core only)	120	11
Factory/office (high-tech) for letting (ground floor shell, first floor offices)	140	13
Factory/office (high tech) for owner occupation (controlled environment, fully finished)	170	16
High tech laboratory workshop centres (air conditioned)	250	23
Warehouses, low bay (6 to 8m high) for letting (no heating)	70	7
Warehouses, low bay for owner occupation (including heating)	75	7
Warehouses, high bay for owner occupation (including heating)	95	9
Cold stores/refrigerated stores	160	15
Administrative and commercial buildings		
Civic offices, non air conditioned	210	20
Civic offices, fully air conditioned	255	24
Offices for letting, 5 to 10 storeys, non air conditioned	200	19
Offices for letting, 5 to 10 storeys, air conditioned	260	24
Offices for owner occupation 5 to 10 storeys, non air conditioned	230	21
Offices for owner occupation 5 to 10 storeys, air conditioned	275	26
Prestige/headquarters office, 5 to 10 storeys, air conditioned	310	29

	Cost m^2 C£	Cost ft^2 C£
Health and education buildings		
General hospitals (348 beds)	390	36
Private hospitals (100 beds)	305	28
Health centres	230	21
Nursery schools	150	14
Primary/junior schools	170	16
Secondary/middle schools	180	17
Recreation and arts buildings		
Sports halls including changing and social facilities	200	19
Swimming pools (international standard) including changing and social facilities	265	25
Swimming pools (schools standard) including changing facilities	205	19
Residential buildings		
Social/economic single family housing (multiple units)	150	14
Private/mass market single family housing 2 storey detached/semidetached (multiple units)	170	16
Purpose designed single family housing 2 storey detached (single unit)	200	20
Social/economic apartment housing, low rise (no lifts)	130	12
Social/economic apartment housing, high rise (with lifts)	140	13
Private sector apartment building (standard specification)	150	14
Private sector apartment buildings (luxury)	180	17
Hotel, 5 star, city centre	355	33
Hotel, 3 star,city/provincial	265	25

Regional variations

The approximate estimating costs are based on projects in Nicosia. For other parts of the country, no adjustments for regional variations are required.

EXCHANGE RATES AND INFLATION

The combined effect of exchange rates and inflation on prices within a country and price comparisons between countries is discussed in section 4.

Exchange rates

The graph below plots the movement of the Cyprus pound against sterling and the US dollar since 1980. The figures used for the graph are quarterly and the method of calculating these, and other related issues are discussed in section 4. The exchange rate at January 1991 was £C 0.83 to the pound sterling and £C 0.43 to the US dollar.

Cyprus Pound against Sterling and the US Dollar

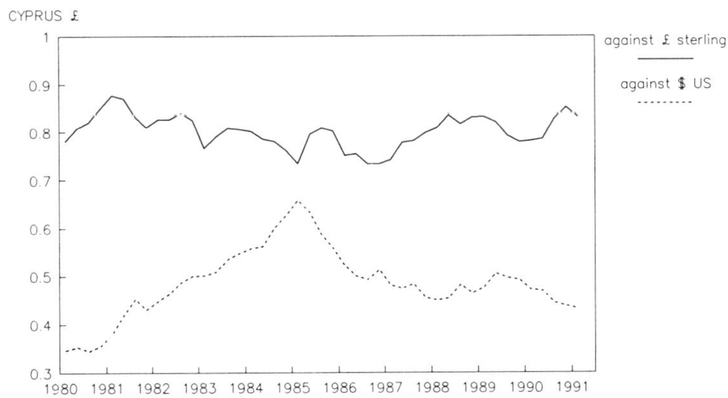

Inflation

The graph below presents a number of indices showing the trends in inflation, prices and costs. All the indices have been rebased to 1985=100. Notes on inflation and the distinction between costs and prices are included in section 4.

Cost and price indices

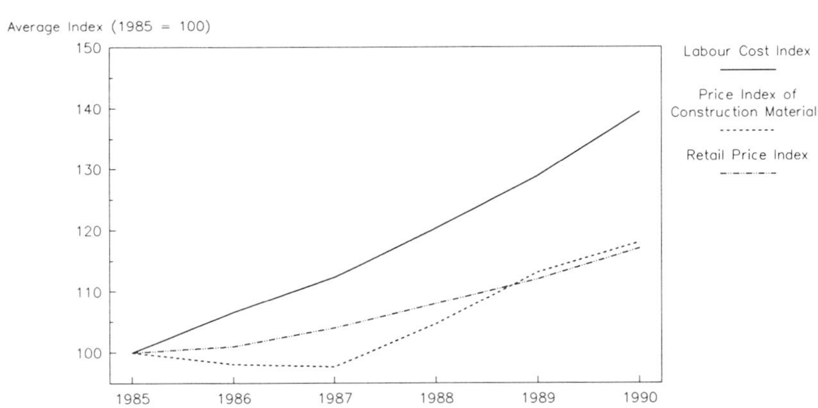

Consumer price inflation

The basis of the figures is the official consumer price index converted to quarterly averages. Inflation has risen steadily since 1985 to an annual rate of 4.5% in 1990.

The table below presents general inflation in Cyprus since 1985. The annual change is the percentage change between the last quarters of consecutive years.

Retail price index 1985 = 100

	1986	1987	1988	1989	1990
average index	101	104	108	112	117
annual change (%)	1.0	3.0	3.8	3.7	4.5

Labour cost index

The table below is a labour cost index compiled by the Ministry of Finance, Department of Statistics and Research.

Labour cost index 1985 = 100

	1986	1987	1988	1989	1990
average index	107	112	120	129	139
annual change (%)	6.6	4.7	7.1	7.5	7.8

Building materials price index

The table below is a building materials price index compiled by the Ministry of Finance, Department of Statistics and Research.

Building materials price index 1985 = 100

	1986	1987	1988	1989	1990
average index	98	98	105	113	118
annual change (%)	-2.0	neg.	7.1	7.6	4.4

USEFUL ADDRESSES

Planning Bureau
Apellis Street
Ay. Omoloyitae
Nicosia
Tel: 010357 2 303275
Fax: 010357 2 366810

Ministry of Communication and Works
Dem. Severis Ave
Nicosia
Tel: 010357 2 303235
Fax: 010357 2 465462

Cyprus Chamber of Commerce and Industry
38 Gr. Dhigenis Ave
PO Box 1455
Nicosia
Tel: 010357 2 449500
Fax: 010357 2 449048

Cyprus Civil Engineers and Architects Association
12 Annis str
Nicosia
Tel: 010357 2 441221

Cyprus Organization of Standards and Control of Quality
Ministry of Commerce and Industry
6 A. Araouzos Street
Nicosia
Tel: 010357 2 303441
Fax: 010357 2 366120

Department of Statistics and Research, Ministry of Finance
13 Byron Avenue
Nicosia
Tel: 010357 2 302349

Czechoslovakia

KEY DATA

Population

Population (1989)	15.6m
Urban population	67%
Population under 15	23.3%
Population over 65	11.6%
Average annual growth rate (1983 to 1988)	0.3%

Geography

Land area	128 000 km^2
Agricultural area	68 260 km^2
Capital city	Prague (population 1 190 576)

Economy

Monetary unit	Koruna (Kcs)
Exchange rate (January 1991 commercial rate) to	
the ECU	Kcs 36.3
the pound sterling	Kcs 51.9
the US dollar	Kcs 26.9
Average annual inflation (1984 to 1989)	n.a.
Gross Domestic Product (GDP)	Kcs 606bn
GDP per capita (1989)	Kcs 38 846
GDP growth (real) (1980 to 1988)	2.1%
Private consumption (% of total)	70.2%
Public consumption (% of total)	8.7%
Investment (% of total)	21.1%

Construction

Total value of construction output (1989)	Kcs 136bn
Construction output per capita	Kcs 8718
Construction output as a percentage of GDP	22%

THE CONSTRUCTION INDUSTRY

Construction output

In 1989 the value of gross construction output in Czechoslovakia was Kcs 136 billion, equivalent to 8.3 - 12.8 billion ECUs depending on the exchange rate used. Indirect indicators suggest it may be nearer to 18 billion ECUs (US$5.1 billion). In addition there is about Kcs 7 billion of DIY construction which is not normally included with construction output. The official figures represent about 22% of GDP - a higher share than in other East European countries and considerably higher than most Western developed countries.

No breakdown of total output is available. However, there is a breakdown for some 65% (Kcs 100 billion) of the work done by building enterprises in 1988 and this is shown below.

Construction output in 1989

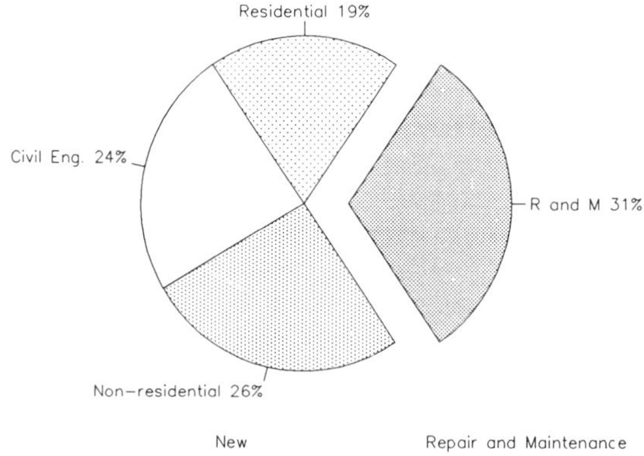

These are the official figures. However, some experts in Czechoslovakia regard the restoration and repair figure as too high. Only about 2.7% of "new" construction work is renovation and modernization. Some 55% to 60% of non-residential building work is in industrial building. The level of output is expected to have decreased in 1990 because of the lack of investment funds in the economic system and this fall is expected to continue in 1991.

The regional distribution of construction work in 1988 compared with that of population is shown below.

Population and construction output 1988: regional distribution

Region	*Population %*	*Construction output %*
Prague	7.7	13.1
Middle Bohemia	7.2	5.7
South Bohemia	4.5	5.9
West Bohemia	5.6	4.7
North Bohemia	7.6	8.3
East Bohemia	8.0	5.7
South Moravia	13.2	9.5
North Moravia	12.6	12.0
Total Czech Republic	66.4	64.9
Bratislava	2.8	4.9
West Slovakia	11.0	12.0
Middle Slovakia	10.3	9.0
East Slovakia	9.5	8.2
Total Slovakian Republic	33.6	35.1
Total CSFR	100.0	100.0

Source: Statistical Yearbook of CSFR, 1989.

The value of construction exports in 1987 was about Kcs 3 billion of which Kcs 2½ billion went to other Eastern European countries. Most exports were from the Ministry of the Building Industry of Czech as opposed to that of Slovakia.

Finance for construction

Investment capital has traditionally been provided by the State. Some 82% of total investment funds (of which construction is a part) are supplied by government, 13% by co-operatives and 5% by the private sector. There are insufficient finances available to maintain the level of demand on the industry. However, the State has slowly diminished finance for housing and is allowing more private initiative. Subsidies for one and two family dwellings are being removed. But the private sector does not yet have adequate finance for housing, although when assets are returned to private ownership, there may be greater possibilities. In addition, inflation and other problems will restrict the amount people are willing to spend on housing.

The structure of the industry

Until recently the contracting industry was state-owned and, in many areas of construction, firms had a monopoly. There were about 230 building organizations and the output of the industry was almost entirely produced by giant organizations. Indeed, only 2.5% of output was produced by firms employing fewer than 500 persons - a greater concentration than in other East European countries.

The building enterprises were especially large organizations with over 5000 employees. They were, in fact, conglomerates where one third of the employees executed so-called non-building activities, (e.g. manufacture) and other activities such as apprentice education and research and development. The administrative apparatus was especially large and accounted for more than one third of the employees. A large part of non-building activities arose from the failure of client-contractor relationships. The share of sub-contracts between the enterprises was low (about 14%), while within the enterprises this was based on a technological division between labour and plant: the main construction work was divided into building construction and civil engineering works and there were separate divisions for finishing work, equipment, and so on.

All this is now changing. The number of construction firms is now increasing rapidly as large organizations split up and new ones are formed. However, construction is also undertaken by other organizations, which together accounted for some 30% of gross output in 1989.

Construction output in 1989: share by organization type

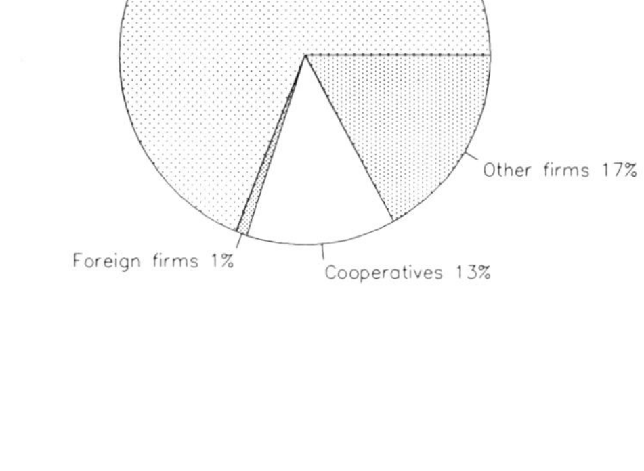

Design offices in the government sector are large organizations of several hundred persons, sometimes connected to large contracting organizations, but often independent. In addition, there are some architects working on their own account on design projects. The qualifications of designers are not controlled but there is a central register of those authorized to undertake design work. There are two types of architect: one is trained only in the *beaux arts* tradition, while the other is part engineer and part architect with the title of engineer/architect. Both engineers and architects have good reputations and Czech designers have won design awards at international exhibitions. This structure, although still formally in place, is now breaking apart. Chief architects, for example, want to buy their organizations but there is as yet no mechanism for them to do so. Architects are leaving their jobs and setting up on their own.

The planning process and change in the construction industry

In Czechoslovakia, as in other Eastern European countries, the last 40 years of centrally planned economies were characterized by the application of the Soviet model of capital construction. Here the state was the client which controlled investment and the execution of work through a nationalized construction industry. This was true of all new construction by building enterprises and repair work, including work by locally controlled building-repair firms. The structure of investment and of building production did not follow from the level of demand, but from subjective decisions, mainly by the political centre. Construction work was assigned according to the state plan; a building market did not exist. But the load placed on the industry could not be met. Therefore, despite high building intensity, unsatisfied needs were growing.

Incentives in the system were geared to the largest possible financial volume of output. Quality was low and there was no competition between enterprises and, therefore, no incentive to improve quality or to reduce prices. Low earnings did not allow the necessary differentiation of wages according to performance and quality, so that the motivational effect of remuneration was near to zero.

With central planning, an artificial over-employment was created by the low productivity of labour and low wage levels, together with the need for a high level of employment of women and pensioners. Unemployment exceeding 8% is expected in construction in 1991, due largely to the lack of investment funds combined with a surplus of construction operatives, who in the past were employed mainly by government bodies.

Construction technology was dominated by large prefabricated concrete elements. Standardization of production, mainly in housing construction was enormous and was detrimental to the quality of the urban environment. As the amount of heavy panel construction declines there will be a surplus of assembly workers and a corresponding shortage of craftsmen for lighter construction methods.

In spite of the lack of incentives and the enforced use of heavy panel construction, the construction industry in Czechoslovakia has higher standards of workmanship than in many of the Eastern European countries. This is because training has been given and because of the tradition of high quality building and the large stock of historic buildings which require sophisticated repair and maintenance.

However, the performance of the industry in terms of output achieved is not good. Completion times can be several times the UK levels and work under construction is about the same as two years output. Costs escalate because of the lack of planning and of the availability of materials in relation to project demands listed in the 5 Year Plan. Once a project was in this plan, it was started irrespective of whether there were adequate resources to undertake it. There was an informal priority system for projects in the Plan. Those with high priority for any reason may have been completed within both a reasonable time span and level of productivity. Others languished, starved of inputs, for years. Some of the reasons for the failure to meet demand that are likely to detract from performance of the construction industry over the years ahead are:

- the high cost of new plant and equipment constitutes a disincentive to buy, even though the old outdated machines are subject to breakdown
- plant is allocated to sites for long periods of time and there is no scope to increase its capacity utilization
- low wages and low differentials produce low motivation. There are no effective incentives to higher productivity
- there is no penalty for delays and long duration of contracts
- there is no research or analysis of work on management in construction
- suppliers have a monopolistic position with no incentive to improve service or product
- absolute shortages of materials make good site management very difficult - if not impossible.

Contractual and procurement arrangements

The arrangements for procurement of a building before the break-up of the Communist system are still largely intact. After a preliminary study by engineers, an approximate cost estimate of the project was prepared based on a comprehensive price book. The investor group in charge of the project would seek funds from the State budget and it would eventually be included in the 5 Year Plan. For larger projects the design would be allocated to a government design group sometimes after a design competition between three or four offices. The final allocation would be determined by the chief architect. Small "artistic" projects could be designed by small private offices specializing in this type of work; they would be supervised by the Foundation for Applied Art. The project would then be allocated to a contracting organization.

The construction process

Currently, instead of being allocated as formerly, the budget for the whole project is given to three or four contractors who put in prices as a percentage up or down on the budget. Usually the lowest price is lower than the budget - perhaps 15% lower. The bids are checked to see that there are no errors and the lowest is accepted.

However, projects tend to cost much more than the initial price. The main reasons are that there are shortages of materials, there are long waits on site for materials to arrive and productivity is very low. When this happens the contracting organization is recompensed; in effect the contractor never loses on a project.

When demand falls, as is forecast for coming years, there should be a better supply of materials which could enable the substantial number of unfinished projects to be completed and construction times to be reduced. There is, however, a tendency to suggest that what is required is more investment to "mop up" the supply of building materials. It would be inefficient to allow the number of new projects started to rise at the expense of finishing projects.

Detailed arrangements for co-operation between the contractor and other parties to the process is contained in the Basic Terms of Delivery of Construction Work (1973).

Development control and standards

The Town and Country Planning and Building By-laws Act of 1976 contains general principles guiding planning and procedures for approval and acceptance of buildings. The law is supplemented by legal regulations which are the Official Announcements relating to technical requirements for construction, documentation for planning applications, certification of products; and authorization of planning, design and engineering activities.

The final building acceptance is given by an authorized state body as stipulated in the 1976 act. Co-operation between the participants in the building process and remedies connected with bad quality or incompetent work are governed by the Economic Code (1970).

Information is published in the Code of Laws and By-Laws, in the Report of the Federal Ministry for the Development of Technology and Investments, the Ministry for Construction, the Ministry of Technology and the Ministry for the Building Industry.

The State Testing Act (1968) and the Technical Standardization Act (1964) control the quality of products and the production pocesses used. The standards are mandatory. Standards are classified according to their validity: state standards - CSN; branch standards - ON; and company standards - PN.

Information may be obtained from the Bulletin of the Authority for Standardization and Measurement.

Research and development

The Federal Ministry for Development of Technique and Investments is responsible for scientific research and development which is financed from the national budget. Other research is undertaken in the enterprises.

Materials

Czechoslovakia has higher material costs in total output than Western developed economies and materials used tend to be heavy and consume much energy in their production. Nevertheless quality is reasonable.

Timber and timber products, although rarely used, are generally easily available because of the large, well distributed area of forest. The volume of imported materials for construction is very low at only 1.6%. Most imports of materials come from Austria and Germany.

There has been a switch away from prefabricated concrete components and there is excess capacity in this area. Generally, however, building materials do not arrive on site on time due perhaps to the distribution system as well as constraints on manufacturing capacity.

Foreign participation and future opportunities

The problems of foreign participation in Czechoslovak enterprises are still substantial: prohibition on ownership of land, inflation, high interest rates, administrative difficulties and so on. Many of these are being resolved. A Law of Foreign Investments is expected. Bilateral international agreements for the support and protection of investments are a significant safeguard. They exist with almost all European developed countries.

A Law of Joint Venture which helps in transferring foreign exchange but also imposes levies of foreign currency on joint ventures is already in place, as is also a Law on Income Tax from Profit. In the building industry there are already many registered joint ventures with companies from Austria, Germany, Poland, UK and UAE.

Further reading

Jiri Hezky, "The Czechoslovak Building Industry, Paper presented at NEDO/Euroconstruct Conference, London, 20 November 1990.

Jiri Hezky, "Das Bauwesen in der Tschecho-Slowakei 1989/91" in Euroconstruct, Baukonjunktur 1990/91 in Westeuropa, IFO Institut für Wirtschaftsforschung, München, Dezember 1990.

Rolf Bollinger, "Chancen und Risken für die Westeuropäische Bauwirtschaft nach den Wende in Osteuropa" in Euroconstruct, Baukonjunktur 1990/91 in Westeuropa, IFO Institut für Wirtschaftsforschung, München, Dezember 1990.

Acknowledgements

The authors would like to thank Ing Klaudius Lepen and Ing Jiri Hezky of the Institute for Economics and Organizations in Building in Bratislava for their substantial contribution to this section.

CONSTRUCTION COST DATA

Cost of labour

The figures below are typical of labour costs in Prague as at January 1991. The "wage rate" is the basis of an employee's income, while the "cost of labour" indicates the cost to a contractor of employing that employee. The difference between the two covers a variety of mandatory and voluntary contributions - a list of items which could be included is given in section 4.

	Wage rate (per month) Kcs	*Cost of labour (per month) Kcs*	*Number of worked hours per year*
Site operatives			
Mason/bricklayer	3500	5950	1892
Carpenter	4000	6800	1892
Plumber	3600	6100	1892
Electrician	3200	5450	1892
Structural steel erector	4500	7650	1892
HVAC installer	3600	6100	1892
Unskilled labourer	2800	4750	1892
Equipment operator	4200	7150	1892
Watchman/security	2000	3400	1892

	Wage rate (per month) Kcs	*Cost of labour (per month) Kcs*	*Number of worked hours per year*
Site supervision			
General foreman	4975	8450	1892
Trades foreman	3925	6650	1892
Clerk of works	2950	5000	1892
Contractors' personnel			
Site manager	5450	9250	1892
Resident engineer	4975	8450	1892
Resident surveyor	3650	6200	1892
Junior engineer	2950	5000	1892
Junior surveyor	2550	4350	1892
Planner	4018	6850	1892
Senior architect	5475	9300	1892
Senior engineer	4775	8100	1892
Senior surveyor	3650	6200	1892
Qualified architect	4975	8450	1892
Qualified engineer	4475	7600	1892
Qualified surveyor	3650	6200	1892

Cost of materials

The figures that follow are the costs of main construction materials, delivered to site in the Prague area, as incurred by contractors in January 1991. These assume that the materials would be in quantities as required for a medium sized construction project and that the location of the works would be neither constrained nor remote.

	Units	*Cost Kcs*
Cement and aggregate		
Ordinary portland cement in 50kg bags	tonne	650
Coarse aggregates for concrete	m^3	100
Fine aggregates for concrete	m^3	130
Ready mixed concrete (mix 350)	m^3	600
Steel		
Mild steel reinforcement	tonne	4500
High tensile steel reinforcement	tonne	4200
Structural steel sections	tonne	12 000
Bricks and blocks		
Common bricks (300 x 150 x 75mm)	1000	1800
Good quality facing bricks	1000	2500
Hollow concrete blocks	1000	350

	Units	Cost Kcs
Solid concrete blocks	1000	1000
Precast concrete cladding units with exposed aggregate finish	m^2	230
Timber and insulation		
Softwood sections for carpentry	m^3	1600
Softwood for joinery	m^3	1400
Hardwood for joinery	m^3	1900
Exterior quality plywood (10 mm)	m^2	64
Plywood for interior joinery (10 mm)	m^2	101
Softwood strip flooring (12 mm)	m^2	85
Chipboard sheet flooring (10 mm)	m^2	120
100 mm thick quilt insulation	m^2	40
100 mm thick rigid slab insulation	m^2	250
Softwood internal door complete with frames and ironmongery	nr	290
Glass and ceramics		
Float glass (4 mm)	m^2	42
Sealed double glazing units	m^2	52
Good quality ceramic wall tiles	m^2	90
Plaster and paint		
Plaster in 50 kg bags	tonne	600
Plasterboard (50mm thick)	m^2	80
Emulsion paint in 5 litre tins	kg	19
Gloss oil paint in 5 litre tins	kg	21
Tiles and paviors		
Clay floor tiles (100 x 100 x 100mm)	m^2	60
Vinyl floor tiles (100 x 100 x 7mm)	m^2	25
Precast concrete paving slabs	m^2	60
Clay roof tiles	1000	2700
Precast concrete roof tiles	1000	4500
Drainage		
WC suite complete	nr	500
Lavatory basin complete	nr	350
100mm diameter clay drain pipes	m	10
150mm diameter cast iron drain pipes	m	110

Unit rates

The descriptions below are generally shortened versions of standard descriptions listed in five languages (English, French, Italian, German and Spanish) in section 6. Where an item has a two digit reference number (e.g. 05 or 33), this relates to the full description against that number in section 6. Where an item has an alphabetic suffix (e.g. 12A or 34B) this indicates that the standard description has been modified. Where a modification is major the complete modified description is included here and the standard description should be ignored; where a modification is minor (e.g. the insertion of a named hardwood) the shortened description has been modified here but, in general, the full description in section 6 prevails.

The unit rates below are for main work items carried out by local contractors in a typical construction project built to local standards in the Prague area in January 1991. The rates include all necessary labour, materials, equipment and, where appropriate, allowances for contractors overheads and profit, preliminary and general items and contractors profit and attendance on specialist trades. Higher rates can be expected for similar work undertaken by international contractors to Western European standards.

		Unit	*Rate Kcs*
Excavation			
01	Mechanical excavation of foundation trenches	m^3	57
02	Hardcore filling making up levels	m^2	50
03	Earthwork support	m^2	3
Concrete work			
04	Plain insitu concrete in strip foundations in trenches	m^3	650
05	Reinforced insitu concrete in beds	m^3	757
06	Reinforced insitu concrete in walls	m^3	1770
07	Reinforced insitu concrete in suspended floor or roof slabs	m^3	2650
08	Reinforced insitu concrete in columns	m^3	2020
09	Reinforced insitu concrete in isolated beams	m^3	2020
10	Precast concrete slab	m^2	290
Formwork			
11	Softwood or metal formwork to concrete walls	m^2	69
12	Softwood or metal formwork to concrete columns	m^2	82
13	Softwood or metal formwork to horizontal soffits of slabs	m^2	88
Reinforcement			
14	Reinforcement in concrete walls	tonne	9470
15	Reinforcement in suspended concrete slabs	tonne	8830
16	Fabric reinforcement in concrete beds	m^2	25
Steelwork			
17	Fabricate, supply and erect steel framed structure	tonne	13 900

		Unit	Rate Kcs
Brickwork and blockwork			
18	Precast lightweight aggregate hollow concrete block walls	m^2	164
19	Solid (perforated) concrete blocks	m^2	126
20	Solid (perforated) sand lime bricks	m^2	164
21	Facing bricks	m^2	88
Roofing			
22	Concrete interlocking roof tiles 430 x 380mm	m^2	151
23	Plain clay roof tiles 260 x 160mm	m^2	107
24	Fibre cement roof slates 600 x 300mm	m^2	82
25	Sawn softwood roof boarding	m^2	252
26	Particle board roof coverings	m^2	278
27	3 layers glass fibre based bitumen felt roof covering	m^2	101
28	Bitumen based mastic asphalt roof covering	m^2	50
29	Glass-fibre mat roof insulation 160mm thick	m^2	76
30	Rigid sheet loadbearing roof insulation 75mm thick	m^2	101
31	Troughed galvanized steel roof cladding	m^2	126
Woodwork and metalwork			
32	Preservative treated sawn softwood 50mm x 100mm	m^2	151
33	Preservative treated sawn softwood 50 x 150mm	m^2	189
34	Single glazed casement window in hardwood, size 650 x 900mm	each	505
35	Two panel glazed door in hardwood, size 850 x 2000mm	each	631
36	Solid core half hour fire resisting hardwood internal flush doors, size 800 x 2000mm	each	1010
37	Aluminium double glazed window, size 1200 x 1200mm	each	1640
38	Aluminium double glazed door, size 850 x 2100mm	each	1890
39	Hardwood skirtings	m	379
40	Framed structural steelwork in universal joist sections	tonne	16 400
41	Structural steelwork lattice roof trusses	tonne	13 900
Plumbing			
42	UPVC half round eaves gutter	m	82
43	UPVC rainwater pipes	m	76
44	Light gauge copper cold water tubing	m	126
45	High pressure plastic for cold water supply	m	88
46	Low pressure plastic pipes for cold water distribution	m	63
47	UPVC soil and vent pipes	m	95
48	White vitreous china WC suite	each	820
49	White vitreous china lavatory basin	each	568
51	Stainless steel single bowl sink and double drainer	each	1830
Electrical work			
52	PVC insulated and copper sheathed cable	m	19
53	13 amp unswitched socket outlet	each	101
54	Flush mounted 20 amp, 1 way light switch	each	88

		Unit	Rate Kcs
Finishings			
55	2 coats gypsum based plaster on brick walls	m^2	82
56	White glazed tiles on plaster walls	m^2	164
57	Red clay quarry tiles on concrete floor	m^2	164
58	Cement and sand screed to concrete floors	m^2	50
59	Thermoplastic floor tiles on screed	m^2	252
60	Mineral fibre tiles on concealed suspension system	m^2	126
Glazing			
61	Glazing to wood	m^2	139
Painting			
62	Emulsion on plaster walls	m^2	5
63	Oil paint on timber	m^2	25

Approximate estimating

The building costs per unit area given below are averages incurred by building clients using local contractors for typical buildings in the Prague area as at January 1991. The equivalent building costs per unit area for international contractors for buildings constructed to Western European standards would be higher.

Approximate estimating costs generally include for mechanical and electrical installations but exclude furniture, loose or special equipment, and external works; they also exclude fees for professional services. The costs shown are for specifications and standards appropriate to Czechoslovakia and this should be borne in mind when attempting comparisons with similarly described building types in other countries. A discussion of this issue is included in section 4. Comparative data for countries covered in this publication, including construction cost data, are presented in Part Three.

Approximate estimating costs must be treated with reserve; they cannot provide more than a rough guide to the probable cost of building.

Note that for industrial buildings, administrative and commercial buildings, health and education buildings and residential buildings, rates given are per m^3.

	Cost m^3 Kcs	Cost ft^2 Kcs
Industrial buildings		
Factories for letting	450	-
Factories for owner occupation (light industrial use)	430	-
Factories for owner occupation (heavy industrial use)	475	-
Factory/office (high-tech) for letting (shell and core only)	1000	-
Factory/office (high-tech) for letting (ground floor shell, first floor offices)	950	-
Factory/office (high tech) for owner occupation (controlled environment, fully finished)	1025	-

	Cost m^3 Kcs	Cost ft^2 Kcs
Warehouses, low bay (6-8m high) for letting (no heating)	410	-
Warehouses, low bay for owner occupation (including heating)	600	-
Warehouses, high bay for owner occupation (including heating)	550	-
Cold stores/refrigerated stores	800	-
Administrative and commercial buildings		
Civic offices, non air conditioned	785	-
Civic offices, fully air conditioned	945	-
Offices for letting, 5-10 storeys, non air conditioned	1040	-
Offices for letting, 5-10 storeys, air conditioned	1260	-
Offices for letting, high rise, air conditioned	1555	-
Offices for owner occupation 5-10 storeys, non air conditioned	960	-
Offices for owner occupation 5-10 storeys, air conditioned	1150	-
Offices for owner occupation high rise, air conditioned	1400	-
Prestige/headquarters office, 5-10 storeys, air conditioned	1440	-
Prestige/headquarters office, high rise, air conditioned	1860	-
Health and education buildings		
General hospitals	945	-
Teaching hospitals	1025	-
Private hospitals	1330	-
Health centres	765	-
Nursery schools	965	-
Primary/junior schools	725	-
Secondary/middle schools	795	-
University (arts) buildings	865	-
University (science) buildings	910	-
Management training centres	780	-

	Cost m^2 Kcs	Cost ft^2 Kcs
Recreation and arts buildings		
Theatres (over 500 seats) including seating and stage equipment	5380	500
Theatres (less than 500 seats) including seating and stage equipment	4490	420
Concert halls including seating and stage equipment	5270	490
Sports halls including changing and social facilities	4710	440
Swimming pools (international standard) including changing and social facilities	5300	490
Swimming pools (schools standard) including changing facilities	3920	365
National museums including full air conditioning and standby generator	6300	585
Local museums including air conditioning	5500	510
City centre/central libraries	5270	490
Branch/local libraries	4580	425

	Cost m^3Kcs	Cost ft^2Kcs
Residential buildings		
Social/economic single family housing (multiple units)	550	-
Private/mass market single family housing 2 storey detached/semidetached (multiple units)	630	-
Purpose designed single family housing 2 storey detached (single unit)	750	-
Social/economic apartment housing, low rise (no lifts)	610	-
Social/economic apartment housing, high rise (with lifts)	685	-
Private sector apartment building (standard specification)	770	-
Private sector apartment buildings (luxury)	830	-
Student/nurses halls of residence	750	-
Homes for the elderly (shared accommodation)	800	-
Homes for the elderly (self contained with shared communal facilities)	880	-
Hotel, 5 star, city centre	1545	-
Hotel, 3 star,city/provincial	1215	-
Motel	815	-

EXCHANGE RATES AND INFLATION

The combined effect of exchange rates and inflation on prices within a country and price comparisons between countries is discussed in section 4.

Exchange rates

The graph below plots the movement of the Czech Koruna against sterling and the US dollar since 1980. The figures used for the graph are quarterly and the method of calculating these, and other related issues are discussed in section 4. The exchange rate at January 1991 was Kcs 51.9 to the pound sterling and Kcs 26.9 to the US dollar.

The Czech Koruna against Sterling and the US Dollar

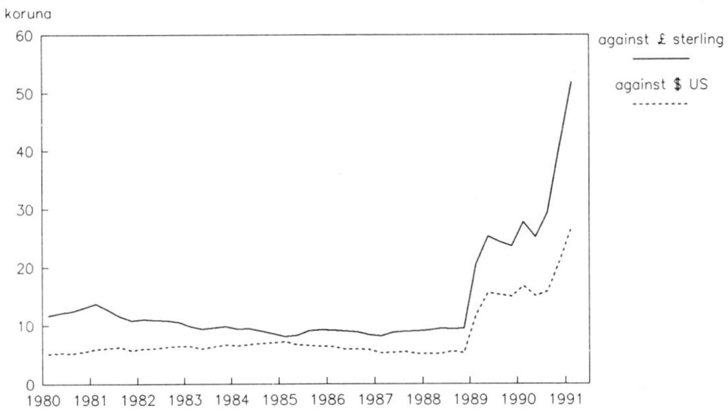

Inflation

The table below presents a number of indices showing the trends in inflation, prices and costs. All the indices have been rebased to 1980=100. Notes on inflation and the distinction between costs and prices are included in section 4.

Cost and prices indices

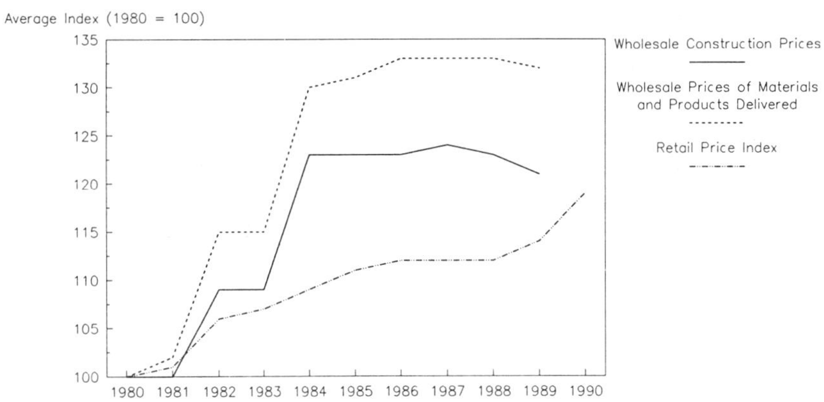

Retail price index

Retail price inflation peaked in 1982. Since then the annual rate has been below 2%, although it rose sharply (to over 4%) in 1990. All of the indices in this section are prepared by the Federal Office of Statistics.

Consumer price inflation 1980 = 100

	1981	1982	1983	1984	1985	1986	1987	1988	1989	1990
average index	101	106	107	109	111	112	112	112	114	119
annual change (%)	1.0	5.0	0.9	1.9	1.8	0.9	0.0	0.0	1.8	4.4

Wholesale construction price index

Wholesale construction price index 1980 = 100

	1981	1982	1983	1984	1985	1986	1987	1988	1989	1990
average index	100	109	109	123	123	123	124	123	121	n.a.
annual change (%)	0.0	9.0	0.0	12.8	0.0	0.0	0.8	-0.8	-1.6	n.a.

Wholesale materials price index

Wholesale price of materials and products delivered index 1980 = 100

	1981	1982	1983	1984	1985	1986	1987	1988	1989	1990
average index	102	115	115	130	131	133	133	133	132	n.a.
annual change (%)	2.0	12.7	0.0	13.0	0.0	0.8	1.5	0.0	-0.8	n.a.

USEFUL ADDRESSES

Ministerstvo vystavby a stavebníctvi Slovenske republiky
(Ministry of Building Industry of the Slovak Republic)
Stefanovicova 5 813 25 Bratislava
Tel: 469 51

Ministerstvo vystavby a stavebníctví Ceske republiky
(Ministry of Building Industry of the Czech Republic)
Na Porícním právu 1, 120 07 Praha 2
Tel: 2135-1111

Vyskumny ústav vystavby a architektúry
(Research Institute for Construction and Architecture)
110 00 Praha 1, Letenská 3

Ustav ekonomiky a organizácie stavebníctva
(Institutc of Economics and Organization of Building Industry)
Ruzová dolina 27, 824 69 Bratislava
Tel: 2016-111
Fax: 67897

Ustav racionalizac ve stavebnictví
(Institute of Rationalization in Building Industry)
Legerova 72 120 57 Praha 1
Tel: 240441-5
Fax: 122852

Zväz stavebnych podnikatel'ov Slovenska
(Union of Building Entrepreneurs of Slovakia)
Zahradnícka 46 824 93 Bratislava
Tel: 64204
Fax: 0766556

Svaz podnikatelú ve stavebníctví
(Union of Building Entrepreneurs)
Národní tr. 10
110 00 Praha 1
Tel: 204 05
Fax: 299832

Interstav, a.s.
Ruzová dolina 27
821 09 Bratislava
Tel: 2016-227

Denmark

KEY DATA

Population

Population (1989)	5.13m
Urban population	86%
Population under 15	17.0%
Population over 65	15.5%
Average annual growth rate (1983 to 1988)	0.1%

Geography

Land area	43 000 km²
Agricultural area	28 000 km²
Capital city	Copenhagen
	(population 467 850)

Economy

Monetary unit	Danish Krone (Dkr)
Exchange rate (January 1991) to	
the ECU	Dkr 7.87
the pound sterling	Dkr 11.3
the US dollar	Dkr 5.82
Average annual inflation (1984 to 1989)	4.3%
Gross Domestic Product (GDP)	Dkr 724bn
GDP per capita (1989)	Dkr 141 131
GDP growth (real) (1980 to 1988)	1.9%
Private consumption (% of total)	53.6%
Public consumption (% of total)	25.9%
Investment (% of total)	17.5%
Central government expenditure as percentage of GDP	57.3%

Construction

Total value of construction output (1989)	Dkr 100.8bn
Construction output per capita	Dkr 19 649
Construction output as a percentage of GDP	13.6%

THE CONSTRUCTION INDUSTRY

Construction output

The value of gross construction output in Denmark in 1989 was DKr 100.8 billion or 12.5 billion ECUs. This represents 13.6% of GDP. An estimate of the breakdown of output by type of work is shown below.

Construction output 1989

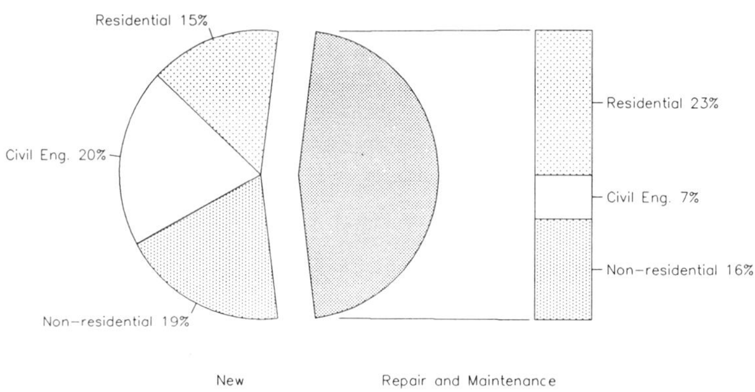

The share of repair and maintenance in total output is very high and residential construction low, which may be expected in a mature economy.

Output was fairly buoyant until 1986 but since then has fallen in most types of work and no recovery is foreseen in the near future.

Characteristics and structure of the industry

The main contracting method used in Denmark is the separate trades contracting system but main contracting and turnkey contracting are also used. The industry has a well developed system of subcontracting and this is reflected in the preponderance of small firms in the industry.

In 1984, there were just over 20 000 contracting firms in Denmark. Most of these (99%) had less than 50 employees, 69% less than 5 employees and only some 0.5% had more than 100 employees. Firms with over 100 employees did, however, employ more than 15% of the total workforce of over 100 000. In 1989 there were 14 Danish contractors in Building magazine's "Top 300 European Contractors" and two in the top 100, most with some export business. Three are in the Engineering News Record's "Top 250 International Contractors". Major contractors are listed below.

Major Danish contractors

Major contractors	*Place in Building's "Top 300 European Contractors"*	*Exports (% of turnover)*	*Main work types*
Hojgaard	88	20-49	
Rasmussen and Schiotz	100	less than 10	
Phonix Contractors	150	10-19	Building Industrial Transport
Superfos Damann Luxol	160	20-49	
Christiani and Nielson	208	over 50	All work types
L and N	210	less than 10	

Source: Building

In 1986 there were about 3000 engineering consultancy firms in Denmark, some with very substantial numbers of employees. Indeed, the greater part of the work is done by firms with more than 100 employees, but there is also a large number of small firms. The major Danish firms are multi-disciplinary and operate in most regions of Denmark; they are also important abroad, especially in Africa, where 45% of their work outside Denmark is carried out.

Engineers in Denmark are well educated and have a broad remit, e.g. almost every Danish engineering firm provides services for the housing sector.

The title of architect is not protected nor is the employment of an architect obligatory.

Clients and finance

The division between public and private housing is complex. Government organizations actually act as clients for very little housing but state subsidized housing accounts for about 70% of all housing construction in terms of the total number of units. Subsidies are a mixture of long-term interest free loans with a state guarantee and rent and repayment contributions. Special attention is devoted to housing for the young and the elderly. It is the unsubsidized sector that is declining rapidly.

Finance of private building construction is generally provided by mortgage institutions and banks, the latter financing the contractors on a short term basis for the period of construction.

Contractual and procurement arrangements

Except for very large projects, designers are not selected in competition. There are about 20 to 30 competitions a year generally open to all and judged mainly by architects. Continuing clients will co-operate with a few firms of consultants and select one of them for each project on the basis of good personal relations and competence for the particular job.

The most widely used method of tendering is selective competitive tendering with 3 to 5 tenderers. However, open tendering is also often used but negotiation is rare. Tendering in the public sector is regulated by legislation and with selective tendering and open tendering the lowest price must be accepted, except that a tender may be rejected if the client believes the contractor is unable to carry out the work.

The principal tender documents comprise a comprehensive specification, drawings, form of agreement, form of tender and general conditions of contract. The building industry's standard form of building contract, *Almindelige Betingelser for Arbejder og Leverancer* (1972) is widely used.

Although cost reimbursement contracts are known, the majority of contracts are let on a lump-sum basis. Until recently, it was standard practice for tenderers to submit details of their proposed tenders to their trade organizations for checking and approval; the trade organization could prevent the submission or acceptance of a tender they considered to be too low and the only recourse of the building client was through a special form of arbitration. However, Parliament has now prohibited tendering contractors from "adjusting" their tenders. Trade organizations are still permitted to hold meetings of tenderers to discuss work, but the building client must be invited and it is prohibited for prices to be discussed at such meetings.

Trade associations are very prominent in Denmark and many of them have prepared standard schedules of prices for their members. Trade contractors prepare their own unit quantities and often base their prices on the standard schedules prepared by their particular association. Several Danish building institutions have advocated the wider use of bills of quantities, but to date they are used only occasionally, mainly in association with advanced techniques of computer analysis and planning. Co-ordination of the various trades on site is usually the responsibility of the supervising architect, but the complexity of planning required on some large schemes has led to the introduction of project supervisors and other special overall planning consultants.

All Government financed and state subsidized building contracts are let on a firm price, fixed time basis. Exceptions to the firm price rule are for expenses incurred by the contractor due to *force majeure* or circumstances occasioned by the authority. Many private contracts are let under the same conditions. Statements for interim payments submitted by the contractor are usually subject to a 10% retention. The final account is prepared by the contractor.

Liability and insurance

For public works projects, liability is regulated by the General Conditions for Technical Consultation and Assistance (ABR from the Danish title). The consultant is liable for errors and omissions in discharging the work but only if fault is proven. The liability of the consultant expires five years after conclusion of the consultancy work. The organizations of clients and professionals have made a *concordat* to the effect that where the conditions of engagement are nationally agreed and where there is more than one defendant, the consultant is liable for a just and equitable proportion only.

According to the ABR, the agreement can stipulate a penalty for exceeding deadlines which the consultant is bound to pay to the client.

Insurance is compulsory for all members of the professional engineering association, the *Foreningen af Raadgivende Ingeniøret* (FRI). Although it is not a legal requirement, most clients also prescribe that the supplier take out insurance cover. Insurance for architects is not obligatory.

A directive for quality assurance in public sector building came into force in 1986. The main idea behind the directive is that prevention is generally less costly than repair. The directive states that clients must impose on their suppliers (contractors, engineers and architects) an obligation to carry out quality assurance for the services they provide. The client must determine a supervision and inspection scheme for the building process. Suppliers are obliged to follow the scheme and to assure the quality of the services rendered. The directive is for public sector building, but has also had a profound effect on private-sector construction. As a consequence, most building at the present time is carried out in accordance with the quality assurance directive.

Development control and standards

Physical planning in Denmark is subject to the following codes and regulations:

- the National and Regional Planning Act defining the principles to be applied in planning
- the Local Authority Planning Act governing the preparation of local district plans
- the Urban and Rural Zoning Act governing building development in the open countryside
- the Building (Principles) Act stipulating the authorities for building control, procedures etc.

There are three legal authorities responsible for physical planning. The National Agency for Physical Planning guides physical planning and development in the preferred direction. Counties are obliged to produce regional plans revised every two years. The municipalities prepare a structure plan and appropriate local plans. Control is exercised through building permits issued by the municipal authorities. The rules allow some discretion to politicians.

In addition to planning laws there are Building Regulations which are national regulations stipulating technical standards for the location of buildings on site, dimensions, thermal, acoustic and moisture insulation, safety and structural requirements, fire safety, sanitation, etc. They refer to various technical standards (see below) some of which are mandatory. Technical, architectural and other provisions of the Building Regulations are, as far as possible, stated as performance specifications.

In principle, the codes and regulations are applicable to building but not to civil engineering and public works, for which regulations are issued by the Ministry of Public Works. In fact, however, many regulations are common to both.

The Danish Building Centre publishes a systematic collection of all building regulations codes of practice, building standards, regulations on town planning and the environment. It is known as *Byggedata* and is regularly updated.

A Danish Standardization Council established by the Ministry of Industry is responsible for Danish standards. There is a scheme for certification of building products after tests conducted by the Danish Standards Institute to ensure that the product conforms to the relevant specifications. After certification, regular checks are made normally based on the manufacturer's own production and quality control procedures.

Most building standards issued by the Danish Standards Institute are recommended, not mandatory. The technical procedures they contain must comply with the Building Regulations. Some standards are referred to in the Regulations with mandatory effect.

There are other separate certification schemes for water and drainage materials under the Ministry of Housing; electrical applicances under the Danish Board for Testing and Approval and the Electrical Equipment Testing Authority (DGP). The last two organizations are under the control of the Ministry of Public Works.

Research and development

The following institutes undertake research for building:

The Danish Building Research Institute
The Danish Standards Association
The Danish Institute for Testing and Verification
The Union of Danish Civil Engineers
The Danish Academy of Technical Sciences and its research centres
The Technological Institute
The Jutland Technological Institute
University research departments

Most of the research for testing and standards of construction and products is state financed. Testing and control necessary for Official Approval Certificates is financed by the material producers.

Materials and labour

Many building materials used in Denmark are imported, due to a shortage of natural resources, although cement, bricks and some timber products are home produced. To attain high productivity, components for system built timber houses are imported from Norway and Sweden. Costs and standards are both high and the trend has been towards industrialized building techniques imported mostly from West Germany. Traditional construction in Denmark follows the pattern of that in North Germany using bricks and mortar rather than timber. Imported materials are not subject to any restriction and no duty is payable on goods imported from the EEC.

Industrialized building systems are popular in Denmark. Initially these were mainly prefabricated timber systems but systems based on other materials, especially precast concrete, are now common.

The labour force in the construction industry is well trained and geared to high productivity. Vocational training starts in the latter stages of schooling and even the "unskilled" receive training under programmes set up by legislation in 1984 and funded by employers contributions. Almost all construction workers belong to the appropriate craft unions which are affiliated to the Federation of Trade Unions. Local negotiations take place regarding wages and working conditions whilst biennial agreements on general wage increases, hours, and holidays are negotiated at national level.

Further reading

Association of Consulting Engineers et al, Engineering Consultancy in the European Community: Denmark. London: Telford 1990.

D Edwards, "Denmark" in Department of the Environment Planning Control in Western Europe, London: HMSO, 1989 pages 79 to 148.

CONSTRUCTION COST DATA

Cost of labour

The figures below are typical of labour costs in Copenhagen as at January 1991. The "wage rate" is the basis of an employee's income, while the "cost of labour" indicates the cost to a contractor of employing that employee. The difference between the two covers a variety of mandatory and voluntary contributions - a list of items which could be included is given in secton 4.

	Wage rate (per hour) Dkr	*Cost of labour (per hour) Dkr*	*Number of hours worked per year*
Site operatives			
Mason/bricklayer	125	163	1628
Carpenter	125	163	1628
Plumber	125	163	1628
Electrician	125	163	1628
Structural steel erector	125	163	1628
HVAC installer	125	163	1628
Semi-skilled worker	125	163	1628
Unskilled labourer	125	163	1628
Equipment operator	100	130	1628
Watchman/security	100	130	1628
Site supervision			
General foreman	135	176	1628
Trades foreman	135	176	1628
Contractors' personnel			
Site manager	135	176	1628

Cost of materials

The figures that follow are the costs of main construction materials, delivered to site in the Copenhagen area, as incurred by contractors in January 1991. These assume that the materials would be in quantities as required for a medium sized construction project and that the location of the works would be neither constrained nor remote.

All the costs in this section exclude value added tax (VAT - see below).

	Units	*Cost Dkr*
Cement and aggregate		
Ordinary portland cement in 50kg bags	tonne	1030
Coarse aggregates for concrete	m^3	65
Fine aggregates for concrete	m^3	100
Ready mixed concrete (10N/mm^2)	m^3	500
Ready mixed concrete (25N/mm^2)	m^3	800
Steel		
Mild steel reinforcement	tonne	5000
High tensile steel reinforcement	tonne	6600
Structural steel sections	tonne	6000

	Units	Cost Dkr
Bricks and blocks		
Common bricks (size 110 x 220 x 50)	1000	2000
Good quality facing bricks (110 x 220 x 50mm)	1000	2500
Hollow concrete blocks (500 x 190 x 150mm)	1000	6000
Solid concrete blocks (590 x 190 x 100mm)	1000	12 000
Precast concrete cladding units 75mm thick	m^2	160
Timber and insulation		
Softwood sections for carpentry	m^3	2380
Softwood for joinery	m^3	3180
Hardwood for joinery	m^3	30 000
Exterior quality plywood (12mm)	m^2	52
Plywood for interior joinery (12mm)	m^2	119
Softwood strip flooring (129 x 22mm)	m^2	299
Chipboard sheet flooring (12 mm)	m^2	32
100 mm thick quilt insulation	m^2	35
100 mm thick rigid slab insulation	m^2	35
Softwood internal door complete with frames	nr	1389
Glass and ceramics		
Float glass (4 mm)	m^2	100
Sealed double glazing units	m^2	1544
Good quality ceramic wall tiles (150 x 150mm)	m^2	50
Plaster and paint		
Plaster in 50 kg bags	tonne	4000
Plasterboard (13mm thick)	m^2	23
Emulsion paint in 5 litre tins	litre	30
Gloss oil paint in 5 litre tins	litre	100
Tiles and paviors		
Clay floor tiles (110 x 240 x 10mm)	m^2	100
Vinyl floor tiles (300 x 300 x 3mm)	m^2	250
Precast concrete paving slabs (210 x 210 x 70mm)	m^2	75
Clay roof tiles	1000	4800
Precast concrete roof tiles	1000	5700
Drainage		
WC suite complete	nr	7000
Lavatory basin complete	nr	2500
100mm diameter clay drain pipes	m	30
150mm diameter cast iron drain pipes	m	75

Unit rates

The descriptions below are generally shortened versions of standard descriptions listed in five languages (English, French, Italian, German and Spanish) in section 6. Where an item has a two digit reference number (e.g. 05 or 33), this relates to the full description against that number in section 6. Where an item has an alphabetic suffix (e.g. 12A or 34B) this indicates that the standard description has been modified. Where a modification is major the complete modified description is included here and the standard description should be ignored; where a modification is minor (e.g. the insertion of a named hardwood) the shortened description has been modified here but, in general, the full description in section 6 prevails.

The unit rates below are for main work items on a typical construction project in the Copenhagen area in January 1991. The rates include all necessary labour, materials, equipment and, where appropriate, allowances for contractors overheads and profit, preliminary and general items and contractors profit and attendance on specialist trades. All the rates in this section exclude value added tax (VAT - see below).

		Unit	*Rate Dkr*
Excavation			
01	Mechanical excavation of foundation trenches	m^3	300
02	Hardcore filling making up levels	m^2	90
Concrete work			
04	Plain insitu concrete in strip foundations in trenches	m^3	1600
05	Reinforced insitu concrete in beds	m^3	1830
06	Reinforced insitu concrete in walls	m^3	1925
07	Reinforced insitu concrete in suspended floor or roof slabs	m^3	2015
08	Reinforced insitu concrete in columns	m^3	2110
09	Reinforced insitu concrete in isolated beams	m^3	2200
Formwork			
11	Softwood or metal formwork to concrete walls	m^2	450
12	Softwood or metal formwork to concrete columns	m^2	800
13	Softwood or metal formwork to horizontal soffits of slabs	m^2	1050
Steelwork			
17	Fabricate, supply and erect steel framed structure	tonne	7100
Brickwork and blockwork			
18	Precast lightweight aggregate hollow concrete block walls	m^2	300
19	Solid (perforated) concrete blocks	m^2	450
20	Solid (perforated) sand lime bricks	m^2	350
21	Facing bricks	m^2	450

		Unit	*Rate Dkr*
Roofing			
22	Concrete interlocking roof tiles 430 x 380mm	m^2	385
23	Plain clay roof tiles 260 x 160mm	m^2	450
24	Fibre cement roof slates 600 x 300mm	m^2	530
26	Particle board roof coverings	m^2	120
27	3 layers glass-fibre based bitumen felt roof covering	m^2	550
28	Bitumen based mastic asphalt roof covering	m^2	650
29	Glass-fibre mat roof insulation 160mm thick	m^2	450
30	Rigid sheet loadbearing roof insulation 75mm thick	m^2	750
31	Troughed galvanised steel roof cladding	m^2	400
Woodwork and metalwork			
32	Preservative treated sawn softwood 50mm x 100mm	m	20
33	Preservative treated sawn softwood 50 x 150mm	m	30
34A	Single glazed casement window in pine softwood, size 650 x 900mm	each	1000
35	Two panel glazed door in hardwood, size 850 x 2000mm	each	2820
36	Solid core half hour fire resisting hardwood internal flush doors, size 800 x 2000mm	each	1500
37	Aluminium double glazed window, size 1200 x 1200mm	each	3100
38	Aluminium double glazed door, size 850 x 2100mm	each	4300
39	Hardwood skirtings	m	32
Plumbing			
42	UPVC half round eaves gutter	m	65
43	UPVC rainwater pipes	m	60
48	White vitreous china WC suite	each	1700
49	White vitreous china lavatory basin	each	2650
50	Glazed fireclay shower tray	each	2220
51	Stainless steel single bowl sink and double drainer	each	1200
Finishings			
55	2 coats gypsum based plaster on brick walls	m^2	92
56	White glazed tiles on plaster walls	m^2	290
57	Red clay quarry tiles on concrete floor	m^2	300
60	Mineral fibre tiles on concealed suspension system	m^2	270

Approximate estimating

The building costs per unit area given below are averages incurred by building clients for typical buildings in the Copenhagen area as at January 1991. They are based upon the total floor area of all storeys, measured between external walls and without deduction for internal walls.

Approximate estimating costs generally include for mechanical and electrical installations but exclude furniture, loose or special equipment, and external works; they also exclude fees for professional services. The costs shown are for specifications and standards appropriate to Denmark and this should be borne in mind when attempting comparisons with similarly described building types in other countries. A discussion of this issue is included in section 4. Comparative data for countries covered in this publication including construction cost data are presented in Part Three.

Approximate estimating costs must be treated with reserve; they cannot provide more than a rough guide to the probable cost of building.

All the rates in this section exclude value added tax (VAT - see below).

	Cost m^2 *Dkr*	*Cost* ft^2 *Dkr*
Industrial buildings	2500	230
Administrative and commercial buildings	7500	700
Health and education buildings		
General hospitals	10 000	930
Teaching hospitals	10 000	930
Private hospitals	10 000	930
Health centres	10 000	930
Nursery schools	5500	930
Primary/junior schools	5500	510
Secondary/middle schools	5500	510
University (arts) buildings	5500	510
University (science) buildings	5500	510
Management training centres	5500	510
Recreation and arts buildings	10 000	930
Residential buildings		
Social/economic single family housing (multiple units)	6000	560
Private/mass market single family housing 2 storey detached/semidetached (multiple units)	6000	560
Purpose designed single family housing 2 storey detached (single unit)	8000	740
Social/economic apartment housing, low rise (no lifts)	8000	740
Social/economic apartment housing, high rise (with lifts)	8000	740
Private sector apartment building	10 000	930
Student/nurses halls of residence	10 000	930
Homes for the elderly	10 000	930
Hotel	10 000	930

Regional variations

The approximate estimating costs are based on projects in Copenhagen. For other parts of the country no adjustment is required for regional variations.

Value added tax

The standard rate of value added tax (VAT) is currently 22%, chargeable on construction work.

EXCHANGE RATES AND INFLATION

The combined effect of exchange rates and inflation on prices within a country and price comparisons between countries is discussed in section 4.

Exchange rates

The graph below plots the movement of the Danish Krone against sterling and the US dollar since 1980. The figures used for the graph are quarterly and the method of calculating these, and other related issues are discussed in section 4. Denmark is a member of the European Monetary System (EMS). The exchange rate at January 1991 was Dkr 11.3 to the pound sterling and Dkr 5.82 to the US dollar.

The Danish Krone against Sterling and the US Dollar

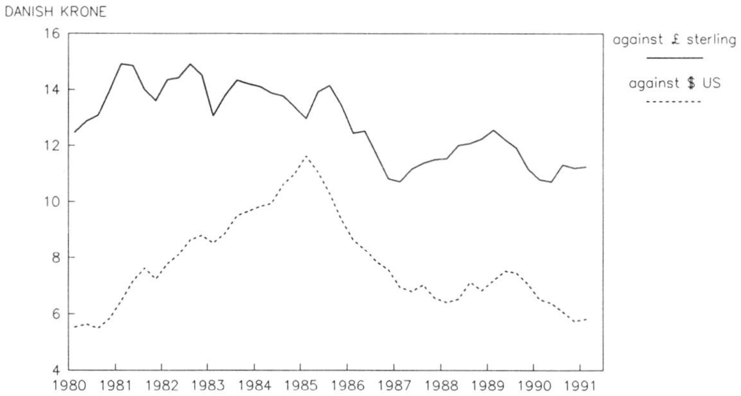

Inflation

The graph and the table below present the consumer price index showing the trend in inflation since 1980. The index has been rebased to 1980=100. Notes on inflation and the distinction between costs and prices are included in section 4.

Consumer price inflation

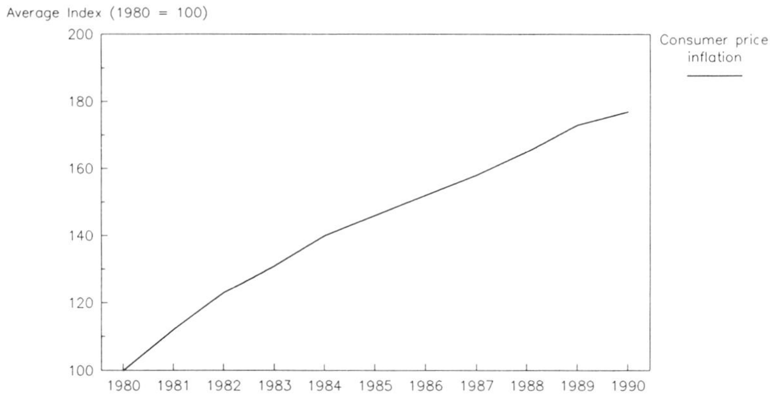

The basis of the figures is the official consumer price index converted to quarterly averages. Inflation peaked in 1981; since then inflation has fallen steadily to less than 3% in 1990.

Consumer price inflation 1980 = 100

	1981	1982	1983	1984	1985	1986	1987	1988	1989	1990
average index	113	123	132	140	146	152	158	165	173	178
annual change (%)	11.7	10.1	6.9	6.3	4.7	3.6	4.0	4.6	4.8	2.6

USEFUL ADDRESSES

Ministry of the Environment
(Miljoministeriet), Slotsholmsgade 12
1216 Kobenhavn K

Ministry of Housing
(Boligministeriet), Slotsholmsgade 12
1216 Kobenhavn K

Danish Contractors Association
(Entreprenorforeningen), Norre Voldgade 106
1358 Kobenhavn K

Association of Consulting Architects
(Praktiserende arkitekters rad (PAR), Bredgade 66
1260 Kobenhavn K

Building Industry Association
(Byggesocietetet), Vimmelskaftet 47
1161 Kobenhavn K

Foreningen af Radgivende Ingeniorer, FRI.
Esplanaden 34, 1263 Kobenhavn K
Tel: (1) 113737

Statens Byggeforskningsinstitut
Dr Meergaardsvej 15, Postboks 119
2970 Horshdm
Tel: (2) 86 55 33

Dansk Standardiseringsraad
Aurehojvej 12, Postbox 77
DK-2900 Hellerup

Finland

KEY DATA

Population

Population (1989)	4.95m
Urban population	60%
Population under 15	19.3%
Population over 65	13.2%
Average annual growth rate (1983 to 1988)	0.4%

Geography

Land area	338 000 km²
Agricultural area	25 000 km²
Capital city	Helsinki (population 485 626)

Economy

Monetary unit	Markka (Fmk)
Exchange rate (January 1991) to	
the ECU	Fmk 4.93
the pound sterling	Fmk 7.05
the US dollar	Fmk 4.93
Average annual inflation (1984 to 1989)	6.6%
Gross Domestic Product (GDP)	Fmk 438bn
GDP per capita (1989)	Fmk 88 485
GDP growth (real) (1980 to 1988)	3.2%
Private consumption (% of total)	53.4%
Public consumption (% of total)	20.5%
Investment (% of total)	26.4%
Central government expenditure as percentage of GDP	35.9%

Construction

Total value of construction output (1989)	Fmk 85bn
Construction output per capita	Fmk 17 171
Construction output as a percentage of GDP	19.4%

THE CONSTRUCTION INDUSTRY

Construction output

In 1989 the value of gross construction output in Finland was about Fmk 85 billion which represents about 19.4% of GDP. The breakdown of the output by type of work is shown below.

Construction output 1989

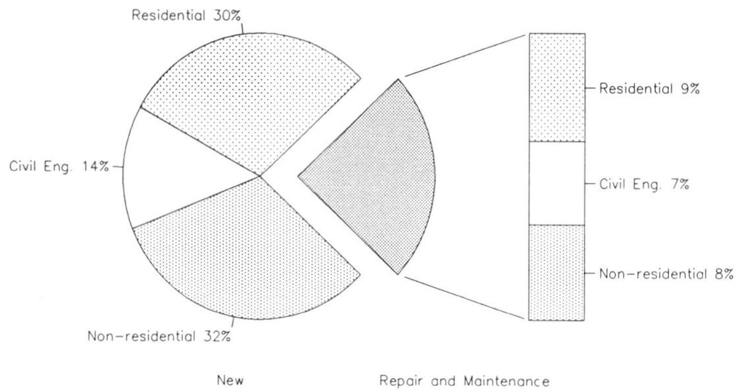

During the mid 1970s construction in Finland accounted for a much higher percentage of GDP and the proportion of output accounted for by housing was particularly high. The share of repair and maintenance has so far been small, but as the sizeable building stock erected in the 1970s needs repair and maintenance around the turn of the century, the situation is going to change. By the year 2010, repair and maintenance is expected to account for about 40% of total production.

There was a boom in construction from 1988 to 1990 but this has now levelled off.

Finnish contractors were relatively late starters in the race for foreign contracts. The real breakthrough in Finnish contracting exports took place in the second half of the 1970s. In 1971, for example, the total foreign invoicing by Finnish construction exporters amounted to only about US$100 million, but reached US$1000 million by the beginning of the 1980s.

The foreign output of Finnish building contractors is equal to 7% of the volume of domestic construction, a high ratio by international standards. By 1988, Finnish contractors had been involved in over 500 projects in 50 countries, a large majority of them in the Soviet Union. The total value of all the foreign construction projects in progress was about Fmk 6100 million at the end of 1988. In international construction markets, Finnish contractors have had a competitive advantage, mainly in the fields of industrial works, public buildings, civil engineering, water supply, sewage treatment, infrastructure and precast element technology.

The share of foreign construction has, however, diminished as the special trade arrangement between Finland and the USSR was abolished and the clearing system was replaced by trade in freely convertible currencies. The internal turmoil in the Soviet Union is a risk factor for Finnish built Soviet projects. The Soviet Union owes Finland a balance of about Fmk 10 billion from bilateral trade and is expected to pay it off by the end of 1992.

Characteristics and structure of the industry

In 1986 there were nearly 400 contractors in Finland, of which eight firms employed more than 1000 persons and did over a third of the work and 16 firms employed over 500 and did 44% of the work. This represents a very high degree of concentration compared with most European countries. There are eight Finnish contractors in Building magazine's "Top 300 European Contractors" and four in the top 100. There are four in the Engineering News Record's "Top 250 International Contractors".

Major Finnish contractors

Major contractors	*Place in Building's "Top 300 European Contractors"*	*Exports (% of turnover)*
Haka Corporation	29	5
Yit Construction Corporation	34	5-10
Polar Construction Corporation	35	less than 5
Puslimatka Oy	47	5-10
Lemminkainen Oy	106	15
Rakennus Ruola Oy	112	n.a.
Hartela Construction	168	n.a.
Terasbetoni Oy	172	15

Source: Building

The most prevalent method of organizing contracting in Finland is the main contractor system where a single contractor bears the overall responsibility for construction.

Another common method is the turnkey system where design is also part of the tender. Arguments in favour of this method claim that it allows the contractor to submit a lower bid as he can adapt design to his production methods. The image of this procurement method has improved during the 1980s.

The separate contracts method is far less common. Under this method the project is divided into several sub-projects. The client, through his consultant, invites separate tenders and enters into contracts directly with various contractors. Most often the contracts are for works by different trades, such as frame construction, prefabricated components, heating, plumbing, air conditioning, electrical services, etc.

In Finland, the architect is the leader of the design team unless the exceptional nature of the undertaking demands other arrangements. The team leader outlines the building project, co-ordinates and supervises the work of other designers. He is also responsible for the development of plans, their implementation and the overall buildability of the design.

The structural design of the building is entrusted to an independent structural engineer. Geotechnical design is treated as a separate field and such services are often needed at the early phases of projects as foundation conditions are considered. Prefabricated component design may either be part of structural design or may be the responsibility of the manufacturer.

Heating, plumbing and sewerage, ventilation and electrical design constitute separate fields.

Generally, design offices are small consisting of teams of a few people. The mid-sized offices employ 30 to 40 designers, and there are only 30 to 40 offices larger than that in all of Finland. Sometimes the client's organization, especially in the public sector, has design know-how, but most often an independent architect and a design office are engaged.

Finnish construction is highly industrialized. The level of prefabrication of structures is high and prefabricated components are widely used.

Selection of design consultants

The client selects qualified designers, usually based on previous experience or recommendations. The scope of the service and fees are agreed afterwards. Clients rarely invite competitive tenders from designers.

Architectural competitions, closed or open, are often arranged when designing large or especially demanding projects. Rules approved by the sector organizations are applied to competitions which are usually judged by a jury. In an open competition several proposals are awarded a prize and the one considered best is usually selected for implementation. The number of proposals submitted to an open competition in Finland can be over 200. In a closed competition, that is, one by invitation only, proposals are normally requested from 3 to 5 groups of designers and the participating designers all receive equal fees, independent of the quality of their proposal. Normally, the winner is recommended to undertake the project.

Contractual arrangements

Normally the client appoints a contractor on the basis of selected competitive tenders. In certain cases only a single contractor is contacted. In the public sector, project tenders must always be invited from at least five and, in large building construction projects, from at least ten contractors. Contractors are notified of this through advertisements in newspapers.

Generally, the tender includes plans and a work specification. Each contractor prepares his own bill of quantities as a basis for his tender price quotation. The bill of quantities is not, however, part of the contract.

A table is often drawn-up to rank the offers. It is used to study, for instance, the effect of technical solutions on price, delivery times and terms, etc. The most advantageous offer is not necessarily the cheapest one, but quite often the cheapest one is selected. The financial standing of the contractor is also established.

Usually the main contract is a lump sum contract, but the method of setting a target price has become more prevalent. If the main contractor can cut total costs, his remuneration grows larger, and if costs exceed the target level, he is liable for them.

The contract documents include the tender itself, the payment programme, other correspondence between the client and contractor, contract negotiation documents and the contract proper on a standard form.

The state has very few detailed rules for its own building projects, but among the most important ones are the General Conditions of Construction Contract and the Building Information File. The latter consists of instructions - not rules - by the National Board of Public Construction that provide clear practical advice, for instance, on labour inputs. The File is intended to facilitate the planning of the works on the basis of data collected from practical experience. The General Conditions of Construction Contract (YSE 1983) have been adopted by both public and private builders. They are not made part of the contract; a mere reference to them suffices.

Liability and insurance

The designer has both a legal and a moral responsibility. The legal responsibility is based on a design contract and the attached general terms of contract of consultancy. That responsibility extends to, for instance, carrying out the tasks of the design contract in the agreed time and liability for any possible errors. In monetary terms the designer's liability is normally limited only to the amount of his design fee. The normal liability period for the designer is 10 years. A product liability law, which is expected to shift some legal liability from the contractor to the designer, goes into effect in the autumn of 1991.

Everyone involved in a project shares in the moral responsibility, the greatest burden belonging to who most can influence the end result. Therefore, the designer's responsibility is more moral in nature and the contractor bears more of the legal responsibility.

Inexperienced clients often have difficulty in evaluating the quality of the design work on the basis of drawings and other documents. Especially for one-off and large projects, clients often hire independent project consultants to ensure the desired end result.

The contractor is liable to the client for any damages resulting from the contractor's failure to satisfy the conditions of contract. In general, the contractor is liable for:

- plans he makes or has made plus information and work performed on the basis of the plans
- measurements he has made
- extra works and modifications
- the fulfilment of contractual obligations by all parties under the main contract.

The contractor is responsible for performing in accordance with the contract for one year in the case of building and electrical work and two years in the case of HVAC work. The defects liability period begins on the date the work passes an acceptance inspection or as the building is taken into use without an inspection.

The contractor is obliged to repair any deficiencies and defects detected during the liability period that he cannot prove to have been caused by factors unrelated to his performance and which are pointed out to him before the end of this period. He is not, however, obliged to perform such repairs that are connected with the regular maintenance of equipment or are required by the faulty operation or neglected or improper maintenance.

To guarantee performance of the works, the contractor must in connection with the acceptance inspection, give a bond, generally equal to 2% of the contract price. The bond is returned to the contractor after the liability period has expired.

The contractor's liability ends at the expiration of the liability period, except for such flaws and deficiencies that the client can prove to be the result of the contractor's gross negligence or intentional breach of contract and which the client could not reasonably have been expected to identify during the acceptance inspection or the defects liability period. The contractor is released from that liability after 10 years have passed from the acceptance inspection.

Designers are increasingly taking out professional indemnity insurance; presently about half of architectural designers carry it. It normally indemnifies losses up to Fmk 1 million with a Fmk 10000 deductible. A further increase in the number of insured designers is likely with their greater legal liability after Autumn 1991.

Development control

A building permit is required for a new building or an extension. Municipalities issue special application forms for that purpose. The application must be accompanied by the following statutory documents:

- proof of the fact that the applicant is in possession of the lot or construction site
- a land register map
- a ground plan of the lot or construction site in triplicate showing buildings to be erected and demolished (in cities this must show green areas, playgrounds, access roads, parking and loading areas and other yard arrangements)
- design documents with explanations in triplicate.

The building committee may also demand that the applicant provide data on specific working and structural drawings, strength calculations, accounts of building materials, quality of foundations, used structures and the soil of the construction site, etc.

In most instances the client obtains the building permit which should be secured before the contract is signed. Exceptions are made in cases of compelling urgency.

Before a decision is made on the permit, the committee must obtain an expert opinion from health, fire or any other authorities as necessary. As building authorities study applications for a building permit, they must ascertain that the building conforms to the ratified plan and site layout plan and see that other rules and regulations of construction are complied with. The Finnish National Building Code specifies the statutory requirements regarding the facts that need to be established before granting a building permit. It includes rules that bind the authorities as well as rules for solutions that conform to the set requirements.

A decision on a building permit is rendered 14 days after presentation of the proposal for public inspection during which period the owners of adjoining properties, industrial buildings and the neighbourhood at large may appeal. A building permit remains in force for three years.

The application process for a building permit lasts 3 to 6 months, depending on the municipality, and about 5% of applications are rejected. A large majority of applications are accepted, as the seeking of a building permit constitutes a negotiation process with the authorities; that is, the applicant knows what may be built on his lot according to the city plan and whether a special permit is possible, etc.

Acknowledgements

The authors are grateful to Tampere University of Technology and particularly to Mr Harri Huru and Miss Mervi Forsblom for the very comprehensive information they have provided on Finland which constitutes the basis for this section.

CONSTRUCTION COST DATA

Cost of labour

The figures below are typical of labour costs in Helsinki as at January 1991. The "wage rate" is the basis of an employee's income, while the "cost of labour" indicates the cost to a contractor of employing that employee. The difference between the two covers a variety of mandatory and voluntary contributions - a list of items which could be included is given in section 4.

	Wage rate (per hour) Fmk	*Cost of labour (per hour) Fmk*	*Number of hours worked per year*
Site operatives			
Mason/bricklayer	78.70	131.40	1920
Carpenter	66.50	111.50	1920
Plumber	68.40	114.20	1920
Electrician	50.40	84.20	1920
Semi-skilled worker	46.00	76.80	1920
Unskilled labourer	45.90	76.70	1920
Site supervision	*(per month)*	*(per month)*	
General foreman	14 700	24 550	
Trades foreman	9900	16 530	
Clerk of works	10 500	17 540	
Contractors' personnel			
Site manager	15 900	26 550	
Resident engineer	14 500	24 200	
Junior engineer	12 400	20 700	
Consultants' personnel			
Senior architect	22 500	37 580	
Senior engineer	22 000	36 740	
Qualified architect	15 000	25 050	
Qualified engineer	15 500	25 900	

Cost of materials

The figures that follow are the costs of main construction materials, delivered to site in the Helsinki area, as incurred by contractors in January 1991. These assume that the materials would be in quantities as required for a medium sized construction project and that the location of the works would be neither constrained nor remote.

	Units	*Cost Fmk*
Cement and aggregate		
Ordinary portland cement in 50kg bags	tonne	475
Coarse aggregates for concrete	m^3	35
Fine aggregates for concrete	m^3	35
Ready mixed concrete (K20)	m^3	350
Ready mixed concrete (K30)	m^3	400
Steel		
Mild steel reinforcement	tonne	3500
High tensile steel reinforcement	tonne	3500
Bricks and blocks		
Common bricks (size 285 x 85 x 85mm)	1000	1120
Good quality facing bricks	1000	4000
Hollow concrete blocks (590 x 125 x 190mm)	1000	6000
Solid concrete blocks (600 x 200 x 200mm)	1000	10 000
Timber and insulation		
Softwood sections for carpentry	m^3	1800
Softwood for joinery	m^3	1800
Exterior quality plywood	m^2	75
Plywood for interior joinery	m^2	75
Softwood strip flooring (32mm)	m^2	125
Chipboard sheet flooring (22mm)	m^2	35
100mm thick quilt insulation	m^2	17
100mm thick rigid slab insulation	m^2	50
Softwood internal door complete with frames and ironmongery	nr	700
Glass and ceramics		
Sealed double glazing units	m^2	1100
Good quality ceramic wall tiles (150 x 150mm)	m^2	85
Plaster and paint		
Plaster in 50 kg bags	tonne	2500
Plasterboard (13mm thick)	m^2	15
Emulsion paint in 5 litre tins	litre	110
Gloss oil paint in 5 litre tins	litre	170

	Units	Cost Fmk
Tiles and paviors		
Clay floor tiles	m^2	125
Vinyl floor tiles	m^2	75
Precast concrete paving slabs	m^2	90
Clay roof tiles	1000	4000
Precast concrete roof tiles	1000	4150
Drainage		
Lavatory basin complete	nr	1200

Unit rates

The descriptions below are generally shortened versions of standard descriptions listed in five languages (English, French, Italian, German and Spanish) in section 6. Where an item has a two digit reference number (e.g. 05 or 33), this relates to the full description against that number in section 6. Where an item has an alphabetic suffix (e.g. 12A or 34B) this indicates that the standard description has been modified. Where a modification is major the complete modified description is included here and the standard description should be ignored; where a modification is minor (e.g. the insertion of a named hardwood) the shortened description has been modified here but, in general, the full description in section 6 prevails.

The unit rates below are for main work items in a typical construction project in the Helsinki area in January 1991. The rates include all necessary labour, materials, equipment and, where appropriate, allowances for contractors overheads and profit, preliminary and general items and contractors profit and attendance on specialist trades.

		Unit	Rate Fmk
Excavation			
01	Mechanical excavation of foundation trenches	m^3	190
02	Hardcore filling making up levels	m^2	55
Concrete work			
04	Plain insitu concrete in strip foundations in trenches	m^3	920
05	Reinforced insitu concrete in beds	m^3	1130
06	Reinforced insitu concrete in walls	m^3	1300
07	Reinforced insitu concrete in suspended floor or roof slabs	m^3	1350
08	Reinforced insitu concrete in columns	m^3	1390
09	Reinforced insitu concrete in isolated beams	m^3	1425
Formwork			
11	Softwood formwork to concrete walls	m^2	250
12	Softwood or metal formwork to concrete columns	m^2	490
13	Softwood or metal formwork to horizontal soffits of slabs	m^2	700

		Unit	Rate Fmk
Reinforcement			
14	Reinforcement in concrete walls	tonne	5500
15	Reinforcement in suspended concrete slabs	tonne	6000
Steelwork			
17	Fabricate, supply and erect steel framed structure	tonne	4000
Brickwork and blockwork			
18	Precast lightweight aggregate hollow concrete block walls	m^2	180
19	Solid (perforated) concrete blocks	m^2	275
21	Facing bricks	m^2	300
Roofing			
22	Concrete interlocking roof tiles 430 x 380mm	m^2	240
23	Plain clay roof tiles 260 x 160mm	m^2	270
24	Fibre cement roof slates 600 x 300mm	m^2	300
26	Particle board roof coverings	m^2	70
27	3 layers glass-fibre based bitumen felt roof covering	m^2	300
28	Bitumen based mastic asphalt roof covering	m^2	400
29	Glass-fibre mat roof insulation 160mm thick	m^2	300
30	Rigid sheet loadbearing roof insulation 75mm thick	m^2	500
31	Troughed galvanized steel roof cladding	m^2	270
Woodwork and metalwork			
32	Preservative treated sawn softwood 50mm x 100mm	m	25
33	Preservative treated sawn softwood 50 x 150mm	m	35
34	Single glazed casement window in hardwood, size 650 x 900mm	each	600
35	Two panel glazed door in hardwood, size 850 x 2000mm	each	1845
36	Solid core half hour fire resisting hardwood internal flush doors, size 800 x 2000mm	each	830
37	Aluminium double glazed window, size 1200 x 1200mm	each	2000
38	Aluminium double glazed door, size 850 x 2100mm	each	2600
39	Hardwood skirtings	m	30
Plumbing			
48	White vitreous china WC suite	each	1200
49	White vitreous china lavatory basin	each	1500
50	Glazed fireclay shower tray	each	1500
51	Stainless steel single bowl sink and double drainer	each	750
Finishings			
55	2 coats gypsum based plaster on brick walls	m^2	60
56	White glazed tiles on plaster walls	m^2	185
57	Red clay quarry tiles on concrete floor	m^2	180

Approximate estimating

The building costs per unit area given below are averages for all of Finland incurred by building clients for typical buildings at January 1991. They are based upon the total floor area of all storeys, measured between external walls and without deduction for internal walls.

Approximate estimating costs generally include for mechanical and electrical installations but exclude furniture, loose or special equipment, and external works; they also exclude fees for professional services. The costs shown are for specifications and standards appropriate to Finland and this should be borne in mind when attempting comparisons with similarly described building types in other countries. A discussion of this issue is included in section 4. Comparative data for countries covered in this publication, including construction cost data are presented in Part Three.

Approximate estimating costs must be treated with reserve; they cannot provide more than a rough guide to the probable cost of building.

	Cost m^2 *Fmk*	*Cost* ft^2 *Fmk*
Industrial buildings	4740	440
Administrative and commercial buildings	5850	545
Health and education buildings	7070	660
Residential buildings	5150	480

Regional variations

The approximate estimating costs are based on average costs for Finland. Adjust these costs by the following factors to take account of regional variations:

Central Finland	: -12% to +12%
Northern Finland	: +38%
Helsinki and environs	: +25% to +50%

EXCHANGE RATES AND INFLATION

The combined effect of exchange rates and inflation on prices within a country and price comparisons between countries is discussed in section 4.

Exchange rates

The graph below plots the movement of the Finnish Markka against sterling and the US dollar since 1980. The figures used for the graph are quarterly and the method of calculating these, and other related issues are discussed in section 4. The exchange rate at January 1991 was Fmk 7.05 to the pound sterling and Fmk 4.93 to the US dollar.

The Finnish Markka against Sterling and the US Dollar

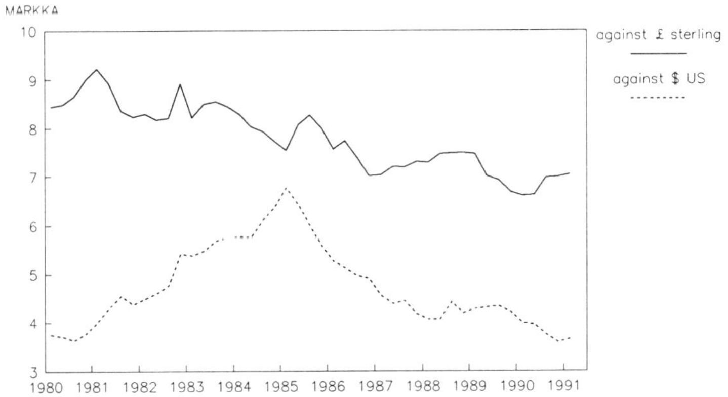

Inflation

The graph and the table below present a building cost index showing the movement in general building costs since 1980. The index has been rebased to 1980=100. Notes on inflation and the distinction between costs and prices are included in section 4.

Building cost index

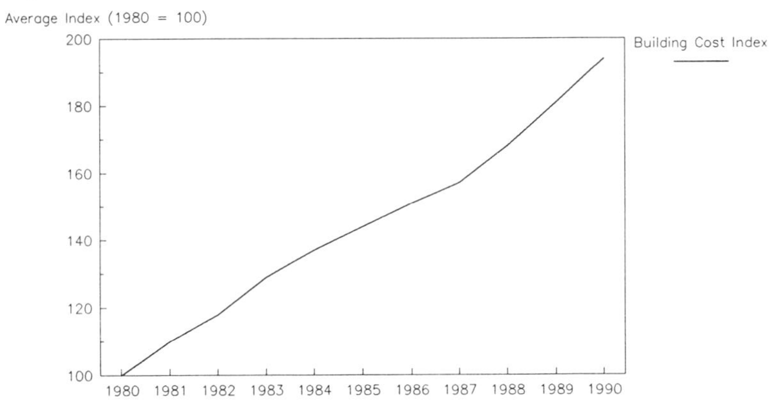

Building cost index

Building cost inflation peaked in 1981 but has, in recent years, returned to levels prevailing in the early 1980s.

Building cost index 1980 = 100

	1981	1982	1983	1984	1985	1986	1987	1988	1989	1990
average index	110	118	129	137	144	151	157	168	181	194
annual change (%)	10.0	7.3	9.3	6.2	5.1	4.9	4.4	7.0	7.7	7.2

USEFUL ADDRESSES

Ympäristöministeriö (Ministry of the Environment)
(building and physical planning)
PO Box 399
SF-00121 Helsinki

Valtiovarainministeriö (Ministry of Finance)
(in charge of financial aspects of works)
Snellmaninkatu 1 A
SF-00170 Helsinki

Rakennushallitus (National Board of Building)
(in charge of works in progress under
the Ministry of Finance)
PO Box 237
SF-00531 Helsinki

Liikenneministeriö (Ministry of Communications - Transport)
(in charge of works in the fields of communications, roads, etc.)
Eteläesplanadi 16
SF-00130 Helsinki

Valtion teknillinen tutkimuskeskus (Technical Research Centre of Finland)
(building research body)
Vuorimiehentie 5
SF-02150 Espoo

Suomen Rakennusurakoitsijaliitto (The Association of General Contractors in Finland)
Eteläranta 10
SF-00130 Helsinki

Suomen Arkkitehtiliitto SAFA (The Finnish Association of Architects)
Eteläesplanadi 22 A
SF-00130 Helsinki

Suomen Rakennusinsinöörienliitto (Association of Finnish Civil Engineers)
Meritullinkatu 16 A 5
SF-00170 Helsinki

Maanmittausinsinöörien Liitto (The Finnish Association of Surveyors)
Nervanderinkatu 5 D 42
SF-00100 Helsinki

Suomen Standardisoimisliitto SFS (Finnish Standards Association SFS)
PO Box 205
SF-00121 Helsinki

Tilastokeskus (Central Statistical Office of Finland)
PO Box 504
SF-00101 Helsinki

France

KEY DATA

Population

Population (1989)	55.95m
Urban population	74%
Population under 15	20.2%
Population over 65	15.6%
Average annual growth rate (1983 to 1988)	0.4%

Geography

Land area	547 000 km²
Agricultural area	313 000 km²
Capital city	Paris (population 2 176 243 town, 8 706 963 suburbs)

Economy

Monetary unit	Franc (FFr)
Exchange rate (January 1991) to	
the ECU	FFr 6.95
the pound sterling	FFr 9.94
the US dollar	FFr 5.14
Average annual inflation (1984 to 1989)	3.5%
Gross Domestic Product (GDP)	FFr 5159bn
GDP per capita (1989)	FFr 92 289
GDP growth (real) (1980 to 1988)	1.9%
Private consumption (% of total)	60.1%
Public consumption (% of total)	18.8%
Investment (% of total)	20.9%
Central government expenditure as percentage of GDP	49.4%

Construction

Total value of construction output (1989)	FFr 557.9bn
Construction output per capita	FFr 9971
Construction output as a percentage of GDP	10.8%

THE CONSTRUCTION INDUSTRY

Construction output

The value of gross construction output in France in 1989 was FFr 557.9 billion or 79.2 billion ECUs. In addition, another FFr 152.8 billion or 21.7 billion ECUs is estimated to be produced by other sectors, DIY or the "black economy". Construction industry output is 10.8% of GDP. After a period of substantial decline, the level of output has been increasing since 1985, but it is expected to level off in the next few years. The breakdown by type of work is shown below.

Construction output 1989

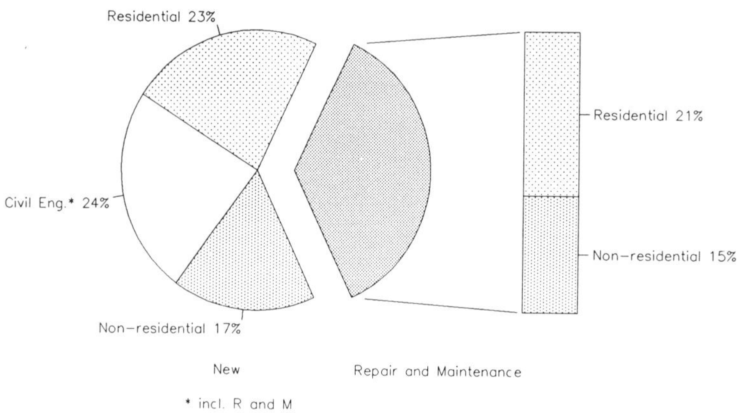

The proportion of civil engineering in total output is very high while non-residential construction is very low.

Characteristics and structure of the industry

Architects have a title protected by law and are required to sign the application for a building permit. Practices are generally small. There are 23 000 architects, 404 per million inhabitants, and the number of building engineers is estimated at about the same. The French construction industry, as indeed many aspects of French life, is dominated by engineers who are prevalent in client organizations and contractors as well as in design organizations in the public and private sector. Many operate in *Bureaux d'Etudes Techniques* which are design organizations which may undertake the whole design, and also operate as subcontractors to architects and contractors. They range from those employing more than 1000 to very small firms. The title of engineer is not protected by law. Quantity Surveyors and Building Surveyors do not exist as such in the French system though there are *économistes* formerly known as *métreurs-vérificateurs*, who are the measures and checkers, but with a status below the engineer and architect.

The traditional system of contracting is the separate trades system which favours smaller specialist trade firms. However, in the post-war period, main contracting has been on the increase, mainly for very large projects and the main contractor in turn subcontracts specialized work. In 1988, there were only 200 firms employing more than 200 persons. There is a large gap between the major firms and the mass of small contractors. Because of mergers, the number of large contractors has been reduced from ten in 1985 to six major groups by January 1991 as shown below.

Major French contractors

Major contractors	*Head office*	*Place in Building's "Top 300 European Contractors"*	*Place in ENR's "Top 250 International Contractors"*	*Exports (% of turnover)*
Bouygues	Guyancourt	1	15	20
SGE	Paris	4	13	35
Dumez	Nanterre	8	8	65
SAE	Paris	9	17	35
Spie Batignoles	Cergy -Pontoise	10	21	30
GTM Entrepose	Nanterre	15	25	25

Source: Building; Engineering News Record

These companies have individual operating companies which are often counted separately.

All these groups undertake a wide range of work covering both building and civil engineering. All of them are very active in export markets. Indeed French contractors received more export work than any other European contractors in 1989, accounting for about a quarter of the output by European contractors abroad, equivalent to some FFr 46 billion.

There were 43 French companies in Building magazine's "Top 300 European Contractors" in 1989 and 16 in the top 100.

Clients and finance

In 1988 public sector clients in France accounted for 24% of building work and 71% of civil engineering work, equivalent to 37% of all new work. Local government is the most important type of public sector client in both building and civil engineering. The role of the state is, in fact, greater than these figures indicate because it often acts as an initiator of projects which may be financed by private money or very often by a combination of private and public funds in *Sociétes Mixtes* (SEMs).

For example, the *Délégation à l'Aménagement du Territoire et à l'Action Régionale* (DATAR), the government agency in charge of regional development, provides grants of up to 25% of investment in land, buildings and equipment purchased during the first three years of operation. These grants are available in most parts of western and southern France, as well as in the traditionally industrial areas of the north and east with redevelopment problems. Only substantial companies can take advantage of the grants. Grants vary according to location but will be between 12% and 25% of the total investment, payable in stages.

Investors not eligible for DATAR grants may receive funds from local authorities or from the European Community.

The majority of business parks and industrial zones in France are owned and operated by municipalities or Chambers of Commerce, which have the status of public authorities and are active in promoting the development of infrastructure. They have a total budget of about FF10 billion, of which about a third goes on infrastructure.

Chambers of Commerce usually sell or lease land and the buildings on it. One of the most popular forms of financing building is a form of leasing known as *crédit bail*. This includes an option to buy the building for FF1 after a 15-25 year period. Municipalities can subsidize up to 25% of new building with their own funds. They may also use their better financial conditions for borrowing to give loans at beneficial rates to companies. In some cases land may be given free to entrepreneurs.

In enterprise zones companies may, under certain conditions, be exempt from corporate income tax for 10 years. In this case no other government grants would be available, but other forms of grant or loans might be. Thus, for industrial development, grants may account for well over 30% of the total finance. Loans on favourable conditions may be over 50%. The investor's share in a new investment may be less than 20% and is often below 30%. Grants are also available for research and development centres and development operations.

The usual sources of other finance for private work, especially commercial or industrial projects, are banks or insurance companies. In France, these companies are closely connected with business and, indeed, with the building and civil engineering industry. Many of the large contractors are partly owned by banks or insurance companies and the *bureau d'etudes* not owned by contractors are often closely linked to financial institutions. This can make the process of getting finance for a project easier because *bureaux d'etudes* may liaise between client and financier. Moreover, French banks have traditionally provided long-term as well as short-term finance. They spread their risks, however, by limiting the size of loans to any one company. As a result many French businesses have accounts at more than one bank to obtain adequate loan funding.

Selection of design consultants

The rules for the selection of the design team in the public sector are very clearly laid down. The team must consist of an architect and an engineer - the type of project determines which of them is the team leader. The architect and/or engineer or the client then appoints the other professionals including a *bureau d'études* and an *économiste*.

The method of selection of the principal designer is by some form of competition *(concours)* which must be announced in the *Bulletin Official des Annonces des Marchés Publics* and in the Press, notably in the weekly magazine *Le Moniteur*. The selection method varies according to the size of project. Below FF450 000 it is on the basis of past performance. From FF450 000 to 900 000 there is a form of simplified competition and submissions of experience are formally examined by a jury. Over FF900 000 a publicly announced competition takes place with plans, elevations, perspective views etc. Fees are paid for submissions.

In the private sector, competitions are relatively rare. Clients may select architects known to them or whose reputation for the specific type of building is high and then commission them directly. They may invite a contractor to participate in the design at an early stage. Alternatively, a limited competition may be adopted. There are few in-house design teams in the private sector.

The role assigned to the architect in the private sector varies and may include both design and site supervision or be much less. In the private sector, as in the public sector, there are different practices as to whether the architect or client appoints the other professionals.

There are a number of different contracts used in the private sector, notably those issued by the *Ordre des Architectes*, the *Union Nationale des Syndicats Francais d'Architects*, the *Syndicat National des Architects Agrées* and *des Mâitre d'Oeuvre en Bâtiment*. In the public sector the *Acte d'Engagement*, the consultants contract, differs according to the extent of the architect's involvement.

Contractual arrangements

As in the case of designer selection, the choice of contractor varies from the public sector where the rules are clearly defined, to the private sector where the client can determine his own method.

A system of classification of contractors, based mainly on the criteria of size and qualification, is administered by the Professional Organization for the Qualification and Classification of the Building Industry and Associated Activities (OPQCB). This body comprises representatives of the construction professions, contractors' federations, and technical control bodies. The classification is recognized nationwide in the French building industry. On presentation of substantiated documentary evidence that the contractor has carried out certain types of project, firms receive a certificate, renewed annually, upon which a "curriculum vitae" is given. This is the basic document which determines the contractors who may be invited to tender for any project.

The traditional form of contracting is separate trades contracting (*lots séparés*) or a grouping of contractors. In the public sector, the selection must be by competition, except in a few special circumstances. The call for tender is known as the *appel d'offres*. However, because in France the detailed design work is usually undertaken by the contractor rather than the consultant designer, the design cannot be finalized at the tender stage. The *Code des Marchés Publics* - the regulating document for the public sector - permits the acceptance of the tender which is "economically the most advantageous".

It can happen, however, that when submissions are studied, none of the tenders is found to be satisfactory and the bids may be called *infructueux* (literally, unfruitful). The tender may be cancelled and the project again put out to tender with a different (more detailed) specification, or it may be negotiated. It is understood that the latter is increasingly becoming the usual practice, and of course the *maître d'ouvrage* and the *maître d'oeuvre* have the advantage of having seen the proposals of all the bidding contractors, so they can take the best ideas from each. Another reason why bids may be declared *infructueux* is that there may be a price ceiling for certain types of work, notably public housing. If no bid is lower than the ceiling, there is again an opportunity for negotiation.

Contractors may also be selected on a combination of design and construction factors. This is known as an *appel d'offres avec concours* and has advantages when technical or aesthetic requirements justify in-depth studies by the designer and contractor working together.

Negotiation is also permitted for a number of specific reasons listed in the *Code des Marchés Publics*, such as for buildings where a security factor is involved, where there is great urgency, or where buildings are erected as part of a research project.

In the private sector clients may use main contractors or separate trades contractors. The normal method is to have an *appel d'offres* of four to six contractors. They may then negotiate on design, price and time and select the most appropriate for the project. Alternatively, they may negotiate directly with a contractor. A negotiated contract is known as *gré à gré*.

The standard form of contract, most commonly used for private work is the *norme PO3.001* published by *l'Association Francaise de Normalisation* and it is not uncommon for architects to draw up their own supplementary clauses. For public work, the building ministries involved in construction have their own forms.

Contract documents for each trade contractor usually comprise the following:

- *série de plans* (drawings)
- *spécification*
  - (a) *le devis descriptif* (for workmanship)
  - (b) *le cahier des préscriptions techniques particulières* (for materials).
  - (c) *préscriptions techniques generales* (National Code of Practice)
- contract conditions, embodying:
  - (a) *cahiers des conditions et charges générales* (Standard Form of Contract)
  - (b) *cahiers des charges particulières* (special conditions of contract)
- *la lettre de commande* (contract agreement)
- *le planning* (the agreed progress chart)
- a detailed break-down of the tender price to provide rates for the purpose of pricing variations. Occasionally this might take the form of a bill of quantities in respect of rates only
- *soumission* (form of tender).

Standard published schedules of rates, known as the *séries des prix* are often incorporated in contracts for measuring variations. There are a number of priced schedules published for various purposes. The most widely used, including by government departments, is that known as *La Série Centrale* prepared by the Society of Architects in conjunction with the building employers. A monthly supplement is issued with coefficients to bring the prices for each trade up to date, for all parts of France.

Although *marché à forfait* (lump sum contracts) are the most popular form of contract in France, re-measurement contracts are also used. In negotiated tenders the rates are often agreed between the parties, but in competitive tendering the contractors quote percentages on or off (usually off) the *série de prix*.

All contracts contain a fluctuations clause and an agreed formula based upon price indices published by the Ministry of Construction is used to arrive at the additional payments due. The price indices are issued monthly and cover individual trades, thus enabling the formula to be applied to interim payments. Interim payments are normally monthly and subject to a 10% retention, except in the case of very large contracts where a maximum retention sum may be fixed. Half of the retention fund is released on practical completion of the contract, the remainder being paid to the builder at the end of the six months maintenance period.

Variations can be authorized to the work described in the contract documents. If possible their value is agreed on the contract rates or on a lump sum basis before the work is put in hand, omissions being valued at 85% of the contract rates. The final account is prepared by the contractor and checked by the architect.

Liability, insurance and inspection

The French have the most far-reaching, comprehensive and formalized system of liability and insurance in Europe and have a related system of building control.

The most interesting aspect of their system is the special decennial responsibility introduced originally by Napoleon in the Civil Code and modified over two centuries. The Spinetta Law of 1978 considers all parties involved in the construction process - contractors, architects, engineers, *bureaux de contrôle*, manufacturers and importers of products and components, even individuals or companies involved in the sale of the finished building - to carry some responsibility. Decennial responsibility commences with the handover of the works to the client which is a recognized formal event in the process. This liability covers solidity of construction and fitness for purpose.

In addition, there is responsibility for perfect achievement for one year from handover; biennial responsibility for perfect functioning from the handover date; third party responsibility under common law starting from the date of damage for 10 years; and of course contractual responsibility for a maximum period of 30 years.

It is a legal requirement that all parties to the process be insured against the risks associated with their liabilities. The client also has to be insured against decennial responsibility. No limit can be attached to this policy so that the client will always be able to obtain the necessary funds to complete the project. The objective is to finance any necessary works while responsibility is allocated amongst the parties to the process.

Linked to this insurance process is a system of inspection for construction projects. A *contrôleur technique*, who must be independent of the parties to the process, is engaged by the client to check all stages of the project: general concept, design, detailed specification, working drawings, choice of materials and products, and construction on site. His employment is obligatory when the building is open to the public but, in any case, this facilitates obtaining insurance cover. It is estimated that about 70% of the market for building construction is covered by *bureaux de contrôle* though only, say, 40% in the case of dwellings.

Development control and standards

Planning in France is based on a two tier system - the overall plan for the 30 year development of urban regions and the *Plan d'occupation des Sols* (POS) which is the local plan. The whole process of detailed planning has been decentralized to the *Commune*.

There is a clearly laid down procedure for obtaining a *permis de construire* which covers both the planning and control of buildings. Normally the *permis de construire* is issued by the *maire* who takes the decision himself after the process of consultation with all interested technical departments and with opportunity for objection from the public. Consultations with the *maire* at an early stage are important to ascertain what is possible, especially as there is very little room for discretion when the POS has been approved. If the application is properly completed and complies with all the rules laid down and with the POS there are very few reasons for refusal of permission. About 95% of all *permis de construire* applications in France are granted.

Mandatory regulations apply to safety, health and energy conservation. In addition there are national regulations applicable to new construction contained in various laws, decrees and orders, amplified where necessary by administrative circulars. The Scientific and Technical Centre for Building *Centre Scientifique et Technique du Bâtiment* (CSTB), publishes model solutions to many regulations.

Standards produced by the French Standards Association, *Association Française de Normalisation* (AFNOR), comprise technical recommendations which builders are advised to follow, though a number are included in the mandatory regulations. In private contracts standards are binding only if the contracts prescribe their application. In public contracts, application is a statutory requirement.

Research and development

The principal bodies active in research are:

- The Scientific and Technical Centre for Building (STB) which is a publicly funded institution
- The Inter-Professional Technical Union and its affiliates including the Experimental Research and Study Centre for Building and Public Works. The equipment of this centre is partly financed by the State
- various technical professional centres under the aegis of the Ministry of Industrial and Scientific Development. They may receive operational subsidies
- other research groups including universities.

Further reading

J L Meikle and P M Hillebrandt, The French Construction Industry : A Guide for UK Professionals, CIRIA, 1989.

RIBA, Architectural Practice in Europe : France, RIBA, 1990.

Association of Consulting Engineers et al, Engineering Consultancy in the European Community : France, London: Telford, 1990.

J V Punter "France" in Development of the Environment Planning Control in Western Europe London: HMSO, 1989, pp 149-252.

CONSTRUCTION COST DATA

Cost of labour

The figures below are typical of labour costs in Paris as at January 1991. The "wage rate" is the basis of an employee's income, while the "cost of labour" indicates the cost to a contractor of employing that employee. The difference between the two covers a variety of mandatory and voluntary contributions - a list of items which could be included is given in section 4.

	Wage rate (per hour) FFr	*Cost of labour (per hour) FFr*	*Annual working hours per year*
Site operatives			
Skilled tradesman	44.97	95.87	1866
Mason/bricklayer	44.97	97.08	1866
Carpenter	44.97	96.78	1866
Plumber	44.97	92.97	1866
Electrician	44.97	91.26	1866
Structural steel erector	44.97	93.91	1866
HVAC installer	44.97	93.57	1866
Semi-skilled worker	39.02	84.56	1866
Unskilled labourer	31.38	70.98	1866
Site supervision			
Trades foreman	50.20	105.81	1866

Cost of materials

The figures that follow are the costs of main construction materials, delivered to site in the Paris area, as incurred by contractors in January 1991. These assume that the materials would be in quantities as required for a medium sized construction project and that the location of the works would be neither constrained nor remote.

All the costs in this section exclude value added tax (VAT - see below).

	Units	*Cost FFr*
Cement and aggregate		
Ordinary portland cement in 50kg bags	tonne	704
Coarse aggregates for concrete	m^3	163
Fine aggregates for concrete	m^3	172
Ready mixed concrete (20N/mm^2)	m^3	528

	Units	*Cost FFr*
Steel		
Mild steel reinforcement	tonne	5280
High tensile steel reinforcement	tonne	5680
Structural steel sections	tonne	6660
Blocks and bricks		
Good quality facing bricks (225 x 115 x 75mm)	1000	2180
Hollow concrete blocks (450 x 225 x 225mm)	1000	2690
Solid concrete blocks (450 x 225 x 100mm)	1000	3790
Timber and insulation		
Softwood sections for carpentry	m^3	1872
Softwood for joinery	m^3	2563
Hardwood for joinery	m^3	5358
Plywood for interior joinery (10 mm)	m^2	61
Softwood strip flooring (19 mm)	m^2	45
Chipboard sheet flooring (19 mm)	m^2	40
100 mm thick quilt insulation	m^2	20
100 mm thick rigid slab insulation	m^2	41
Softwood internal door complete with frames and ironmongery (in a 75mm thick wall)	nr	605
Glass and ceramics		
Float glass (4mm)	m^2	109
Sealed double glazing units (4-6-4)	m^2	251
Good quality ceramic wall tiles (110 x 110mm)	m^2	84
Plaster and paint		
Plaster in 50 kg bags	tonne	609
Plasterboard (10mm thick)	m^2	18
Emulsion paint in 5 litre tins	litre	13
Gloss oil paint in 5 litre tins	litre	25
Tiles and paviors		
Clay floor tiles (150 x 150 x 16mm)	m^2	114
Vinyl floor tiles (300 x 300 x 3mm)	m^2	95
Precast concrete paving slabs (500 x 500 x 3mm)	m^2	61
Clay roof tiles (240 x 300mm)	1000	4580
Precast concrete roof tiles (10Nr/m^2)	1000	5160
Drainage		
WC suite complete	nr	1643
Lavatory basin complete with brackets and mixer taps	nr	880
100mm diameter clay drain pipes	m	13
150mm diameter cast iron drain pipes	m	187

Unit rates

The descriptions below are generally shortened versions of standard descriptions listed in five languages (English, French, Italian, German and Spanish) in section 6. Where an item has a two digit reference number (e.g. 05 or 33), this relates to the full description against that number in section 6. Where an item has an alphabetic suffix (e.g. 12A or 34B) this indicates that the standard description has been modified. Where a modification is major the complete modified description is included here and the standard description should be ignored; where a modification is minor (e.g. the insertion of a named hardwood) the shortened description has been modified here but, in general, the full description in section 6 prevails.

The unit rates below are for main work items on a typical construction project in the Paris area in January 1991. The rates include all necessary labour, materials, equipment and, where appropriate, allowances for contractors overheads and profit, preliminary and general items and contractors profit and attendance on specialist trades. All the rates in this section exclude value added tax (VAT - see below).

		Unit	*Rate FFr*
Excavation			
01	Mechanical excavation of foundation trenches	m^3	80
02	Hardcore filling making up levels	m^2	68
03	Earthwork support	m^2	69
Concrete work			
04	Plain insitu concrete in strip foundations in trenches	m^3	961
05	Reinforced insitu concrete in beds	m^3	1033
06	Reinforced insitu concrete in walls	m^3	1060
07	Reinforced insitu concrete in suspended floor or roof slabs	m^3	1028
08	Reinforced insitu concrete in columns	m^3	1177
09	Reinforced insitu concrete in isolated beams	m^3	1151
10	Precast concrete slab	m^2	330
Formwork			
11	Softwood formwork to concrete walls	m^2	148
12	Softwood or metal formwork to concrete columns	m^2	425
13	Softwood or metal formwork to horizontal soffits of slabs	m^2	177
Reinforcement			
14	Reinforcement in concrete walls	tonne	10 020
15	Reinforcement in suspended concrete slabs	tonne	10 400
16	Fabric reinforcement in concrete beds	m^2	38

		Unit	Rate FFr
Brickwork and blockwork			
18	Precast lightweight aggregate hollow concrete block walls	m^2	216
19	Solid (perforated) concrete blocks	m^2	218
Roofing			
22A	Concrete interlocking roof tiles 10Nr/m^2	m^2	412
23A	Plain clay roof tiles 170 x 240mm	m^2	434
24	Fibre cement roof slates 600 x 300mm	m^2	204
25A	Sawn softwood roof boarding	m^2	147
26A	Particle board roof coverings	m^2	153
27	3 layers glass-fibre based bitumen felt roof covering	m^2	185
29	Glass-fibre mat roof insulation 160mm thick	m^2	72
30A	Rigid sheet loadbearing roof insulation 100mm thick	m^2	116
31	Troughed galvanized steel roof cladding	m^2	213
Woodwork and metalwork			
32	Preservative treated sawn softwood 50mm x 100mm	m	37
33	Preservative treated sawn softwood 50 x 150mm	m	51
34A	Single glazed casement window in Tropical hardwood, size 600 x 950mm	each	864
35A	Two panel glazed door in Tropical hardwood, size 800 x 2150mm	each	2786
36A	Solid core half hour fire resisting hardwood internal flush doors, size 830 x 2040mm	each	1044
37A	Aluminium double glazed window, size 1500 x 2150mm	each	4157
38	Aluminium double glazed door, size 850 x 2100mm	each	3271
39	Hardwood skirtings	m	45
40	Framed structural steelwork in universal joist sections	tonne	13 100
Plumbing			
42	UPVC half round eaves gutter	m	155
43	UPVC rainwater pipes	m	131
44	Light gauge copper cold water tubing	m	109
45	High pressure plastic for cold water supply	m	39
46	Low pressure plastic pipes for cold water distribution	m	17
47	UPVC soil and vent pipes	m	112
48	White vitreous china WC suite	each	1390
49	White vitreous china lavatory basin	each	1220
50	Glazed fireclay shower tray	each	132
51	Stainless steel single bowl sink and double drainer	each	880
Electrical work			
52	PVC insulated and copper sheathed cable	m	19
53	13 amp unswitched socket outlet	each	229
54	Flush mounted 20 amp, 1 way light switch	each	198

		Unit	Rate FFr
Finishings			
55	2 coats gypsum based plaster on brick walls	m^2	72
56	White glazed tiles on plaster walls	m^2	230
57	Red clay quarry tiles on concrete floor	m^2	261
58	Cement and sand screed to concrete floors	m^2	130
59	Thermoplastic floor tiles on screed	m^2	97
60	Mineral fibre tiles on concealed suspension system	m^2	187
Glazing			
61	Glazing to wood	m^2	200
Painting			
62	Emulsion on plaster walls	m^2	40
63	Oil paint on timber	m^2	60

Approximate estimating

The building costs per unit area given below are averages incurred by building clients for typical buildings in the Paris area as at January 1991. They are based upon the total floor area of all storeys, measured between external walls and without deduction for internal walls.

Approximate estimating costs generally include for mechanical and electrical installations but exclude furniture, loose or special equipment, and external works; they also exclude fees for professional services. The costs shown are for specifications and standards appropriate to France and this should be borne in mind when attempting comparisons with similarly described building types in other countries. A discussion of this issue is included in section 4. Comparative data for countries covered in this publication, including construction cost data are presented in Part Three.

Approximate estimating costs must be treated with caution; they cannot provide more than a rough guide to the probable cost of building.

All the rates in this section exclude value added tax (VAT - see below).

	Cost m^2 FFr	Cost ft^2 FFr
Industrial buildings		
Factories for letting	1950	180
Factories for owner occupation (light industrial use)	3400	320
Factories for owner occupation (heavy industrial use)	5550	520
Factory/office (high-tech) for letting (shell and core only)	3450	320
Factory/office (high tech) for owner occupation (controlled environment, fully finished)	6950	650

	Cost m^2 FFr	Cost ft^2 FFr
High tech laboratory workshop centres (air conditioned)	15 000	1390
Warehouses, low bay (6 to 8m high) for letting (no heating)	1790	170
Warehouses, low bay for owner occupation (including heating)	2550	230
Warehouses, high bay for owner occupation (including heating)	2900	270
Administrative and commercial buildings		
Civic offices, non air conditioned	6960	650
Civic offices, fully air conditioned	8340	775
Offices for letting, 5 to 10 storeys, non air conditioned	5950	550
Offices for letting, 5 to 10 storeys, air conditioned	7500	700
Offices for letting, high rise, air conditioned	9650	900
Offices for owner occupation 5 to 10 storeys, non air conditioned	7070	660
Offices for owner occupation 5 to 10 storeys, air conditioned	9150	850
Offices for owner occupation high rise, air conditioned	14 950	1390
Prestige/headquarters office, 5 to 10 storeys, air conditioned	10 530	980
Prestige/headquarters office, high rise, air conditioned	15 000	1390
Health and education buildings		
General hospitals	7500	700
Teaching hospitals	6350	590
Private hospitals	8350	780
Health centres	5200	480
Primary/Secondary schools	5400	500
University (arts) buildings	6050	560
University (science) buildings	7150	660
Recreation and arts buildings		
Theatres (over 500 seats) including seating and stage equipment	10 130	940
Theatres (less than 500 seats) including seating and stage equipment	8000	740
Concert halls including seating and stage equipment	13 230	1230
Sports halls including changing and social facilities	4750	440
Swimming pools (international standard) including changing and social facilities	7730	720
Swimming pools (schools standard) including changing facilities	6000	560
National museums including full air conditioning and standby generator	19 670	1830
Local museums including air conditioning	7800	730
City centre/central libraries	7500	700

	Cost m²FFr	Cost ft² FFr
Residential buildings		
Social/economic single family housing (multiple units)	3000	280
Private/mass market single family housing 2 storey detached/semidetached (multiple units)	5090	470
Purpose designed single family housing 2 storey detached (single unit)	3600	330
Social/economic apartment housing, low rise (no lifts)	3850	360
Social/economic apartment housing, high rise (with lifts)	4250	390
Private sector apartment building (standard specification)	4010	370
Private sector apartment buildings (luxury)	7000	650
Homes for the elderly (self contained with shared communal facilities)	3840	350
Hotel, 5 star, city centre	11 670	1080
Hotel, 3 star, city/provincial	9200	850
Motel	6500	600

Regional variations

The approximate estimating costs are based on projects in Paris. Adjust these costs by the following factors for regional variations:

Alsace	: -2%	Lorraine	: -3%
Aquitane	: -3%	Midi Pyrénées	: -6%
Auvergue	: -6%	Nord pas de Calais	: -3%
Bourgogue	: -4%	Basse Normandie	: -2%
Bretague	: -5%	Haute Normandie	: -5%
Centre	: -5%	Pays de Loire	: -5%
Champaque-Ardennes	: 0%	Picardie	: -2%
Corse	: -8%	Poitou-Charentes	: -4%
Franche-Comté	: -4%	Provence-Alpes -Côte D'Azur	: -8%
Languedoc-Roussil	: -5%	Rhône-Alpes	: 0%
Limousin	: -6%		

Value added tax (VAT)

The standard rate of value added tax (VAT) is currently 18.60%, chargeable on general building work.

EXCHANGE RATES AND INFLATION

The combined effect of exchange rates and inflation on prices within a country and price comparisons between countries is discussed in section 4.

Exchange rates

The graph below plots the movement of the French Franc against sterling and the US dollar since 1980. The figures used for the graph are quarterly and the method of calculating these, and other related issues are discussed in section 4. France joined the European Monetary System (EMS) at its inception in 1979. The exchange rate at January 1991 was FFr 9.94 to the pound sterling and FFr 5.14 to the US dollar.

The French Franc against Sterling and the US Dollar

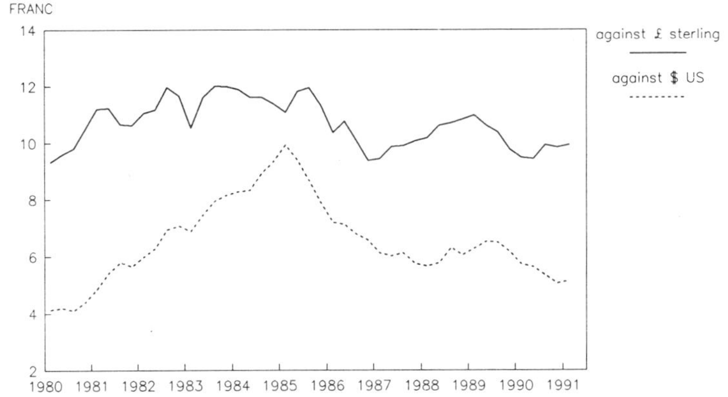

Inflation

The graph and the table below present the consumer price index showing the trend in inflation since 1981. The index has been rebased to 1981=100. Notes on inflation and the distinction between costs and prices are included in section 4.

Consumer price inflation

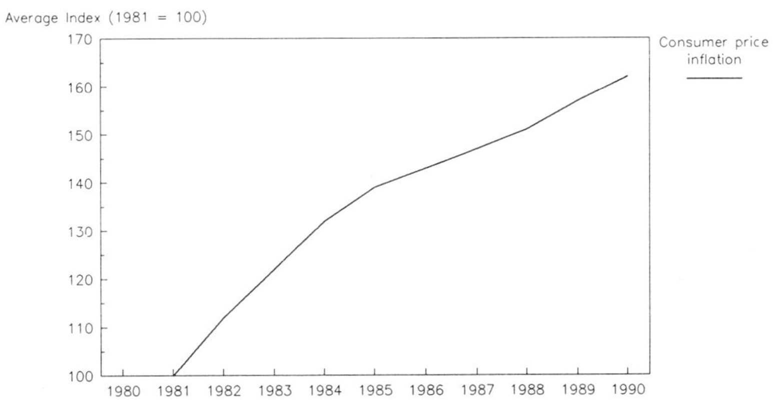

The basis of the figures is the official consumer price index converted to quarterly averages. Inflation peaked in 1981. Progress in reducing inflation has reduced the annual rate to below 5% since 1985.

Consumer price inflation 1981 = 100

	1982	1983	1984	1985	1986	1987	1988	1989	1990
average index	112	122	132	139	143	147	151	157	162
annual change (%)	11.8	9.6	7.4	5.8	2.7	3.1	2.7	3.6	3.4

USEFUL ADDRESSES

Ministère de Industrie et de l'Aménagement du Territoire
(Ministry of Industry and Territorial Development),
101 rue de Grenelle
75700 Paris

Ministère de Culture, de la communication et des Grands Travaux
(Ministry of Culture and communications and Major Public Works)
3, rue de Valois, Cedex 01
75042 Paris

Centre Scientifique et Technique du Bâtiment
Avenue du Recteur Poincare, Cedex 16
75782 Paris
Tel: (1) 4524 4302

Fédération Nationale du Bâtiment (FNB)
7 Rue la Perouse
75116 Paris

Conseil National de l'ordre des Architects
78 Avenue Raymond Poincare
75116 Paris

Chambre des Ingénieurs - Conseils de France, CICF
Maison de l'Ingeniere, 3 Rue Leon Bonnat
75016 Paris
Tel: (1) 524 43 53

Syndication National Professionnel des
Metreurs-Verificateurs-Prescripteurs Specialistes
24 Boulevard Beaumarchais
Paris Lleme

Association Française de Normalisation (AFNORS)
Tour Europe, Cedex 7
92080 Paris La Défense

Groupement Professionnel des Verificateurs, Reviseurs et
Controleurs des travaux d'architecture des bâtiments
civils, palais nationaux et monuments historiques
9 rue Vulpian
75103 Paris

Ministère de l'Amenagement du Territoire de l'Equipement
du Tourism et des Transports (MATELT)
244-246 Boulevard St Germaine
75007 Paris

International European Construction Federation
9 Rue la Perouse, Cedex 16
75784 Paris

Germany (West)

KEY DATA

Population

Population (1989)	61.2m
Urban population	86%
Population under 15	14.9%
Population over 65	15.4%
Average annual growth rate (1983 to 1988)	-0.1%

Geography

Land area	249 000 km^2
Agricultural area	120 000 km^2
Capital city	Bonn * (population 276 500)

Economy

Monetary unit	Deutschemarks (DM)
Exchange rate (January 1991) to	
the ECU	DM 2.05
the pound sterling	DM 2.93
the US dollar	DM 1.51
Average annual inflation (1984 to 1989)	1.3%
Gross Domestic Product (GDP)	DM 2122bn
GDP per capita (1989)	DM 34 670
GDP growth (real) (1980 to 1988)	1.7%
Private consumption (% of total)	54.5%
Public consumption (% of total)	19.4%
Investment (% of total)	20.3%
Central government expenditure as percentage of GDP	43.1%

Construction

Total value of construction output (1989)	DM 260bn
Construction output per capita	DM 4250
Construction output as a percentage of GDP	12.3%

* Now Berlin (total population 3 300 000)

Germany (East)

KEY DATA

Population

Population (1989)	16.7m
Urban population	77%
Population under 15	19.8%
Population over 65	13.1%
Average annual growth rate (1983 to 1988)	n.a.

Geography

Land area	108 000 km²
Agricultural area	n.a.
Capital city	Berlin (East)* (population 1 284 500)

Economy

Monetary unit	Ostmarks (EM) (Deutschemark (DM) after July 1990)
Exchange rate (official rates: end 1989) to	
the ECU	EM 2.05
the pound sterling	EM 2.93
the US dollar	EM 1.51
Average annual inflation (1984 to 1989)	n.a.
Gross Domestic Product (GDP)	EM 268bn
GDP per capita (1989)	EM 16050
GDP growth (real) (1980 to 1988)	4.0%
Private consumption (% of total)	72.1%
Public consumption (% of total)	6.4%
Investment (% of total)	21.5%

Construction

Total value of construction output (1989)	EM 51bn
Construction output per capita	EM 3050
Construction output as a percentage of GDP	19%

* Now Berlin (total population 3 300 000)

NATIONAL CONTEXT

Introduction

In October 1990 the former two Germanies - the German Democratic Republic (GDR or East Germany) and the Federal Republic of Germany (FRG or West Germany) - were reunited for the first time since 1945. It is too soon to describe in detail the characteristics of the enlarged state, thus what follows inevitably needs on occasion to refer to the old states with the terminology of "old GDR" and "old FRG".

Geography and population

The united Germany has an area of 356 755 square kilometres of which 248 577 were formerly in the old FRG and 108 178 in the old GDR. The population is 77.8 million of which 61.1 million is from the old FRG. Thus 30% of the land of the present Germany but 21% of the population was originally the old GDR.

Economy and industry

The old FRG had a very stable economy with the highest GNP per capita in the European Community. The old GDR was successful in terms of the Comecon bloc, but, in the opinion of German economists, lagged behind the FRG's economy by about 25 years. The differences in standards of living between the old FRG and the old GDR were substantial. With open competition between the two, the level of unemployment in the East, which was negligible before unification, has risen sharply to about 6% in November 1990, equivalent to an extra 1 million persons in one year. In addition about 20% of employment in the old GDR is on enforced part-time working. The causes include outmoded capital stock, poor management and problems of adapting to a free market economy. By contrast unemployment in the old FRG is the lowest for some time. In 1985 it was about 9% but by October 1990 it had fallen to 6%. At the same time emigration of workers from the old GDR to the old FRG has been vast. Between the last quarter of 1989 and the second quarter of 1990 alone, some 815 000 refugees entered FRG from the old GDR. The unemployment situation will inevitably pose problems for the united Germany and will, in the next few years, lower the overall performance of Germany in Europe.

THE CONSTRUCTION INDUSTRY

Construction output

The value of construction output in 1989 in the old FRG was some DM 260 billion, equivalent to 125.5 billion ECUs; it is difficult to assess the equivalent figure for the old GDR. In 1989 it was 51 billion Ostmarks. The exchange conversion rate fixed for Ostmarks and Dmarks was 1:1, but the purchasing power is not the same. However, on this basis, the old GDR has less than 19% of building work compared with 21% of the population of the united Germany but the real output of construction is probably considerably less.

In the old FRG gross construction output was about 12% of GNP while in the old GDR it was about 19%. The breakdown of output in the old FRG is shown below.

Construction output 1989 (FRG)

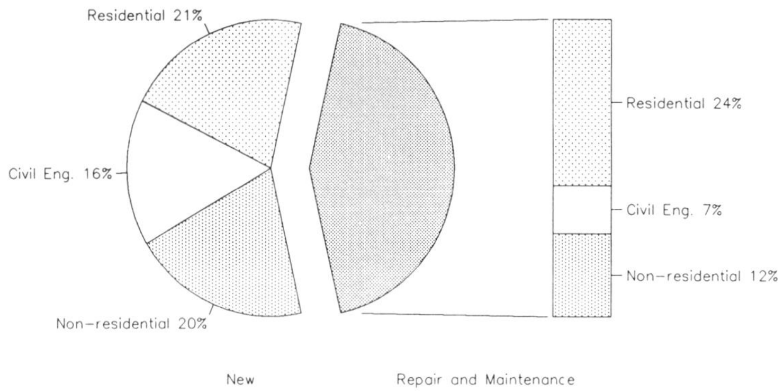

The proportion of repair and maintenance in total output is high representing a country with high standards. It is expected that output will increase overall but particularly in new housing in response to the migration taking place from the East.

It is difficult to break down the output in the old GDR. There is said to be a high proportion of repair and maintenance work in the old GDR in comparison with most Eastern European countries, but there is little agreement on the exact proportion, though it may be as high as 40%. There has been a major programme of housing renovation. New building work, including housing, probably accounts for about 47% and civil engineering accounts for 13% giving a total for new work of 60% of gross output.

There is a great need for investment in the old GDR in the physical and social infrastructure, and in industry. Despite this, there has been a marked fall in the amount of construction work carried out in 1990 as compared to 1989. This was because of disruption due to uncertainty regarding ownership of land, the economic situation and the migration of construction workers.

Characteristics and structure of the industry

The unification of Germany makes a description of the construction industry difficult. Some data have necessarily to apply either to the old FRG or GDR while it has been possible to merge other data. The Federal nature of Germany is very evident in the construction industry since each *Länder* makes its own interpretation of, and amendments to the Federal laws and rules. To some extent this makes it easier to regard the Eastern *Länder* as simply an extension of the old FRG system.

The majority of work is either let via trade contracts or general contractors. In the separate trades system, site co-ordination is the responsibility of the architect or independent project manager. Where general contracting is used it is common for several contractors to form a joint company to tender for and construct large contracts. About one-third of all construction work in the old FRG is undertaken as joint-ventures. Many clients favour the system as it helps to share financial responsibility - if one member of the venture goes into liquidation the other members will usually support the venture through to completion.

Most contractors operate in relatively local markets and the differences in regulations from *Länder* to *Länder* means that it is a major decision to expand over the border. Hence there are relatively few large contractors. In 1988 there were 88 contractors employing more than 500 employees in the FRG and these undertook 9% of total construction output. However, there are a few large contractors; the old FRG in 1989 accounted for 27 of Building magazine's "Top 300 European Contractors" and 12 in the top 100. There are 16 German contractors in the Engineering News Record's "Top 250 International Contractors".

Major German contractors (FRG)

Major contractors	*Place in Building's "Top 300 European Contractors"*	*Place in ENR's "Top 250 International Contractors"*	*Exports (% of turnover)*	*Main work types*
Philipp Holzmann	7	12	40	All types.
Hochtief	16	22	35	Mainly building, civil eng, power and all other.
Bilfinger und Berger	24	18	55	Building, civil eng and all other.
Strabag Bau	31	-	n.a.	Building, civil eng and all other.
Dyckerhoff und Widman	32	98	10	Building, civil eng and all other.
Walter Thosti Boswau	48	-	10	Building and most civil eng.
Wayss und Freytag	50	128	n.a.	Building, civil eng and power.
Ed Züblin	59	108	15	Building, civil eng and all other.

Source: Building; Engineering News Record

The industry in the old GDR was predominantly organized in construction "combines" - public organizations, employing more than 10 000 persons and involved in a number of processes. There were three main types: housebuilders, heavy construction organizations and repair and maintenance organizations. These have now been split up and formed into private limited companies. The average number of employees has fallen to around half the previous level.

The combines undertook most construction - firms employing 500 employees or more undertook over 60% of the work. However, down the scale, there were relatively few small firms and these did a small proportion of the work. The new organizations are still too large in relation to their western counterparts and have too many staff and administrative workers: 30% to 35% compared with around 20% in the west.

The systems of education in the East and West of Germany are substantially the same. The two main professions in the construction and building field are architecture and building engineering. Each have completely different education programmes.

There are two main streams of education:

- the universities or technical universities (*Technische Hochschule*)
- or the *Fachhochschule* (comparable to the Polytechnics).

Students of both the University and the *Fachhochschule* finalise with the degree of *Diplom Ingenieur*.

After two years practical experience in an architect's office an architect registers at the Chamber of Architects in the state in which he lives. The title "Architect" is protected by law. There are a large number of professionals in West Germany in relation to the population. There are over 60 000 architects, 990 for every million inhabitants and an estimated 100 000 building engineers, 1639 for every million inhabitants - much higher than in most European countries. Architects operate mainly in small practices of 2 or 3 architects. Engineers have larger practices than architects.

On the design side, organizations are very large in the East. Each of the large combines had its own planning and design organization operating as separate divisions and employing 300 to 700 persons. This organizational structure is not suitable for the new competitive market. Architects offices with 100 employees are the norm in the old GDR whereas in the old FRG they are the exception.

There are a number of arrangements for co-operation between East and West German firms and this is expected to increase. A feature of the old FRG contracting industry is the high participation of banks and financial institutions in the ownership of construction firms which facilitates German construction firms' access to capital. Frequently there are cross shareholdings between contractors.

Clients and finance

In the old FRG, the public sector work accounts for about a quarter of all construction output. Of this, the communes are clients for slightly more than half (51%) followed by the *Länder* (13%) the Federal government (11%) Railways (7%) Post (9%) and Defence and NATO (9%).

Housebuilding is mainly undertaken by small local builders who sell "off plan" and the buyers make stage payments during the construction process. Development of non-residential buildings is, to a large extent, undertaken by the organization wishing to occupy the building using a mixture of cash and debt. They may later sell their building to a property investor. However, banks often employ a developer, but they provide the land and finance themselves and retain ownership of the development. Thus banks build up a portfolio of properties. Insurance companies are also important investors in property. Pension funds are of minor importance. Generally, investors prefer to purchase a building which is complete and already let. Large property developers, such as exist in the UK, are as yet unknown, although speculative development is increasing.

Clients in the old FRG demand a high standard of construction and it seems to be generally agreed that the end product is of high quality. The laws and regulations for construction are detailed and strictly enforced. In the old GDR standards are much lower.

Selection of design consultants

The normal procedure in the public sector is that the client requests three to ten firms to submit proposals. In most cases personal meetings precede the written proposals and sometimes this is the stage at which the designer is chosen, though not formally so. Important criteria for selection are quality of work and special know-how, good references, reliability, a regional presence and personal contacts. Price is not a key element. In any case fees have to be calculated on the basis of *Honorarordnung für Architekten und Ingenieure* (HOAI). The normal procedure for private clients is less formal but follows similar lines.

Contractual arrangements

In the old FRG for Government schemes, the majority of contracts are placed as a result of public advertisement, but for private schemes it is more usual to invite tenders from a limited number of contractors, followed by further negotiation with the lowest tenderer. It is expected that the same procedures will be followed for the old GDR.

Turnkey contracts are the exception in Germany.

Germany has published rules covering contract procedure for all building work, the *Verdingungsordnung für Bauleistungen* (VOB) (Regulations for Building Contracts). This document, published in three parts, is the main compendium of regulations and a good practice guide for the industry. VOB is the basis of most building contracts.

- Part A details a method of tendering for government contracts, and is mandatory for public buildings.

- Part B lists general conditions of contract for the execution of the works. It contains clauses concerning renumeration, execution of the works, termination, payment, liabilities of the parties, and so on.

- Part C comprises 49 separate DIN standards covering all the major building trades arranged in a similar format. These give guidelines for the measurement of quantities and ensure some uniformity of tender documents. The items of work are properly described, spelling out the specification, the unit and the quantity, and the tenderer has to fill in the unit price and the overall price for each item. In a number of contractors the tenderer is, in addition, required to give details of wage costs and material costs. The impact of the general conditions on price are not priced separately but are incorporated in the unit rates.

The specification, together with a bill of quantities, necessary drawings and the VOB (Parts B and C) and any other special conditions, are collected together to form the tender documents; these are supplemented by an exchange of letters placing and accepting the contract. The conditions of contract may be as VOB Part B, but these are frequently supplemented with amended or additional clauses to suit the particular circumstances of each scheme.

The type of contract most commonly entered into is a remeasurement contract based on a specification and bills of quantities (subject to remeasurement on completion). Most contracts are "firm" price though alterations to rates are usually permitted when there are quantity adjustments in excess of 10%. Interim payments are usually paid to a contractor upon receipt of a detailed interim account from him for the work executed at that time.

Authorized variations are permitted providing they do not change the scope of the work; the contractor prepares details of the cost of the variation for agreement with the architect, based as far as possible on rates in the contract documents, but new rates are agreed when necessary. It is normal for the cost of variations to be agreed before the work is undertaken on site.

Most contracts make provision for retention, usually 5% to 10% of the contract sum, which is released upon agreement of the final account. Each contractor, however, is liable for the maintenance of his part of the work for anything between two and five years after completion. It is the duty of the contractor to prepare the final account which in the case of Government work is subject to audit by the Federal Auditor. The audit must be carried out within five years and the account may be re-opened within this time.

Liability, insurance and inspection

In the German system the formal acceptance of work is crucial. The responsibility for building failure rests with the contractor. However, the relevant building authority appoints an independent assessor (*Prüfungsingenieur*) - often an academic or independent consultant - to certify that the building is structurally sound and that the overall requirements have been achieved. He then accepts responsibility and is obliged to have professional indemnity insurance. Owners' rights to claim damages are limited to five years and then only for damage caused directly by the fault of the builder. In the case of design faults, a five year limit applies, but owners also have the right to claim for consequential loss.

The builder according to his contract is, with some exceptions, liable only for two years from handover. Thus there is no statutory guarantee period for architects, engineers or contractors. Some latent defects insurance is available, notably for new housing.

Development control and standards

The system of development control will follow that of the old FRG for the whole of Germany. The old FRG system is described below.

Planning law comes from both the Federal Government and the *Länder* and procedures are based on a system of statutory development plans. Building regulation approval is covered at the same time as development control by the municipality building permit system. Applications must be submitted to the municipality which are then passed on as necessary to the next superior level. Authorities have no discretion over applications which accord with approved land-use plans; they must be approved. Failure to do so may result in legal action by the applicant in the administrative courts. No time limit is specified for giving a decision but, after three months, application may be made to the administrative courts on the grounds of undue delay so that, in practice, this is taken as representing adequate time. Typically 20% of applications comply fully with the regulations, 78% deviate to some extent and the authority will prescribe changes to be made. The remaining 2% deviate so greatly that permission is refused. In practice a wide range of informal consultation takes place as well as the formal procedures.

In addition to VOB, West Germany has a comprehensive range of standards covering technical details and information on installation for a wide range of products used in construction. These are published and kept up to date by DIN (*Deutsches Institut für Normung*), which also has a materials testing division. DIN also maintain a databank of all standards, technical rules, and regulations - DITR (*Deutsches Informationszentrum für technische Regeln*) which is available in a variety of forms - by microfiche, on-line computer connection, or as a telephone service, available to subscribers. Individual *Länder* each have their own set of Building Regulations which are based on a set of national "model bylaws" The variations between individual *Länder* are generally minor. Indeed 90% of the regulations are now harmonised.

West Germany also has an *Agrément* Institute, which is solely concerned with the safety of materials.

The standards are not legally binding but they are generally regarded as stating good practice. However, deviations from standard are permitted subject to approval of the building control authority, though the issue of a certificate or consent given in individual cases.

Both the legal conditions and the building standards apply to both building and civil engineering construction, though there are additional regulations for certain types of construction, e.g. bridges. In the case of public works the building control is exercised by the relevant competent authority.

Labour

Not only is there a problem in the old GDR with the productivity of labour but there is also the problem that the trades do not correspond to likely future demand because of the dominance of large concrete panel construction and other cement intensive structures with only basic finishing. It has been estimated that the mix of trades required is quite different from that available. In future, a far greater number of bricklayers, plasterers, carpenters and plumbers than currently available will be needed. The skills of workers who were assembling the heavy systems and who previously constituted nearly a quarter of the labour force will no longer be needed.

Research and development

Research and development work in building and civil engineering in more than 30 governmental and private institutions is being co-ordinated by the recently established Association for Building Research (*Arbeitsge meinschaft für Bauforschung*) at the Federal Ministry for Regional Planning, Building and Urban Development. Most funds are spent on buiding materials, components and building processes.

In the old GDR the most important research organization was the Building Academy of the German Democratic Republic.

Further reading

Biggs, W D, Betts, M and Cottle, M J, The West German Construction Industry: A Guide for UK Professionals. London: CIRIA Special Publication 68, 1990.

Association of Consulting Engineers et al, Engineering Consultancy in the European Community: West Germany. London: Telford 1990.

A J Hooper "Federal Republic of Germany" in Department of the Environment, Planning Control in Western Europe London: HMSO, 1989. pp 253-335.

Otto Dienemann "Perspektiven für den DDR Baumarkt im Rahmen des E.G. - Binnenmarktes" in Euroconstruct, Baukonjunktur 1990/91 in Westeuropa, IFO Institut für Wirtschafts-forschung, München, Dezember 1990.

Rolf Bollinger, "Chancen und Risken für die Westeuropäische Bauwirtschaft nach den Wende in Osteuropa" in Euroconstruct, Baukonjunktur 1990/91 in Westeuropa, IFO Institut für Wirtschaftsforschung, München, Dezember 1990.

Acknowledgements

The authors are grateful to Professor Otto Dienemann, Building Academy of the former GDR, for helpful discussions.

CONSTRUCTION COST DATA

Cost of labour

The figures below are typical of labour costs in Bonn as at January 1991. The "wage rate" is the basis of an employee's income, while the "cost of labour" indicates the cost to a contractor of employing that employee. The difference between the two covers a variety of mandatory and voluntary contributions - a list of items which could be included is given in section 4.

	Wage rate (per hour) DM	*Cost of labour (per hour) DM*
Site operatives		
Mason/bricklayer	21.00	23.50
Carpenter	21.53	24.12
Plumber	21.11	23.52
Electrician	19.63	22.00
Structural steel erector	18.13	20.30
HVAC installer	21.02	23.50

Cost of materials

The figures that follow are the costs of main construction materials, delivered to site in the Bonn area, as incurred by contractors in January 1991. These assume that the materials would be in quantities as required for a medium sized construction project and that the location of the works would be neither constrained nor remote.

All the costs in this section exclude value added tax (VAT - see below).

	Units	*Cost DM*
Cement and aggregate		
Ordinary portland cement in 50kg bags	tonne	236.88
Coarse aggregates for concrete	m^3	70.15
Fine aggregates for concrete	m^3	70.15
Ready mixed concrete (20N/mm^2)	m^3	180.00
Ready mixed concrete (30N/mm^2)	m^3	240.00
Steel		
Mild steel reinforcement	tonne	1800.00
Bricks and blocks		
Common bricks	1000	700.00
Hollow concrete blocks	1000	730.00
Timber and insulation		
Softwood sections for carpentry 100/100	m	8.86
Exterior quality plywood	m^2	17.00
Chipboard sheet flooring (19mm)	m^2	13.46
Softwood internal door complete with frames and ironmongery	nr	379.00
Glass and ceramics		
Float glass (4mm)	m^2	93.12
Sealed double glazing units	m^2	142.80
Plaster and paint		
Plasterboard	m^2	6.25
Emulsion paint in 10 litre tins	litre	3.60
Gloss oil paint in 25 litre tins	litre	21.60
Tiles and paviors		
Precast concrete paving slabs	m^2	22.32
Clay roof tiles	1000	2400.00
Drainage		
WC suite complete	nr	187.20
Lavatory basin complete	nr	396.00
100mm diameter clay drain pipes	m	26.54
150mm diameter cast iron drain pipes	m	9.13

Unit rates

The descriptions below are generally shortened versions of standard descriptions listed in five languages (English, French, Italian, German and Spanish) in section 6. Where an item has a two digit reference number (e.g. 05 or 33), this relates to the full description against that number in section 6. Where an item has an alphabetic suffix (e.g. 12A or 34B) this indicates that the standard description has been modified. Where a modification is major the complete modified description is included here and the standard description should be ignored; where a modification is minor (e.g. the insertion of a named hardwood) the shortened description has been modified here but, in general, the full description in section 6 prevails.

The unit rates below are for main work items in a typical construction project in the Bonn area in January 1991. The rates include all necessary labour, materials, equipment and, where appropriate, allowances for contractors overheads and profit, preliminary and general items and contractors profit and attendance on specialist trades.

All the rates in this section exclude Value added tax (VAT - see below).

		Unit	*Rate DM*
Excavation			
01	Mechanical excavation of foundation trenches	m^3	50.00
02	Hardcore filling making up levels	m^2	15.50
Concrete work			
04	Plain insitu concrete in strip foundations in trenches	m^3	170.00
05	Reinforced insitu concrete in beds	m^3	180.00
06	Reinforced insitu concrete in walls	m^3	195.00
07	Reinforced insitu concrete in suspended floor or roof slabs	m^3	190.00
08	Reinforced insitu concrete in columns	m^3	220.00
09	Reinforced insitu concrete in isolated beams	m^3	200.00
Formwork			
11	Softwood or metal formwork to concrete walls	m^2	75.50
12	Softwood or metal formwork to concrete columns	m^2	117.00
13	Softwood or metal formwork to horizontal soffits of slabs	m^2	90.00
Reinforcement			
15	Reinforcement in suspended concrete slabs	tonne	2860.00
16A	Fabric reinforcement in concrete beds	tonne	2574.00
Brickwork and blockwork			
18	Precast lightweight aggregate hollow concrete block walls	m^2	40.00
19	Solid (perforated) concrete blocks	m^2	74.00
20	Solid (perforated sand lime bricks	m^2	85.00
21	Facing bricks	m^2	180.00

		Unit	Rate DM
Roofing			
22	Concrete interlocking roof tiles 430 x 380mm	m^2	50.00
23	Plain clay roof tiles 260 x 160mm	m^2	80.50
24	Fibre cement roof slates 600 x 300mm	m^2	133.00
25	Sawn softwood roof boarding	m^2	30.00
26	Particle board roof coverings	m^2	35.00
28	Bitumen based mastic asphalt roof covering	m^2	78.70
Woodwork and metalwork			
32	Preservative treated sawn softwood 50 x 100mm	m	3.40
33	Preservative treated sawn softwood 50 x 150mm	m	5.00
34	Single glazed casement window in hardwood, size 650 x 900mm	each	436.00
36	Solid core half hour fire resisting hardwood internal flush doors, size 800 x 2000mm	each	1280.00
37	Aluminium double glazed window, size 1200 x 1200mm	each	1122.00
38	Aluminium double glazed door, size 850 x 2100mm	each	1100.00
39	Softwood skirtings	m	5.60
Plumbing			
42	Metal half round eaves gutter	m	57.00
43	Metal rainwater pipes	m	50.50
47	UPVC soil and vent pipes	m	31.60
48	White vitreous china WC suite	each	750.00
Finishings			
55	2 coats gypsum based plaster on brick walls	m^2	28.00
56	White glazed tiles on plaster walls	m^2	85.00
58	Cement and sand screed to concrete floors	m^2	35.00
59	Thermoplastic floor tiles on screed	m^2	35.00
60	Mineral fibre tiles on concealed suspension system	m^2	101.40
Glazing			
61	Glazing to wood	m^2	80.00
Painting			
62	Emulsion on plaster walls	m^2	12.10
63	Oil paint on timber	m^2	25.00

Approximate estimating

The building costs per unit area given below are averages incurred by building clients for typical buildings in the Bonn area as at January 1991. They are based upon the total floor area of all storeys, measured from the external face of external walls and without deduction for internal walls.

Approximate estimating costs generally include for mechanical and electrical installations but exclude furniture, loose or special equipment, and external works; they also exclude fees for professional services. The costs shown are for specifications and standards appropriate to Germany and this should be borne in mind when attempting comparisons with similarly described building types in other countries. A discussion of this issue is included in section 4. Comparative data for countries covered in this publication, including construction cost data are presented in Part Three.

Approximate estimating costs must be treated with reserve; they cannot provide more than a rough guide to the probable cost of building.

All the rates in this section exclude value added tax (VAT - see below).

	Cost *m^2 DM*	*Cost* *ft^2 DM*
Industrial buildings		
High tech laboratory workshop centres (air conditioned)	3810	360
Warehouses, low bay (6 to 8m high) for letting (no heating)	1420	130
Warehouses, low bay for owner occupation (including heating)	1770	164
Warehouses, high bay for owner occupation (including heating)	2160	200
Administrative and commercial buildings		
Civic offices, non air conditioned	2000	185
Civic offices, fully air conditioned	2500	230
Offices for letting, 5 to 10 storeys, non air conditioned	2440	225
Offices for letting, 5 to 10 storeys, air conditioned	2685	250
Offices for letting, high rise, air conditioned	4900	455
Offices for owner occupation 5 to 10 storeys, non air conditioned	2700	250
Offices for owner occupation 5 to 10 storeys, air conditioned	3100	290
Offices for owner occupation high rise, air conditioned	5500	510
Prestige/headquarters office, 5 to 10 storeys, air conditioned	4200	390
Prestige/headquarters office, high rise, air conditioned	5940	552
Health and education buildings		
Nursery schools	2040	190
Primary/junior schools	2015	185
Secondary/middle schools	2190	205
University (science) buildings	3430	320
Management training centres	2400	225

	Cost m^2 DM	Cost ft^2 DM
Recreation and arts buildings		
Sports halls including changing and social facilities	2270	210
Swimming pools (international standard) including changing and social facilities	4460	415
Swimming pools (schools standard) including changing facilities	4685	435
Local museums including air conditioning	4630	430
City centre/central libraries	3500	325
Branch/local libraries	2615	242
Residential buildings		
Social/economic single family housing (multiple units)	1165	110
Private/mass market single family housing 2 storey detached/semidetached (multiple units)	1240	115
Purpose designed single family housing 2 storey detached (single unit)	1450	135
Social/economic apartment housing, low rise (no lifts)	1500	140
Social/economic apartment housing, high rise (with lifts)	1800	170
Private sector apartment building (standard specification)	1900	180
Private sector apartment buildings (luxury)	2400	220
Student/nurses halls of residence	2230	210
Homes for the elderly (shared accommodation)	2520	235
Homes for the elderly (self contained with shared communal facilities)	2520	235

Regional variations

The approximate estimating costs are based on projects in the Bonn area. Adjust these costs by the following factors to take account of regional variations:

Baden-Württemberg	:+ 3%	Niedersachsen	:-12%
Bayern	:+15%	Nordrhein-Westfalen	: - 3%
Berlin (West)	:+32%	Rheinland-Pfalz	: - 1%
Brandenburg	: - 5%	Saarland	:+7%
Bremen	: - 8%	Aachen	: - 5%
Hamburg	: - 1%	Sachen-Anhalt	: - 5%
Hessen	: - 4%	Schleswig-Holstein	: - 8%
Mecklenburg-Vorpommern	: -10%	Thüringen	: - 5%.

Value added tax (VAT)

The standard rate of value added tax (VAT) is currently 14%, chargeable on general building work.

EXCHANGE RATES AND INFLATION

The combined effect of exchange rates and inflation on prices within a country and price comparisons between countries is discussed in section 4.

Exchange rates

The graph below plots the movement of the German Deutschemark against sterling and the US dollar since 1980. The figures used for the graph are quarterly and the method of calculating these, and other related issues are discussed in section 4. Germany joined the European Monetary System (EMS) at its inception in 1979. The exchange rate at January 1991 was DM 2.93 to the pound sterling and DM 1.51 to the US dollar.

The German Deutschemark against Sterling and the US Dollar

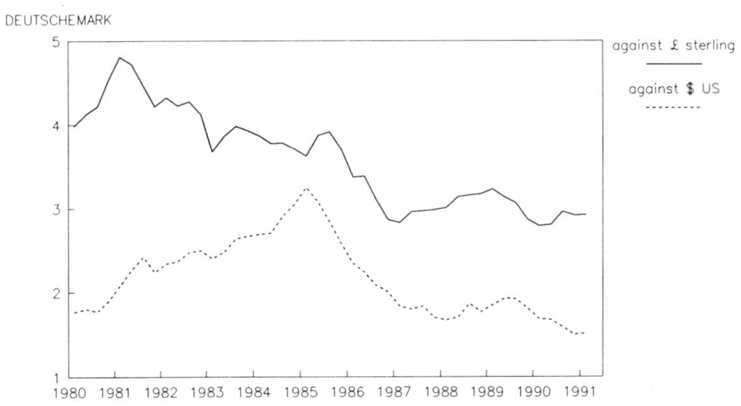

Inflation

The graph below presents a number of indices showing the trends in inflation, prices and costs. All the indices have been rebased to 1980=100 with the exception of the retail price index, which is 1985=100. Notes on inflation and the distinction between costs and prices are included in section 4.

Cost and price indices

Consumer price inflation

The basis of the figures is the official consumer price index converted to quarterly averages. The average annual rate of inflation has been below the average for the EC since the mid 1980s.

Retail price index 1985 = 100

	1985	1986	1987	1988	1989
average index	100	100	100	101	104
annual change (%)	-	0.0	0.0	1.0	3.0

Building cost indices

Commercial Buildings 1980 = 100

	1981	1982	1983	1984	1985	1986	1987	1988	1989	1990
average index	108	114	116	119	122	119	116	118	121	n.a.
annual change (%)	8.0	5.6	1.8	2.6	2.5	-2.5	-2.5	1.7	2.5	n.a.

Multi-storey steel frame buildings 1980 = 100

	1981	1982	1983	1984	1985	1986	1987	1988	1989	1990
average index	107	118	118	117	116	120	123	123	128	n.a.
annual change (%)	7.0	10.9	0.0	-0.8	-0.9	3.4	2.5	0.0	4.1	n.a.

Residential buildings 1980 = 100

	1981	1982	1983	1984	1985	1986	1987	1988	1989	1990
average index	106	109	111	114	115	116	119	121	126	n.a.
annual change (%)	6.0	2.8	1.8	2.7	0.9	0.9	2.6	1.7	4.1	n.a.

USEFUL ADDRESSES

Bundesbaudirektion (BBD) (Federal Constuction Directorate)
1000 Berlin 12, Fasanenstrasse 87
Tel: (030) 31 81

Arbeitsgemeinschaft der für das Bau-, Wohnungs - und Siedlungswesen Zuständigen Minister/Senatoren der Länder (ARGEBAU)
(Consortium of the State Ministers/Senators
Responsible for Building, Housing and Development), 5300 Bonn 12
Dahlmannstrasse 2
Tel: (02221) 22 85 24

Institut für Bautechnik (Anstalt des offentlichen Rechts) (IfBT)
(Institute for Building Technology)
1000 Berlin 30, Reichpietschufer 72/76
Tel: (0311) 25 03 1
Telex: 185413

Bund Deutscher Baumeister, Architekten und Ingenieure e.V. (BDB)
(Federation of German Contractors, Architects and Engineers)
5300 Bonn, 2, Kennedyallee
Tel: (0228) 376784

Bund Deutscher Architekten (BDA)
(Federation of German Architects)
5300 Bonn, Ippendorfer Allee 14b
Tel: (02221) 63 13 81

Verband Beratender Ingenieure e.V. (VBI)
(Association of Consulting Engineers)
4300 Essen, Zweigertstrasse 37-41

Hauptverband der Deutschen Bauindustrie e.V.
(Central Federation for the German Construction Industry)
6200 Wiesbaden Postfach 2966, Abraham Lincoln Strasse 30
Tel: 06121 7220

DIN (Deutsches Institut für Normung),
Burggrafenstrasse 4-10, Postfach 1107
D-1000 Berlin 30

Veerein Deutscher Ingenieure,
VDI - Geseller Technische Gedaudeausrustung,
Graf Reche Strasse 84, D4000 Düsseldorf
Tel: 62 14 266

Bund Deutscher Architechen
Ippendorfer ALLEE 14B, 5300 Bonn
Bauzentrum Rein-Main, Bau-Munster-Hans OHG
61 Darnstadt, Platz Der Deutschen, Einheit 2
Postfach 4091

Greece

KEY DATA

Population

Population (1989)	10.0m
Urban population	62%
Population under 15	13.7%
Population over 65	18.4%
Average annual growth rate (1983 to 1988)	0.3%

Geography

Land area	132 000 km²
Agricultural area	92 000 km²
Capital city	Athens (population 3 027 331)

Economy

Monetary unit	Drachma (Dr)
Exchange rate (January 1991) to	
the ECU	Dr 216
the pound sterling	Dr 309
the US dollar	Dr 160
Average annual inflation (1984 to 1989)	13.8%
Gross Domestic Product (GDP)	Dr 7446bn
GDP per capita (1989)	Dr 744 600
GDP growth (real) (1980 to 1988)	1.4%
Private consumption (% of total)	67.2%
Public consumption (% of total)	20.7%
Investment (% of total)	19.0%
Central government expenditure as percentage of GDP	43.9%

Construction

Total value of construction output (1989)	Dr 923bn
Construction output per capita	Dr 92 300
Construction output as a percentage of GDP	12.4%

THE CONSTRUCTION INDUSTRY

Construction output

The value of gross output of new construction work in Greece in 1989 was 923 billion drachmas, equivalent to about 5.2 billion ECUs. Of this, 44% was housing, 23% other buildings and 33% other construction, i.e. mainly civil engineering. The value of repair and maintenance and rehabilitation work is not known though it is thought to be expanding at a higher rate than new building. This new work represented about 12.4% of GDP. However all these figures must be treated with caution as it has been estimated that the unofficial economy provides the equivalent of 35% of GDP. Construction in Greece was depressed in the 1980s. Over 40% of new construction work in the period 1966 to 1980 was in Athens.

Characteristics and structure of the industry

The engineer is extremely important in Greece; the architect on the other hand is secondary for many types of work. There is a requirement for Greek engineers to register with the Technical Chamber of Greece. The Chamber has approximately 50 000 professional members, of whom about 7500 are officially registered as individual consultants, the others being either employed by public bodies or elsewhere in the construction (or other) industries. The average consultancy is medium-sized, having, say, 2-3 partners and four or five engineers. There are also a number of small firms and a very few large ones. In the late 1980s a new type of organization was established: the public sector consultancy which is publicly owned but operates much as a private firm. However, these public sector consultancies have significant advantages over private firms. They may enter into particular agreements, called programmed or policy agreements with any public authority whereas private consultancies are not permitted to do this. There are about 15 public sector companies, some with several hundred employees.

The title of architect is protected by law (No 4663 of 1930) for all members of the Technical Chamber of Greece which is responsible for granting licences to practice. The employment of architects is not obligatory.

According to official sources there were 425 registered construction companies in 1985, of which thirty had over 100 employees. Most of the rest had less than 50, though in many companies the numbers of employees fluctuate dramatically according to the number of contracts being undertaken. In addition, there are large numbers of very small, unregistered jobbing builders and tradespersons who hire themselves out to registered companies and undertake small local contracts. Only one Greek company is in Building magazine's "Top 300 European Contractors": Eteka (no. 282). Another important company is Biex Inc. which specializes in industrial projects. Consolidated Contractors Intl. Co SAL of Athens is number 48 in the Engineering News Record's "Top 250 International Contractors". It only works abroad where it undertakes building, water, industrial and transport works.

Contractors must hold the appropriate licence for the class of work involved to obtain public sector work. To be licenced, a contractor must have on his staff sufficient numbers of suitably qualified and licenced engineering and architectural personnel, and experience of one of four specific classes of work.

Clients and Finance

Greece has a Five Year Development Plan 1989 to 1993 which focuses principally on regional development. Expenditure is budgeted at Dr 5.6 billion. The Government expects to receive Dr 1400 billion from the European Community: approximately 54% from the regional Development Fund; 19% from the Social Fund and 27% from the Agricultural Fund. The most important funding source for construction is the Regional Development Fund. Greece will also borrow from the European Investment Bank. The low standard of the infrastructure by comparison with other European Community members has led to Greece being one of the main beneficiaries of Structure Funds.

The old system of development grants has been replaced under decree 1892/1990 by tax allowances for investment in productive areas by new and existing industries. Commercial investment such as offices and retail premises is not regarded as productive and does not receive these allowances.

About 80% of housing in Greece is privately owner occupied, the remainder being let by the public or private sector. Housing is financed by private funds.

About half non-residential private building is of hotels and other tourist facilities by organizations including the Greek construction companies.

Selection of design consultants

The selection procedures are very closely regulated. In the public sector there is a distinction between design and consultancy. Design fees are related to the cost of the project. Consultancy is the giving of advice unrelated to the cost of the project. A designer must be registered with the Government Design Register and a consultant in the Consultants' Register. It is not possible to be registered in both.

All government agencies or state-controlled enterprises such as foundations, banks, public utilities, etc. are obliged by law publicly to announce most of their needs for design work and to allot them to professionals registered in the Government Design Register. The Government then organizes a competition for contracts. The open procedure is the most common in Greece. This procedure is a form of price competition. In announcements for some projects, design and construction are combined. However, for many public sector projects, especially in the municipalities, the public engineering companies often obtain work without competition. In a few large projects the Government contacts a company direct. In these situations, foreign companies are often used because of the need for specific experience. The projects are often combined with financial arrangements, e.g. in a joint venture between a foreign and a Greek company.

In the private sector, registration with the Technical Chamber of Greece is required. There is no compulsory legal requirement for projects in the private sector to be announced as there is in the public sector. However, some clients procure projects more or less on the basis of the same guidelines as hold good for public authorities. Very few, and usually only highly specialized projects, are advertised in daily newspapers or technical periodicals.

The best way for engineering consultants to receive information about projects is by knowing people. Only invited consulting engineers are allowed to bid for a project as a general rule. In a typical private project the designer is employed to provide both architectural and engineering services. This includes briefing, definition of the scope of the project, primary design, sketch design, working drawings, final planning, supervision and construction.

Some of these arrangements will change to bring them in line with European Community directives.

Contractual arrangements

Tenders for government contracts in Greece are usually invited by public advertisement. The rules governing public works contracts are set out in Law 1286/6.10.72, Presidential Decree 475/76 and amendment 724/79. Tender documents usually comprise:

- drawings
- general specifications
- technical specifications or technical description
- work item description and unit price list
- schedule of quantities or cost estimate
- form of tender.

Usually, in the case of building contracts, prices included in the unit price list are determined by the tendering contractor, whilst for civil engineering projects the contractor states only his required percentage adjustment to prices already established by the client. The submitted tenders are subject to the uniformity of pricing test, *Omalotita Timon*, which restricts the acceptable adjustments to prices for each category of work to within 0.80 to 1.10 times the average percentage adjustment for the overall project. For public works contracts, fluctuations in prices are governed by procedures included in Law 889/79. Quarterly indices for these purposes are published by the Ministry of Public Works, the base quarter being the calendar quarter in which the contract is signed.

Tenders received are considered by a committee comprising the architects, design engineers and the client's representatives who will assess the technical and the cost implications of the bid.

Private building contracts are not subject to the same degree of regulation as public works. For private work many contracts are negotiated, contractors need not be licenced and the documents supplied to tenderers may vary considerably from those regarded as typical for public works. In some cases the documentation is very limited. Negotiations will normally take place with the tenderers so the contract is not automatically awarded to the tenderer who submits the lowest initial bid.

All construction work must be supervised by a licenced engineer, whose duties also include the recording of the condition and extent of any work which is subsequently concealed. The supervising engineer also has the authority to make interim and final payments. Quantities of "as built" work are ascertained, and used by the contractor to prepare final cost sheets which are subject to the approval of the supervising engineer. In the event of variations creating new items of work, prices are generally calculated in accordance with a set of rules issued by the Ministry of Public Works for the costing of building works.

A performance bond is usually required and the client normally retains 10% of the contract sum until the works are complete. After completion of the works a warranty guarantee bond is held until the expiry of the maintenance period (usually 12 months) during which time the contractor makes good any defects.

Liability

Liabilities are defined by the Civil Code, by law 696 of 1974 and decree no 723 of 1979. Liability for architects is limited to 10 years after handover. Insurance is not obligatory for architects.

Development control and regulations

Current planning legislation is covered by three main laws 1923, 1979 and 1983, each of which extends and amends the previous laws. Since 1983 the planning programme comprises:

- General Town Plans and Implementation Town Plans
- Master Plans for Athens and Salonika
- planning and rehabilitation of main coastal resort areas
- defining of boundaries and building regulations for about 10 000 rural settlements
- planning and extension of 1800 rural settlements
- reports for 51 prefectures.

Planning may be done by central government or local authorities and may be assigned by them to private planning offices. In practice, facilities in the local authorities are insufficient to cope with the requirements which include consultation.

The new programmes are slowly being implemented but people are not used to the system or to the consultation procedures. There are substantial penalties for illegal development but nevertheless there is still some illegal housing construction.

Three compulsory codes are issued by the Government:

- a general building code - a code of practice that is applicable nationwide
- general ordinances for electro-mechanical installations - these are instructions and standards
- general ordinances for fire-fighting and prevention - these are regulations issued by government agencies.

Greek standards are almost entirely in accordance with the European and American standards, such as, DIN, VDE, British standards, and American standards. Some public companies (e.g. Astrofos engineering) have their own standards which are a mixture of the above.

Further reading

Association of Consulting Engineers et al, Engineering Consultancy in the European Community: Greece, London: Telford 1990.

CONSTRUCTION COST DATA

Cost of labour

The figures below are typical of labour costs in Athens as at January 1991. The "wage rate" is the basis of an employee's income, while the "cost of labour" indicates the cost to a contractor of employing that employee. The difference between the two covers a variety of mandatory and voluntary contributions - a list of items which could be included is given in section 4.

	Wage rate (per hour) Dr	*Cost of labour (per hour) Dr*	*Number of hours worked per year*
Site operatives			
Mason/bricklayer	1476	2471	1820
Carpenter	1357	2286	1899
Plumber	1419	2219	1992
Electrician	1276	2124	1992
Structural steel erector	1857	2886	1904
HVAC installer	1457	2357	1999
Semi-skilled worker	1157	1952	1820
Unskilled labourer	1029	1733	1820
Equipment operator	1193	1825	2044
Watchman/security	917	1410	1992

	Wage rate (per hour) Dr	*Cost of labour (per hour) Dr*	*Number of hours worked per year*
Site supervision			
General foreman	1924	2969	1992
Trades foreman	1671	2519	1992
Clerk of works	943	1325	2044
Contractors' personnel			
Site manager	4024	5071	1992
Resident engineer	2643	3417	1992
Resident surveyor	1910	2471	1992
Junior engineer	1129	1531	1992
Junior surveyor	990	1348	1992
Planner	1671	2305	1992
Consultants' personnel			
Senior architect	3250	4357	2044
Senior engineer	3369	4536	2044
Senior surveyor	1871	2600	2044
Qualified architect	2262	3125	2044
Qualified engineer	2262	3125	2044
Qualified surveyor	1793	2243	2044

Cost of materials

The figures that follow are the costs of main construction materials, delivered to site in the Athens area, as incurred by contractors in January 1991. These assume that the materials would be in quantities as required for a medium sized construction project and that the location of the works would be neither constrained nor remote.

All the costs in this section exclude value added tax (VAT - see below).

	Units	*Cost Dr*
Cement and aggregate		
Ordinary portland cement in 50kg bags	tonne	10 961
Coarse aggregates for concrete	m^3	910
Fine aggregates for concrete	m^3	1245
Ready mixed concrete (B225)	m^3	8110
Ready mixed concrete (B300)	m^3	8365
Steel		
Mild steel reinforcement	tonne	68 940
High tensile steel reinforcement	tonne	80 500
Structural steel sections	tonne	93 500

	Units	*Cost Dr*
Bricks and blocks		
Common bricks (6 x 9 x 19)	1000	12 700
Good quality facing bricks (9 x 12 x 19)	1000	40 500
Hollow concrete blocks (19 x 19 x 39)	1000	69 000
Solid concrete blocks (15 x 19 x 39)	1000	60 000
Precast concrete cladding units with exposed aggregate finish	m^2	4600
Timber and insulation		
Softwood sections for carpentry	m^3	53 630
Softwood for joinery	m^3	61 470
Hardwood for joinery	m^3	78 265
Exterior quality plywood (8mm)	m^2	2525
Plywood for interior joinery (18mm)	m^2	1845
Softwood strip flooring (20mm)	m^2	3775
Chipboard sheet flooring (22mm)	m^2	2010
100mm thick quilt insulation	m^2	910
100mm thick rigid slab insulation	m^2	1240
Softwood internal door complete with frames and ironmongery	nr	45 850
Glass and ceramics		
Float glass (5mm)	m^2	1910
Sealed double glazing units	m^2	6000
Plaster and paint		
Good quality ceramic wall tiles (20 x 20cm)	m^2	2870
Plaster in 50 kg bags	tonne	46 800
Plasterboard (12.5mm thick)	m^2	650
Emulsion paint in 5 litre tins	kg	280
Gloss oil paint in 5 litre tins	kg	545
Tiles and paviors		
Clay floor tiles (10 x 20 x 2)	m^2	2370
Vinyl floor tiles (33 x 32 x 2)	m^2	1380
Precast concrete paving slabs (50 x 50 x 2)	m^2	1070
Clay roof tiles	1000	67 970
Drainage		
WC suite complete	nr	32 360
Lavatory basin complete	nr	14 910
100mm diameter clay drain pipes	m	490
150mm diameter cast iron drain pipes	m	4890

Unit rates

The descriptions below are generally shortened versions of standard descriptions listed in five languages (English, French, Italian, German and Spanish) in section 6. Where an item has a two digit reference number (e.g. 05 or 33), this relates to the full description against that number in section 6. Where an item has an alphabetic suffix (e.g. 12A or 34B) this indicates that the standard description has been modified. Where a modification is major the complete modified description is included here and the standard description should be ignored; where a modification is minor (e.g. the insertion of a named hardwood) the shortened description has been modified here but, in general, the full description in section 6 prevails.

The unit rates below are for main work items on a typical construction project in the Athens area in January 1991. The rates include all necessary labour, materials, equipment and, where appropriate, allowances for contractors overheads and profit, preliminary and general items and contractors profit and attendance on specialist trades. All the rates in this section exclude value added tax (VAT - see below).

		Unit	*Rate Dr*
Excavation			
01	Mechanical excavation of foundation trenches	m^3	1470
02	Hardcore filling making up levels	m^2	1550
03	Earthwork support	m^2	1400
Concrete work			
04	Plain insitu concrete in strip foundations in trenches	m^3	14 770
05	Reinforced insitu concrete in beds	m^3	16 670
06	Reinforced insitu concrete in walls	m^3	21 170
07	Reinforced insitu concrete in suspended floor or roof slabs	m^3	22 170
08	Reinforced insitu concrete in columns	m^3	22 670
09	Reinforced insitu concrete in isolated beams	m^3	22 000
10	Precast concrete slab	m^2	7500
Formwork			
11	Softwood formwork to concrete walls	m^2	3100
12	Softwood or metal formwork to concrete columns	m^2	3270
13	Softwood or metal formwork to horizontal soffits of slabs	m^2	3570
Reinforcement			
14	Reinforcement in concrete walls	tonne	135 000
15	Reinforcement in suspended concrete slabs	tonne	130 670
16	Fabric reinforcement in concrete beds	m^2	400
Steelwork			
17	Fabricate, supply and erect steel framed structure	tonne	410 000

		Unit	Rate Dr
Brickwork and blockwork			
18	Precast lightweight aggregate hollow concrete block walls	m^2	2700
19A	Solid (perforated) concrete blocks	m^2	3800
21	Facing bricks	m^2	9000
Roofing			
22	Concrete interlocking roof tiles 430 x 380mm	m^2	6000
23	Plain clay roof tiles 260 x 160mm	m^2	9500
24	Fibre cement roof slates 600 x 300mm	m^2	8500
25	Sawn softwood roof boarding	m^2	5000
27	3 layers glass-fibre based bitumen felt roof covering	m^2	6000
28	Bitumen based mastic asphalt roof covering	m^2	3250
29	Glass-fibre mat roof insulation 160mm thick	m^2	2100
30	Rigid sheet loadbearing roof insulation 75mm thick	m^2	2250
31	Troughed galvanized steel roof cladding	m^2	2530
Woodwork and metalwork			
32	Preservative treated sawn softwood 50 x 100mm	m	1800
33	Preservative treated sawn softwood 50 x 150mm	m	2760
34A	Single glazed casement window in Niagon hardwood, size 650 x 900mm	each	70 000
35A	Two panel glazed door in Niagon hardwood, size 850 x 2000mm	each	200 000
36	Solid core half hour fire resisting hardwood internal flush doors, size 800 x 2000mm	each	52 500
37	Aluminium double glazed window, size 1200 x 1200mm	each	91 600
38	Aluminium double glazed door, size 850 x 2100mm	each	142 500
39	Hardwood skirtings	m	1650
40	Framed structural steelwork in universal joist sections	tonne	416 600
41	Structural steelwork lattice roof trusses	tonne	425 000
Plumbing			
42	UPVC half round eaves gutter	m	2167
43	UPVC rainwater pipes	m	2430
44	Light gauge copper cold water tubing	m	1900
45	High pressure plastic for cold water supply	m	1130
6	Low pressure plastic pipes for cold water distribution	m	1180
47	UPVC soil and vent pipes	m	2360
48	White vitreous china WC suite	each	46 700
49	White vitreous china lavatory basin	each	34 300
50	Glazed fireclay shower tray	each	32 000
51	Stainless steel single bowl sink and double drainer	each	30 000

		Unit	Rate Dr
Electrical work			
52	PVC insulated and copper sheathed cable	m	550
53	13 amp unswitched socket outlet	each	5000
54	Flush mounted 20 amp, 1 way light switch	each	4980
Finishings			
55	2 coats gypsum based plaster on brick walls	m^2	3900
56	White glazed tiles on plaster walls	m^2	9000
57	Red clay quarry tiles on concrete floor	m^2	7400
58	Cement and sand screed to concrete floors	m^2	2100
59	Thermoplastic floor tiles on screed	m^2	4000
60	Mineral fibre tiles on concealed suspension system	m^2	16 500
Glazing			
61	Glazing to wood	m^2	2600
Painting			
62	Emulsion on plaster walls	m^2	1160
63	Oil paint on timber	m^2	1870

Approximate estimating

The building costs per unit area given below are averages incurred by building clients for typical buildings in the Athens area as at January 1991. They are based upon the total floor area of all storeys, measured between external walls and without deduction for internal walls.

Approximate estimating costs generally include for mechanical and electrical installations but exclude furniture, loose or special equipment, and external works; they also exclude fees for professional services. The costs shown are for specifications and standards appropriate to Greece and this should be borne in mind when attempting comparisons with similarly described building types in other countries. A discussion of this issue is included in section 4. Comparative data for countries covered in this publication, including construction cost data are presented in Part Three.

Approximate estimating costs must be treated with reserve; they cannot provide more than a rough guide to the probable cost of building.

All the rates in this section exclude value added tax (VAT - see below).

	Cost m^2 Dr	Cost ft^2 Dr
Industrial buildings		
Factories for letting	75 000	7000
Factories for owner occupation (light industrial use)	80 000	7400
Factories for owner occupation (heavy industrial use)	82 000	7600

	Cost m^2 Dr	Cost ft^2 Dr
Factory/office (high-tech) for letting (shell and core only)	98 000	9100
Factory/office (high-tech) for letting (ground floor shell, first floor offices)	100 000	9300
Factory/office (high tech) for owner occupation (controlled environment, fully finished)	175 000	16 300
High tech laboratory workshop centres (air conditioned)	175 000	16 300
Warehouses, low bay (6 to 8m high) for letting (no heating)	70 000	6500
Warehouses, low bay for owner occupation (including heating)	75 000	6990
Warehouses, high bay for owner occupation (including heating)	80 000	7500
Cold stores/refrigerated stores	160 000	14 900
Administrative and commercial buildings		
Civic offices, non air conditioned	140 000	13 000
Civic offices, fully air conditioned	170 000	15 800
Offices for letting, 5 to 10 storeys, non air conditioned	143 000	13 300
Offices for letting, 5 to 10 storeys, air conditioned	165 000	16 300
Offices for letting, high rise, air conditioned	190 000	17 650
Offices for owner occupation 5 to 10 storeys, non air conditioned	150 000	14 300
Offices for owner occupation 5 to 10 storeys, air conditioned	200 000	18 600
Offices for owner occupation, high rise, air conditioned	210 000	19 500
Prestige/headquarters office, 5 to 10 storeys, air conditioned	215 000	20 000
Prestige/headquarters office, high rise, air conditioned	240 000	22 300
Health and education buildings		
General hospitals (100 beds)	180 000	16 700
Teaching hospitals (100 beds)	180 000	16 700
Private hospitals (100 beds)	180 000	16 700
Health centres	150 000	14 000
Nursery schools	180 000	16 700
Primary/junior schools	150 000	14 000
Secondary/middle schools	155 000	14 400
University (arts) buildings	180 000	16 700
University (science) buildings	173 000	16 100
Management training centres	180 000	16 700
Recreation and arts buildings		
Theatres (over 500 seats) including seating and stage equipment	200 000	18 600
Theatres (less than 500 seats) including seating and stage equipment	200 000	18 600
Concert halls including seating and stage equipment	250 000	23 300
Sports halls including changing and social facilities	130 000	12 100
Swimming pools (international standard) including changing and social facilities (each)	250m	each

	Cost m^2 Dr	Cost ft^2 Dr
National museums including full air conditioning and standby generator	225 000	20 900
Local museums including air conditioning	170 000	15 800
City centre/central libraries	180 000	16 700
Branch/local libraries	160 000	14 900
Residential buildings		
Social/economic single family housing (multiple units)	140 000	13 000
Private/mass market single family housing 2 storey detached/semidetached (multiple units)	160 000	14 900
Purpose designed single family housing 2 storey detached (single unit)	205 000	19 100
Social/economic apartment housing, low rise (no lifts)	120 000	11 200
Social/economic apartment housing, high rise (with lifts)	150 000	13 950
Private sector apartment building (standard specification)	132 000	12 300
Private sector apartment buildings (luxury)	197 000	18 300
Student/nurses halls of residence	150 000	13 950
Homes for the elderly (shared accommodation)	160 000	14 900
Homes for the elderly (self contained with shared communal facilities)	170 000	15 800
Hotel, 5 star, city centre	200 000	18 600
Hotel, 3 star, city/provincial	155 000	14 000

Regional variations

The approximate estimating costs are based on projects in Athens. Adjust these costs by the following factors to take account of regional variations:

Kreta	:+10%	Thesaloniki	: 0%
Kos	: 0%	Arta	: - 10%
Peloponisos	: - 5%	Korfou	:+20%
Kozani	: - 10%	Chios	:+10%
Limnos	:+20%		

Value added tax (VAT)

The standard rate of value added tax (VAT) is currently 18%.

EXCHANGE RATES AND INFLATION

The combined effect of exchange rates and inflation on prices within a country and price comparisons between countries is discussed in section 4.

Exchange rates

The graph below plots the movement of the Greek Drachma against sterling and the US dollar since 1980. The figures used for the graph are quarterly and the method of calculating these, and other related issues are discussed in section 4. The exchange rate at January 1991 was Dr309 to the pound sterling and Dr160 to the US dollar.

Greek Drachma against Sterling and US Dollar

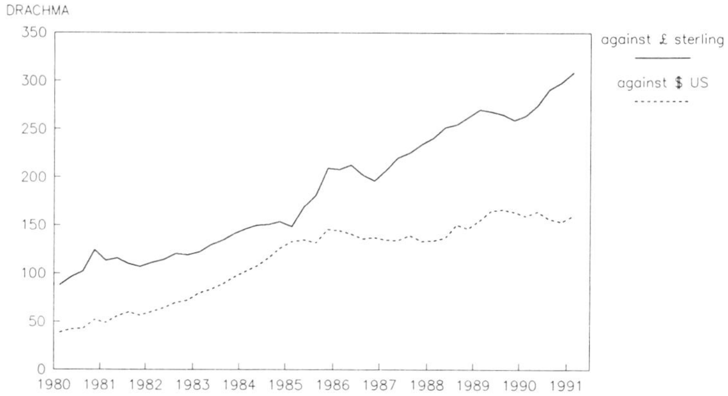

Inflation

The table below presents a number of indices showing the trends in inflation, prices and costs. All the indices have been rebased to 1980=100. Notes on inflation and the distinction between costs and prices are included in section 4.

Cost and price indices

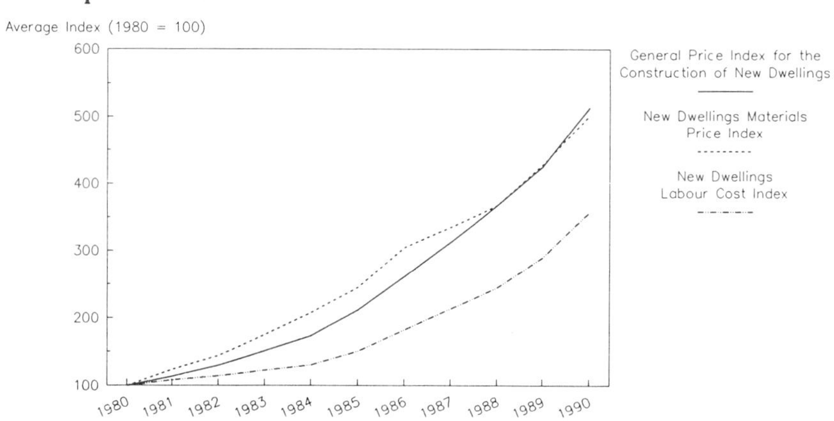

Consumer price inflation

The basis of the figures is the official consumer price index converted to quarterly averages. Retail price inflation peaked in 1986. Since then the annual rate has been between 15% and 20%, considerably higher than average for the rest of the EC. All of the indices in this section are prepared by the National Statistical Service of Greece.

Retail price index 1980 = 100

	1981	1982	1983	1984	1985	1986	1987	1988	1989	1990
average index	114	127	152	174	212	262	313	367	425	513
annual change (%)	13.7	11.4	19.4	14.8	21.9	23.5	19.5	17.3	15.7	20.8

Building price inflation

Two construction price indices are presented below. The materials price index outlines the price movements in a "basket" of construction materials since 1980. The labour price index indicates that construction wage inflation has lagged behind price increases in materials over the period.

Material price index 1980 = 100

	1981	1982	1983	1984	1985	1986	1987	1988	1989	1990
average index	124	144	175	208	246	305	335	367	428	501
annual change (%)	23.7	16.7	21.5	18.5	18.2	24.2	9.8	9.6	16.5	17.0

Labour price index 1980 = 100

	1981	1982	1983	1984	1985	1986	1987	1988	1989	1990
average index	108	114	122	130	151	183	214	245	290	358
annual change (%)	8.2	5.7	6.9	6.6	15.7	21.1	17.1	14.6	18.3	23.3

USEFUL ADDRESSES

Ministry of Environment, Planning and Public Works
17 Amaliados str
Athens

General Secretariat of Public Works, Department of Building Construction, 1 P. Tsaldari
Kallithea
Athens

National Statistical Service of Greece (ESYE)
14-16 Lykourgou str
Athens

Association of Greek Architects
3 Ipitov Street
Athens

Technical Chamber of Greece
4 Karageorgi Servias Street
Athens 10248
Tel: 324 4842

Hellenic Organization for Standardization
Didotou 15, 106 80
Athens

Hellenic Association of Consulting Firms
106 Themistocleavs str
10557 Athens
Tel: 361 6008
Fax: 361 5043

Hungary

KEY DATA

Population

Population (1989)	10.6m
Urban population	60%
Population under 15	19.9%
Population over 65	13.4%
Average annual growth rate (1983 to 1988)	-0.2%

Geography

Land area	93 000 km^2
Agricultural area	n.a.
Capital city	Budapest (population 2 072 000)

Economy

Monetary unit	Forint (Ft)
Exchange rate (January 1991) to	
the ECU	Ft 95.2
the pound sterling	Ft 136
the US dollar	Ft 70.4
Average annual inflation (1984 to 1989)	8.9%
Gross Domestic Product (GDP)	Ft 1411bn
GDP per capita (1989)	Ft 133 113
GDP growth (real) (1980 to 1988)	1.8%
Private consumption (% of total)	61.0%
Public consumption (% of total)	11.3%
Investment (% of total)	24.9%
Central government expenditure as percentage of GDP	n.a.

Construction

Total value of construction output (1989)	Ft 260bn
Construction output per capita	Ft 24 528bn
Construction output as a percentage of GDP	18.4%

THE CONSTRUCTION INDUSTRY

Construction output

In 1989 the value of gross construction output in Hungary was 260 billion forints equivalent to 4 billion ECUs. According to indirect indicators it may be nearer to 7 billion ECUs. Construction output is 18.4% of GDP, based on official figures.

It is reported that repair and maintenance, which had been rising steadily through the 1970s and 1980s and accounted for some 30% of output in 1987, rose to over 40% in 1989. Of new work, residential, which includes hotels, was 24%, representing a large increase from previous years (3 times), possibly because of hotels. Other buildings accounted for 47% and civil works 29%.

Construction output as a percentage of GDP has been falling and so also has the percentage of the workforce engaged in construction. The outlook for the next year or so is for a continued weakness in construction demand as a consequence of the poor economic situation.

Characteristics and structure of industry

There are several types of contracting organizations in Hungary: state building companies and co-operatives which in the early 1980s did over half the work but now probably do only a quarter to a third; other organizations outside the construction sector which undertake about a third of work; associations and partnerships which were very small in the early 1980s but now account for perhaps a sixth of output; and private craft/DIY operations which were always important but now are growing towards taking a quarter of output.

In 1987, according to UN data, there were about 1000 contractors in Hungary, of which over half employed fewer than 50 employees. Some 57 firms employed more than 1000 persons and their total employment was 55% of the work force of construction firms. These large firms undertook over 60% of the work done although other statistics suggest that the industry is much less biased towards large firms. Certainly the situation is changing.

There is one Hungarian company in the Engineering News Record's "Top 250 International Contractors": Transinvest Ltd ranked at 209, specialises in water, sewerage and transport and had only export business.

Foreign companies are already firmly established in Hungary. Over 1000 joint ventures with foreign firms have been registered. Over 100 US construction companies are represented in Hungary. In 1988, 728 companies competed in tenders for communal projects, including 12 foreign contractors. The advent of foreign companies has been facilitated by a number of acts to encourage foreign investment, notably: Act VI (1988), the Companies Act; Act XXVI (1988), the Foreign Investment Act; Act XIII (1989) the Transformation Act on the transformation of economic organizations and economic associations; and Act IX (1988), the Entrepreneurs Profit Tax Act.

Clients and finance

Hungary was the first Eastern bloc country to encourage small scale private enterprise. In 1990 this process was dramatically accelerated. The state as the dominant client of the construction industry is being replaced by privatized organizations and newly established enterprises.

In the past, central government was client for about a quarter of construction work, local councils another quarter and public sector companies about a half. It is expected that work from the companies sector will increase and both central and local government will decline in importance as clients of the industry.

It is expected that for small projects, for example, housing, small hotels, shops, extensions and rehabilitation, private investment will grow. For infrastructure works, government will increasingly support investments by others rather than being the major financier. In the future the construction of major civil engineering projects is likely to be dependent on foreign financing - a 15 year World Bank loan including US$53.8 million for road building is an example of this. Hungary will increasingly be looking to joint ventures with foreign institutions both for finance and construction. There is an International Financial Appeal for development programmes in infrastructure and tourism linked to the World Expo in Budapest in 1995.

According to UN figures, only 10% of housing was public sector in 1988. However, the state has subsidized much private sector housing construction in one way or another. It has been possible to borrow money for private house construction or purchase of housing from the state at fixed rates of interest, sometimes as low as 1% p.a. Government is attempting to reduce expenditure on housing subsidies of various types.

Contractual arrangements

Since 1983 competitive bidding has been compulsory for all public sector construction work. It is thought that this, together with the fall in demand is the reason why, for the first time in many years, construction prices did not increase in the first half of 1990, though general inflation was 30% to 35%.

Development control and standards

Building law in Hungary is concerned with regional planning, the preservation of historic monuments, the protection of recreational areas, the authorization and supervision of the building activities of official building authorities, building design and the construction of buildings.

Building designs are controlled in the first instance by the quality control department of the design enterprise. A separate building permit is granted for each building and the procedure is the same throughout the country.

The National Building Regulations are uniform for the whole country. There are additional regulations for special regions, e.g. Lake Batalon. The regulations cover urban and community planning as well as building. The regulations for civil engineering and public works correspond to the building regulations.

The subjects for standardization include technical requirements (of jointing etc), test methods, packing, transportation and storage of products. National standards are issued by the President of the Hungarian Office for Standardization, and Branch Standards by the Minister of Building and Urban Development. The latter may not conflict with the former. Some standards are mandatory; deviations from others are permitted by agreement. Building sites are regularly supervised by the technical inspector of the builder. Designs are checked on a random sampling method, by the Institute for the Quality Control of Building which also issues or proposes the issue of certificates of fitness for new products.

Information on laws, regulations and standards concerning building are published in gazettes known as *Magyar Közlöny* (Hungarian Gazette), *Épitésügyi Értesitö* (Construction Bulletin) and *Szabványügyi Közlöny* (Gazette of Standards). In addition, there are lists of documents on technical regulations in building (*Épitésügyi Müszakj Szabályozási Kiadványok Jegyzéke*) published regularly by the Information Centre for Building (ETK) together with the Institute for Quality Control of Building (ÉMI). The lists contain the titles of the technical documents (regulations, standards, technical guidelines, type designs, technical-economic norms, permissions for applications, certificates of serviceability, order).

Approximately every five years, the Ministry of Building and Urban Development publishes a collection of the building regulations and instructions in force and the Hungarian Office for standardization publishes the list of State Standards (*Allami Szabványok Jegyzéke*).

Research and development

The Institute for Building Science and the Central Research and Design Institute undertake research for the Ministry of Building and Urban Development. The Institute for the Quality Control of Building carries out research concerning testing methods and the development of regulations.

Research is also carried out by certain departments of the Technical University of Budapest and other institutions of technical education. The Technical Development Fund is formed from the contributions of enterprises, funds central research and, to a lesser extent, technical development work for enterprises.

Foreign participation including joint ventures

The Government acknowledges that it needs to attract technical know-how and skills and has simplified the procedures for foreign investors. Companies are free to establish 50:50 joint ventures with Hungarian firms without any formalities while 100% foreign ownership is also permitted, subject to governmental approval. The Act on Foreign Investment of 1989 guarantees the repatriation of profits and the protection of capital and introduces various tax incentives to encourage investment.

The World Expo 1995 is to be hosted Budapest and will require substantial investment in infrastructure of around US$2 billion with possible further work of US$6 billion.

Further reading

Anna Gáspár, "Hungarian Construction Market 1990", paper for NEDO/Euro-construct conference, London, 20 November 1990.

Rolf Bollinger, "Chancen und Risken für die Westeuropäische Bauwirtschaft nach den Wende in Osteuropa" in Euroconstruct, Baukonjunktur 1990/91 in Westeuropa, IFO Institut für Wirtschaftsforschung, München, Dezember 1990.

EXCHANGE RATES AND INFLATION

The combined effect of exchange rates and inflation on prices within a country and price comparisons between countries is discussed in section 4.

Exchange rates

The graph below plots the movement of the Hungarian Forint against the American dollar and sterling since 1979. The figures used for the graph are quarterly and the method of calculating these, and other related issues are discussed in section 4. The exchange rate at January 1991 was Ft 136 to the pound sterling and Ft 70.4 to the US$.

The Hungarian Forint against Sterling and the US Dollar

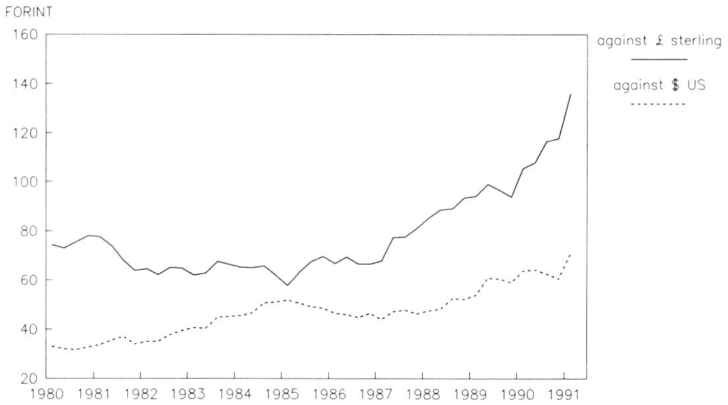

Inflation

The graph and the table below present the consumer price index showing the trend in inflation since 1981. The index has been rebased to 1981=100. Notes on inflation and the distinction between costs and prices are included in section 4.

Consumer price inflation

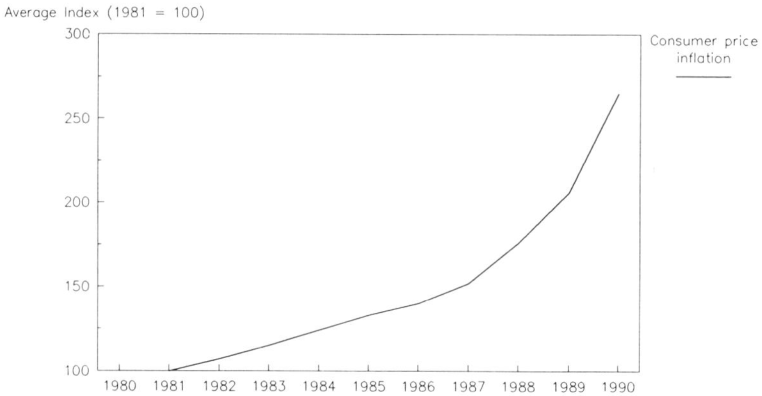

The basis of the figures is the official consumer price index converted to quarterly averages. Inflation has been rising steadily (from a low of 5.3%) since the mid 1980s.

Consumer price inflation 1981 = 100

	1982	1983	1984	1985	1986	1987	1988	1989	1990
average index	107	115	124	133	140	152	176	206	265
annual change (%)	6.9	7.3	8.3	7.0	5.3	8.6	15.7	17.0	28.9

Ireland

KEY DATA

Population

Population (1989)	3.5m
Urban population	58%
Population under 15	27.7%
Population over 65	10.3%
Average annual growth rate (1983 to 1988)	0.2%

Geography

Land area	69 000 km^2
Agricultural area	57 000 km^2
Capital city	Dublin (population 502 749)

Economy

Monetary unit	Punt (IR£)
Exchange rate (January 1991) to	
the ECU	IR£ 0.77
the pound sterling	IR£ 1.10
the US dollar	IR£ 0.57
Average annual inflation (1984 to 1989)	3.7%
Gross Domestic Product (GDP)	IR£ 21bn
GDP per capita (1989)	IR£ 6000
GDP growth (real) (1980 to 1988)	1.7%
Private consumption (% of total)	58.0%
Public consumption (% of total)	17.0%
Investment (% of total)	16.0%
Central government expenditure as percentage of GDP	50.0%

Construction

Total value of construction output (1989)	IR£ 2.59bn
Construction output per capita	IR£ 740
Construction output as a percentage of GDP	12.3%

THE CONSTRUCTION INDUSTRY

Construction output

The total value of construction gross output in Ireland in 1989 was IR£2592 million. This represents about 12.3% of GDP. The breakdown by type of work is shown below.

Construction output 1989

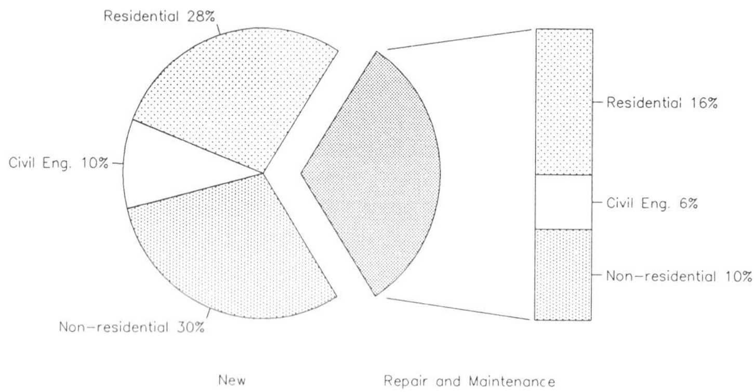

The proportion of civil engineering in total output is low and non-residential building is high.

Between 1980 and 1989 construction output fell steadily. Output in 1990 was probably higher than that in 1989, but thereafter a pause is expected although help from the European Community may enable a continuing increase.

Characteristics and structure in the industry

The main engineering organization is the Institution of Engineers of Ireland (IEI) with a membership of about 6000, of whom half are chartered engineers or fellows. Engineers occupy key positions of authority in central and local government and in industry. About 40% of engineers are employed in the public sector. Due to the fall in construction output in the 1980s, the number of consultancy firms and the numbers employed by them have fallen. By 1989 there were 148 firms employing 569 engineers. Many engineers have emigrated. Some 40% to 50% of Irish engineering graduates seek employment abroad.

The organization for architects is the Royal Institution of the Architects of Ireland (RIAI).

Almost all quantity surveyors are members of the Society of Chartered Surveyors in Ireland which offers them full membership of the Royal Institution of Chartered Surveyors in the UK. Many professionals have trained or gained experience in the UK or USA.

The majority of building contractors in Ireland are medium and small sized with working proprietors who also act as site agents and general foremen. In the whole country there are probably eight to ten contractors with the resources and experience to manage and construct major projects in excess of IR£2.0 million. These contractors have shed directly employed tradesmen in recent years and retained key management personnel. This has led to a growth of subcontracting firms in all trades. Generally, roadbuilding is carried out by local authorities employing direct labour, and relying on plant hire for the provision of equipment, in order to avoid tying up capital during slack periods. Much of the construction equipment available is old and in need of replacement.

Irish contractors do not feature in the top 300 European building contractors.

Clients and finance

Some 55% to 60% of the output of the Irish construction industry is privately funded. Government expenditure declined steadily from 1982 to 1989 but is increasing again with funds from the European Community.

Over 95% of housing is financed privately. The proportion built by local authorities has been falling rapidly. In 1983 it was 25% but by 1988 it had fallen to 5%. However, the stock of local authority housing requires a larger expenditure by local authorities whose maintenance work accounts for around 25% of all housing maintenance.

Central Government also plays a part in private housing, with new housing grants and water and sewerage grants. In 1988, 7% of mortgages were local authority funded compared with 59% by building societies and 33% by banks.

Since joining the European Community in 1973, Ireland has benefited considerably from European grants and loans. Ireland submitted the National Development Plan 1989 to 1993 to the European Commission in 1989. The Commission responded with a Community Support Framework for Ireland which provides for Structural Fund assistance of IR£2.9 billion (3672 million ECUs), about IR£500 million less than requested. European Community Structural Funds will account for about a third of expenditure under the plan with Government funds over a third and the private sector only about a quarter. The Government plans to meet its obligations under the plan partly by a loan from the European Investment Bank.

Selection of consultants

Procedures for appointment of consultants for public work were revised in January 1988. For projects over IR£1 million in value at least three consultancy firms are invited to make a submission on the basis of an outline brief and technical approach. One consultant is selected and then negotiation on fees takes place. If negotiation is inconclusive the second choice consultant is asked to negotiate and so on until agreement is reached.

Consultants are normally appointed directly by the client for private sector work on the basis of experience and personal contacts.

A form of design competition known as "developer competition" is increasingly becoming the main method of procuring large scale building projects in Ireland, particularly in the public sector for health projects and government offices. Bids are invited from consultants and contractors usually on the basis of an outline brief and the submission usually involves considerable design, costing and other technical input for which consultants are not remunerated unless awarded the project. The professional institutions are opposed to this method of procurement.

Contractual arrangements

Tendering and contract procedures in Ireland are similar to those in the UK. The Liaison Committee, which consists of representatives of Architects, Chartered Surveyors and Contractors, draft conditions of contract, with or without quantities, which are published by the RIAI. Public contracts use the GDLA conditions of contract which are similar to those of the RIAI. There are no nationally agreed conditions for management or design and construct contracts and special forms or amended JCT forms are used.

Tendering procedures are governed by a Code of Procedure published by the Liaison Committee. Most major contracts are based on Bills of Quantities issued to a selected list of 4 to 6 contractors. Bills are submitted with tenders. Tenders are usually fluctuating but fixed price contracts are often negotiated after receipt of tenders.

Development control and standards

Statutory development plans are prepared for the 87 local authorities. Planning, including the granting of permission, takes place at local level. The state advises and issues guidelines and directives as well as undertaking regional coordination.

There is a National Code for Building Regulations which specifies performance or functional requirements. The regulations incorporate examples of technical solutions fulfilling the relevant requirements. There are no local regional variations.

The Institute for Industrial Research and Standards (IIRS) draws up and promulgates standard specifications for building products, and it is normal for manufacturers to comply. Architects and other building professionals generally have regard to Irish Standards and Codes of Practice issued by *Eolas* or, in their absence, British Standards or Codes of Practice. The Construction Industry Federation operates a National Housebuilding Guarantee Scheme, which guarantees houses built by contractors registered under the scheme for six years against major structural defects.

Further reading

Association of Consulting Engineers et al, Engineering Consultancy in the European Community: Republic of Ireland, London: Telford, 1990.

CONSTRUCTION COST DATA

Cost of labour

The figures below are typical of labour costs in Dublin as at January 1991. The "wage rate" is the basis of an employee's income, while the "cost of labour" indicates the cost to a contractor of employing that employee. The difference between the two covers a variety of mandatory and voluntary contributions - a list of items which could be included is given in section 4.

	Wage rate (per hour) IR£	*Cost of labour (per hour) IR£*	*Number of hours worked per year*
Site operatives			
Mason/bricklayer	4.17	6.93	1900
Carpenter	4.17	6.93	1900
Plumber	4.17	6.93	1900
Electrician	4.65	7.70	1900
Structural steel erector	4.27	7.01	1900
HVAC installer	4.17	6.93	1900
Semi-skilled worker	4.00	6.50	1900
Unskilled labourer	3.82	5.79	1900
Equipment operator	3.92	4.04	1900
Watchman/security	2.50	3.50	1900
Site supervision			
General foreman	5.00	8.00	1900
Trades foreman	4.60	7.50	1900
Clerk of works	5.00	8.00	1900
Contractors' personnel			
Site manager	13.00	19.50	1900
Resident engineer	11.00	16.50	1900
Resident surveyor	9.00	13.50	1900
Junior engineer	7.50	11.25	1600
Junior surveyor	7.50	11.25	1600
Planner	8.75	13.00	1600
Consultants' personnel			
Senior architect	16.00	26.00	1400
Senior engineer	18.00	29.00	1400
Senior surveyor	18.00	29.00	1400
Qualified architect	14.00	21.00	1400
Qualified engineer	15.00	22.50	1400
Qualified surveyor	15.00	22.50	1400

Cost of materials

The figures that follow are the costs of main construction materials, delivered to site in the Dublin area, as incurred by contractors in January 1991. These assume that the materials would be in quantities as required for a medium sized construction project and that the location of the works would be neither constrained nor remote.

All the costs in this section exclude value added tax (VAT - see below).

	Units	*Cost IR£*
Cement and aggregate		
Ordinary portland cement in 50kg bags	tonne	73.00
Coarse aggregates for concrete	m^3	10.65
Fine aggregates for concrete	m^3	12.60
Ready mixed concrete (20N/mm^2)	m^3	45.00
Ready mixed concrete (30N/mm^2)	m^3	55.00
Steel		
Mild steel reinforcement	tonne	305.00
High tensile steel reinforcement	tonne	315.00
Structural steel sections	tonne	385.00
Bricks and blocks		
Good quality facing bricks (225 x 115 x 75mm)	1000	650.00
Hollow concrete blocks (450 x 225 x 225mm)	1000	610.00
Solid concrete blocks (450 x 225 x 100mm)	1000	340.00
Precast concrete cladding units with exposed aggregate finish	m^2	120.00
Timber and insulation		
Softwood sections for carpentry	m^3	265.00
Softwood for joinery	m^3	525.00
Hardwood for joinery	m^3	950.00
Exterior quality plywood (20mm)	m^2	8.00
Plywood for interior joinery (20mm)	m^2	6.50
Softwood strip flooring (12mm)	m^2	9.00
Chipboard sheet flooring (12mm)	m^2	8.00
100mm thick quilt insulation	m^2	2.00
100mm thick rigid slab insulation	m^2	2.50
Softwood internal door complete with frames and ironmongery	nr	300.00
Glass and ceramics		
Float glass (6mm)	m^2	22.00
Sealed double glazing units	m^2	50.00

	Units	Cost IR£
Plaster and paint		
Good quality ceramic wall tiles (100 x 100mm)	m^2	12.00
Plaster in 50 kg bags	tonne	123.00
Plasterboard (6mm thick)	m^2	3.00
Emulsion paint in 5 litre tins	litre	3.50
Gloss oil paint in 5 litre tins	litre	4.50
Tiles and paviors		
Clay floor tiles (150 x 150mm)	m^2	20.00
Vinyl floor tiles (200 x 200 x 2.5mm)	m^2	6.00
Precast concrete paving slabs (600 x 600 x 50mm)	m^2	10.00
Clay roof tiles	1000	490.00
Precast concrete roof tiles	1000	400.00
Drainage		
WC suite complete	nr	35.00
Lavatory basin complete	nr	30.00
100mm diameter clay drain pipes	m	3.70
150mm diameter cast iron drain pipes	m	3.80

Unit rates

The descriptions below are generally shortened versions of standard descriptions listed in five languages (English, French, Italian, German and Spanish) in section 6. Where an item has a two digit reference number (e.g. 05 or 33), this relates to the full description against that number in section 6. Where an item has an alphabetic suffix (e.g. 12A or 34B) this indicates that the standard description has been modified. Where a modification is major the complete modified description is included here and the standard description should be ignored; where a modification is minor (e.g. the insertion of a named hardwood) the shortened description has been modified here but, in general, the full description in section 6 prevails.

The unit rates below are for main work items on a typical construction project in the Dublin area in January 1991. The rates include all necessary labour, materials, equipment and, where appropriate, allowances for contractors overheads and profit, preliminary and general items and contractors profit and attendance on specialist trades. All the rates in this section exclude value added tax (VAT - see below).

		Unit	Rate IR£
Excavation			
01	Mechanical excavation of foundation trenches	m^3	2.03
02	Hardcore filling making up levels	m^2	2.04
03	Earthwork support	m^2	3.75

		Unit	Rate IR£
Concrete work			
04	Plain insitu concrete in strip foundations in trenches	m^3	64.20
05	Reinforced insitu concrete in beds	m^3	65.30
06	Reinforced insitu concrete in walls	m^3	69.60
07	Reinforced insitu concrete in suspended floor or roof slabs	m^3	69.60
08	Reinforced insitu concrete in columns	m^3	82.40
09	Reinforced insitu concrete in isolated beams	m^3	82.40
10	Precast concrete slab	m^2	40.70
Formwork			
11	Softwood formwork to concrete walls	m^2	12.80
12	Softwood or metal formwork to concrete columns	m^2	21.40
13	Softwood or metal formwork to horizontal soffits of slabs	m^2	16.10
Reinforcement			
14	Reinforcement in concrete walls	tonne	514.00
15	Reinforcement in suspended concrete slabs	tonne	514.00
16	Fabric reinforcement in concrete beds	m^2	1.61
Steelwork			
17	Fabricate, supply and erect steel framed structure	tonne	1500.00
Brickwork and blockwork			
18	Precast lightweight aggregate hollow concrete block walls	m^2	8.56
19	Solid (perforated) concrete blocks	m^2	10.70
21	Facing bricks	m^2	58.90
Roofing			
22	Concrete interlocking roof tiles 430 x 380mm	m^2	19.30
23	Plain clay roof tiles 260 x 160mm	m^2	26.80
24	Fibre cement roof slates 600 x 300mm	m^2	23.50
25	Sawn softwood roof boarding	m^2	19.30
26	Particle board roof coverings	m^2	16.10
27	3 layers glass-fibre based bitumen felt roof covering	m^2	25.70
28	Bitumen based mastic asphalt roof covering	m^2	42.80
29	Glass-fibre mat roof insulation 160mm thick	m^2	4.28
30	Rigid sheet loadbearing roof insulation 75mm thick	m^2	3.85
31	Troughed galvanized steel roof cladding	m^2	18.20
Woodwork and metalwork			
32	Preservative treated sawn softwood 50mm x 100mm	m	2.41
33	Preservative treated sawn softwood 50 x 150mm	m	3.64
34A	Single glazed casement window in Iroko hardwood, size 650 x 900mm	each	85.60

		Unit	Rate IR£
35A	Two panel glazed door in Iroko hardwood, size 850 x 2000mm	each	278.00
36	Solid core half hour fire resisting hardwood internal flush doors, size 800 x 2000mm	each	257.00
37	Aluminium double glazed window, size 1200 x 1200mm	each	278.00
38	Aluminium double glazed door, size 850 x 2100mm	each	321.00
39	Hardwood skirtings	m	4.71
40	Framed structural steelwork in universal joist sections	tonne	1610.00
41	Structural steelwork lattice roof trusses	tonne	1610.00
Plumbing			
42	UPVC half round eaves gutter	m	11.80
43	UPVC rainwater pipes	m	12.80
44	Light gauge copper cold water tubing	m	11.80
45	High pressure plastic for cold water supply	m	6.42
46	Low pressure plastic pipes for cold water distribution	m	7.49
47	UPVC soil and vent pipes	m	13.90
48	White vitreous china WC suite	each	85.60
49	White vitreous china lavatory basin	each	69.60
50	Glazed fireclay shower tray	each	161.00
51	Stainless steel single bowl sink and double drainer	each	96.30
Electrical work			
52	PVC insulated and copper sheathed cable	m	3.69
53	13 amp unswitched socket outlet	each	24.60
54	Flush mounted 20 amp, 1 way light switch	each	17.20
Finishings			
55	2 coats gypsum based plaster on brick walls	m^2	8.56
56	White glazed tiles on plaster walls	m^2	32.10
57	Red clay quarry tiles on concrete floor	m^2	37.50
58	Cement and sand screed to concrete floors	m^2	6.96
59	Thermoplastic floor tiles on screed	m^2	13.40
60	Mineral fibre tiles on concealed suspension system	m^2	15.30
Glazing			
61	Glazing to wood	m^2	19.30
Painting			
62	Emulsion on plaster walls	m^2	3.21
63	Oil paint on timber	m^2	3.75

Approximate estimating

The building costs per unit area given below are averages incurred by building clients for typical buildings in the Dublin area as at January 1991. They are based upon the total floor area of all storeys, measured between external walls and without deduction for internal walls.

Approximate estimating costs generally include for mechanical and electrical installations but exclude furniture, loose or special equipment, and external works; they also exclude fees for professional services. The costs shown are for specifications and standards appropriate to Ireland and this should be borne in mind when attempting comparisons with similarly described building types in other countries. A discussion of this issue is included in section 4. Comparative data for countries covered in this publication, including construction cost data are presented in Part Three.

Approximate estimating costs must be treated with reserve; they cannot provide more than a rough guide to the probable cost of building.

All the rates in this section exclude value added tax (VAT - see below).

	Cost m^2 *IR£*	*Cost* ft^2 *IR£*
Industrial buildings		
Factories for letting	300	28
Factories for owner occupation (light industrial use)	320	30
Factories for owner occupation (heavy industrial use)	340	32
Factory/office (high-tech) for letting (shell and core only)	400	37
Factory/office (high-tech) for letting (ground floor shell, first floor offices)	420	39
Factory/office (high tech) for owner occupation (controlled environment, fully finished)	500	46
High tech laboratory workshop centres (air conditioned)	500	46
Warehouses, low bay (6 to 8m high) for letting (no heating)	270	25
Warehouses, low bay for owner occupation (including heating)	300	28
Warehouses, high bay for owner occupation (including heating)	320	30
Cold stores/refrigerated stores	350	33
Administrative and commercial buildings		
Civic offices, non air conditioned	800	74
Civic offices, fully air conditioned	900	83
Offices for letting, 5 to 10 storeys, non air conditioned	600	56
Offices for letting, 5 to 10 storeys, air conditioned	700	65
Offices for owner occupation 5 to 10 storeys, non air conditioned	800	74
Offices for owner occupation 5 to 10 storeys, air conditioned	900	83
Prestige/headquarters office, 5 to 10 storeys, air conditioned	1200	111

	Cost m^2 IR£	Cost ft^2 IR£
Health and education buildings		
General hospitals (100 beds)	1100	102
Teaching hospitals (100 beds)	1300	121
Private hospitals (100 beds)	1400	130
Health centres	600	56
Nursery schools	580	54
Primary/junior schools	580	54
Secondary/middle schools	580	54
University (arts) buildings	860	80
University (science) buildings	1080	100
Management training centres	860	80
Recreation and arts buildings		
Theatres (over 500 seats) including seating and stage equipment	1200	111
Theatres (less than 500 seats) including seating and stage equipment	1300	121
Concert halls including seating and stage equipment	1500	139
Sports halls including changing and social facilities	600	56
Swimming pools (international standard) including changing and social facilities	1200	111
Swimming pools (schools standard) including changing facilities	1000	93
National museums including full air conditioning and standby generator	1000	93
Local museums including air conditioning	900	84
City centre/central libraries	900	84
Branch/local libraries	850	79
Residential buildings		
Social/economic single family housing (multiple units)	400	37
Private/mass market single family housing 2 storey detached/semidetached (multiple units)	300	28
Purpose designed single family housing 2 storey detached (single unit)	500	46
Social/economic apartment housing, low rise (no lifts)	500	46
Social/economic apartment housing, high rise (with lifts)	600	56
Private sector apartment building (standard specification)	480	45
Private sector apartment buildings (luxury)	550	51
Student/nurses halls of residence	700	65
Homes for the elderly (shared accommodation)	600	56
Homes for the elderly (self contained with shared communal facilities)	700	65
Hotel, 5 star, city centre	1300	121
Hotel, 3 star, city/provincial	1200	111
Motel	1000	93

Regional variations

The approximate estimating costs are based on projects in Dublin. Adjust these costs by the following factors for regional variations:

Cork	:	+2% to +5%
Galway	:	- 3% to - 2%
Limerick	:	+3% to +5%
Midlands	:	- 3% to - 2%

Value added tax (VAT)

The standard rate of value added tax (VAT) is currently 23%. VAT is chargeable on general building work at the reduced rate of 10%. Fittings attract the standard rate of 23%.

EXCHANGE RATES AND INFLATION

The combined effect of exchange rates and inflation on prices within a country and price comparisons between countries is discussed in section 4.

Exchange rates

The graph below plots the movement of the Irish punt against sterling and the US dollar since 1980. The figures used for the graph are quarterly and the method of calculating these, and other related issues are discussed in section 4. Ireland joined the European Monetary System (EMS) at its inception in 1979. As the UK did not join the EMS the one-for-one link between the Irish pound and sterling was abandoned in that year. The exchange rate at January 1991 was IR£1.10 to the pound sterling and IR£0.57 to the US dollar.

The Irish Punt against Sterling and the US Dollar

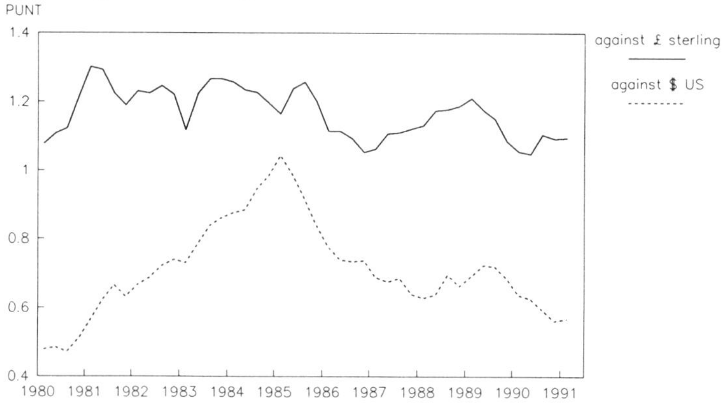

Inflation

The graph below presents a number of indices showing the trends in inflation, prices and costs. All the indices have been rebased to 1980=100. Notes on inflation and the distinction between costs and prices are included in section 4.

Cost and price indices

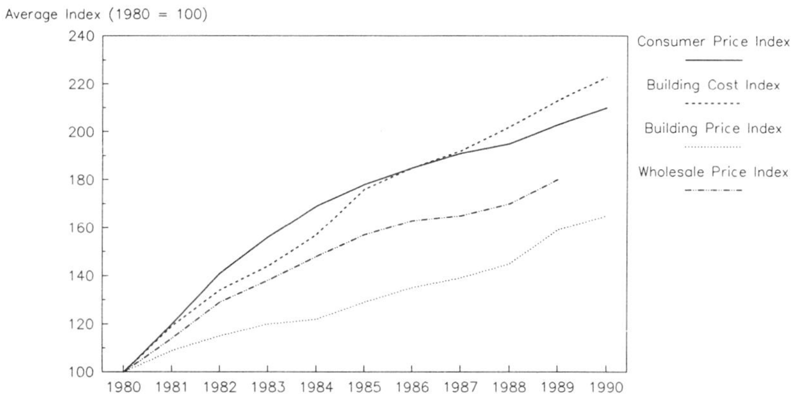

Consumer price inflation

The basis of the figures is the official consumer price index converted to quarterly averages. Inflation peaked in 1981; since then progress in reducing inflation has lagged behind the OECD area as a whole.

The table below presents general inflation in Ireland since 1980. The annual change is the percentage change between the last quarters of consecutive years.

Consumer price inflation 1980 = 100

	1981	1982	1983	1984	1985	1986	1987	1988	1989	1990
average index	120	141	156	169	178	185	191	195	203	210
annual change (%)	20.0	17.5	10.6	8.3	5.3	3.9	3.2	2.1	4.1	3.4

Building cost index

The table below is a cost index compiled by the Society of Chartered Surveyors in Dublin.

Building cost index 1980 = 100

	1981	1982	1983	1984	1985	1986	1987	1988	1989	1990
average index	119	134	144	157	176	185	192	202	213	223
annual change (%)	19.0	12.6	7.5	9.0	12.1	5.1	3.8	5.2	5.4	4.7

Building price index

The table below is a building price index compiled by Patterson Kempster and Shortall (orginally produced by An Foras Forbartha). When this table and the one above are considered together, they indicate that while both costs and prices have increased, the rate of increase in costs has been significantly greater than that in prices testifying to a depressed period for the construction industry.

Building Price Index 1980 = 100

	1981	1982	1983	1984	1985	1986	1987	1988	1989	1990
average index	109	115	120	122	129	135	139	145	159	165
annual change (%)	9.0	5.5	4.3	1.7	5.7	4.7	3.0	4.3	9.7	3.8

Wholesale price index

The table below is an index for wholesale building and construction materials compiled by the Central Statistics offices.

Wholesale Price Index 1980 = 100

	1981	1982	1983	1984	1985	1986	1987	1988	1989	1990
average index	114	129	138	148	157	163	165	170	180	n.a.
annual change (%)	14.0	13.2	7.0	7.2	6.1	3.8	1.2	3.0	5.9	n.a.

USEFUL ADDRESSES

Royal Institute of the Architects of Ireland
8 Merrion Square
Dublin 2

Association of Consulting Engineers Ireland ACE
3 Haddington Road
Dublin 4
Tel: (1) 600374

The Construction Industry Federation
Federation House, Canal Road
Dublin 6
Tel: (1) 977487

Union of Construction Allied Trades and Technicians
56 Parnell Square West
Dublin 1

The Society of Chartered Surveyors in the Republic of Ireland
5 Wilton Place
Dublin 2

Department of the Environment
Custom House
Dublin 1

National Building Agency Ltd
Richmond Avenue
South Dublin 6

National Standards Authority of Ireland
Ballymun Road
Dublin 9

Central Statistics Office
Ardee Road, Rathmines
Dublin 6
Tel: (1) 977144

Italy

KEY DATA

Population

Population (1989)	57.44m
Urban population	68%
Population under 15	17.1%
Population over 65	14.2%
Average annual growth rate (1983 to 1988)	0.2%

Geography

Land area	301 000 km^2
Agricultural area	171 000 km^2
Capital city	Rome (population 2 821 420)

Economy

Monetary unit	Lira (L)
Exchange rate (January 1991) to	
the ECU	L 1538
the pound sterling	L 2198
the US dollar	L 1137
Average annual inflation (1984 to 1989)	6.2%
Gross Domestic Product (GDP)	L 1079 000bn
GDP per capita (1989)	L 18.78bn
GDP growth (real) (1980 to 1988)	2.2%
Private consumption (% of total)	61.7%
Public consumption (% of total)	17.4%
Investment (% of total)	21.4%
Central government expenditure as percentage of GDP	46.1%

Construction

Total value of construction output (1989)	L 115 000bn
Construction output per capita	L 2m
Construction output as a percentage of GDP	10.7%

THE CONSTRUCTION INDUSTRY

Construction output

In 1989 the value of gross construction output in Italy, including new work and renovation and modernization, was Lire 115 000 billion, equivalent to 76 billion ECUs. This represents 10.7% of GDP. Output has been very stable in real terms in recent years and is expected to grow slowly in future. The breakdown of output is shown below.

Construction output 1989

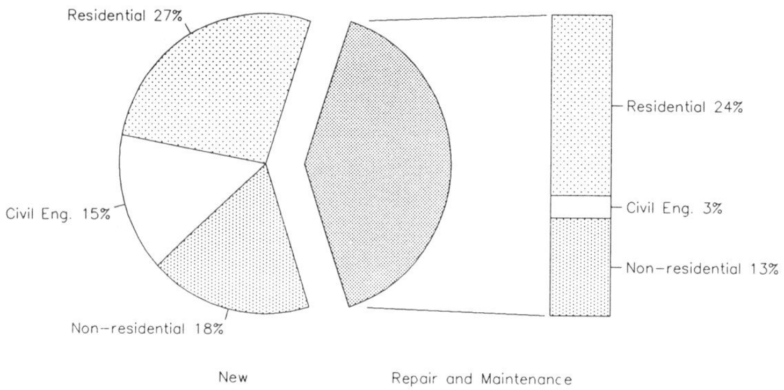

The proportion of non-residential building in total output is low and of this 43% is industrial building. The total amount of construction may in fact be rather more than these figures indicate because there is a large amount of unrecorded output in Italy undertaken by the informal sector. Some, but probably insufficient, account of this is taken in the preparation of official statistics.

Construction investment is spread fairly evenly over the country, but the proportion accounted for by the public sector tends to be high in relation to population in the south of the country (the *Mezzogiorno*), whereas private construction expenditure tends to be concentrated in the industrial north.

In Italy there is a clear distinction between civil works and industrial engineering construction. The former includes building and civil works; the latter includes petrochemicals, heavy industries and power engineering works.

Characteristics and structure of the industry

The State plays a major role in the construction industry in Italy as a client for public works and as a regulator of the process including planning, registration of contractors, standards etc. However, administration of the intentions of government and of laws is not always adequate. It has been estimated that only 50% of budgets for public construction are actually spent and many planning and other regulations are not strictly enforced.

The dominant professions in the construction industry are engineers and architects and one or other of these or both must be involved in all major construction projects. Practising architects must be members of the National Society of Architects (*Ordine Nazionale Architetti*) and engineers must be in the Engineers National Register (*Ordine Nazionale Ingegneri*). The third important consultant is the *geometra* who is more of a technician, but his role is nevertheless significant. Originally he was a land survcyor but now is rcsponsiblc for a largc proportion of small scalc construction work. The quantity surveying functions are undertaken by architects, engineers and *geometri*. Architects operate in studios (which are usually fairly small), in public sector offices, in contracting firms and in commercial consultancy companies. Only about 18% engineers are in private practices. The remainder are in the public service or occupy managerial positions in the commercial sector. The *geometra* works mainly in private practice as a sole practitioner but may also work in architectural practices or with contractors.

There are many small to medium sized contracting firms and a few large firms. Only 500 firms employ over 100 people and probably around 20 to 30 employ over 1000. Most firms are private but there are a few large public sector companies and co-operative companies. Italian contractors have operated overseas for many years and have a broad spread of operations in geographical terms as well as types of work. They account for over a quarter of work done abroad by European Contractors and for 51 of Building magazine's "Top 300 European Contractors", of whom seven are in the top 100. They are very important as exporters. Some 38 of Engineering News Record's "Top 250 International Contractors" are Italian firms - more than any other country - though they do less work than the French and the British. Many Italian firms are very specialized.

Major Italian contractors

Major contractors		*Type of company*	*Place in Building's "Top 300 European Contractors"*	*Place in ENR "Top 250 International Contractors"*	*Exports (% of turnover)*
Italstat	)Gruppo	Public	)		10-20
	)IRI-		) 11		
Condotte	)Italstat	Public	)		20-30
Impresit	] Gruppo	Private	]	16	60
	] Fiat		] 30		
Cogefar	] Impresit		]		
Lodigiani		Private	69	132	30
Impregilo*		Private	72	54	100
Astaldi		Private	86	85	55
Vianini Lavori/Caltagirone		Private	91	214	20-30
CMC		Co-operative			

Notes: * *owned by Fiat, Fiat Impresit, Girola and Lodigiani.*
Source: Building; Engineering News Record

Contractors must be registered with the *Albo Nazionale dei Costruttori* (the National Contractors Register) to be permitted to carry out public work. The classification covers twenty types of work and ten value ranges.

Clients and finance

In 1989 the clients for public works were as follows (with the percentage of value of work in brackets): Provinces and Communes (39%), State Railways (13%), Agency for the Mezzogiorno (13%), Central Government (13%), State Roads Authority (10%), Regions (6%), Postal Authority (4%) and Health (3%). Approximately 5% of new homes are funded directly by the public sector which also contributes in some way to a further 15%. Public housing finance is essentially a regional matter.

A feature of the Italian system is the use of *concessioni* for the execution and maintenance of public works projects such as motorways and airports. Separate companies are established, which may have a state, regional or municipal shareholding, which bid for the right to run the particular *concessioni*.

There is not a well developed speculative property market with major developers as in the UK. Generally, private sector projects are for individual clients' own use or for sale. Over half the office stock is owner occupied. The remainder is owned by insurance companies and banks, and to a lesser extent by pension funds and other private organisations or individuals.

The retail sector, dominated by small, often family run firms, has effectively prevented the construction of major out-of-town shopping centres. Manufacturing industry also has a number of small family firms who own their own factories and often use package deals for procurement of new premises.

Historically, Italy has been a major recipient of financial assistance from the European Community's regional structural funds, receiving 46% of the total in 1975 and 34% in 1980 when Italy's share of the Community's population was only 22%. After the accession of Greece, Spain and Portugal it received 20% of regional funds (still above its share of population: 18%). Italy is the largest recipient of European Investment Bank loans, receiving 39% of the total in 1987. Between 40% and 70% of funding from the Community usually goes on construction investment.

Selection of design consultants

In the public sector the appointment of professional advisers is regulated by Government decrees. Selection of the architect and/or engineer is made directly or by design competition which may be for an outline design or for detailed proposals including a cost estimate. Influence and appropriate contacts often play an important part in awarding design contracts. The winner of a competition is not necessarily commissioned, though his design becomes the property of the public authority. The majority of public works design is undertaken by outside consultants. In-house technical departments concentrate more on checking and approval of designs.

In the private sector appointments are regulated by the provisions of the *Codice Civile* and there is a statutory fee scale. Contracts must be signed by a registered professional; non compliance voids all contracts and on discovery will result in legal proceedings. Signing on behalf of someone who is not registered does occur but brings with it full legal responsibility.

Contractual arrangements

The normal form of contracting is with the employment of a main contractor though clients may seek separate tenders for portions of the work, normally specialist work. Main contractors in turn subcontract most of the work.

In the public sector contracts are usually let by open tender or selective tender. In the private sector some contracts may be negotiated. Overall it is estimated that about 80% of projects go to open tender, 15% to selective tender and 5% to negotiation.

For public works, tenderers are often asked to quote a discount or increase on a base price which is usually an estimate based on pre-priced bills. The winning tender may be the lowest, the average or something else, calculated on a pre-arranged formula. The number of tenderers for open tenders may be substantial; public road contracts up to say 10 billion lire (6.6 million ECUs) may attract 150 to 200 tenderers.

For private sector work only, contractors may, prior to submitting tenders, have to pay a deposit of 2.5% to 5.0% to the client which is refunded, less a nominal tax, to unsuccessful tenderers.

A Decree of 16.7.62 established general contract conditions for public works and the *Associazione Nazionale Costruttori Edili* (ANCE), the building contractors federation, publishes a contract for the private sector.

Contract documents normally comprise the following:

- drawings
- general contract
- specification clauses
- special specification clauses, applicable to the particular contract
- form of tender.

Three copies of the contract documents are made; one each is held by client and contractor and the third is deposited with the Public Registrar. Contracts may be let to a general contractor who will be responsible to the client for the whole project whether or not he employs sub-contractors. Contracts are also sometimes let on a trade-by-trade basis with the client's consultants being responsible for co-ordination.

Most contracts are firm price. In some instances fluctuations are permitted, with a formula for valuing any adjustment set out in the special specification clauses, usually related to a published national index.

Variations are normally permitted and, providing the cost of these falls within certain limits, they are valued at rates pro-rata to the contract. The final account is prepared jointly by the contractor and the employer. Interim payments are made to the contractor and are subject to a retention of 10%, this being held until the satisfactory completion of the contract. Any disputes arising out of the contract are referred to a panel of three arbitrators; one being appointed by the client, one by the contractor and one by the Courts or a similar body, as provided for in the contract documents. An arbitrator's majority decision is binding on both parties.

Liability and insurance

Liability is determined by the provisions of the civil and criminal codes which cover obligations assumed under contract and to third parties both by professionals and contractors. All construction is subject to a ten year liability period for both the contractor and the professionals involved which commences following the issue of the *certificato di abitabilità* (the building occupancy permit).

Professional indemnity insurance is not obligatory and is rarely taken out. The construction industry is not prone to litigation, a reflection of both the cost and the lengthy judicial procedures. Problems tend to be resolved by agreement between the parties concerned, unless there is clear evidence of gross negligence on the part of contractor or consultant.

Construction insurance may be taken out for the period of construction, the maintenance period following completion (usually 12 months, with a maximum of 24) and the ten year period of liability. All insurance premiums are subject to a tax of 21.25%. A main contractor insures the construction phase on the basis of his assessement of the risks involved and this usually includes subcontractors who rarely carry separate cover. In the case of small contractors working at a domestic scale, insurance other than third party cover will often not be held. All risks policies (*assicurazione* - CAR) are increasingly being used and currently account for about 10% of the market. Major contractors may hold general cover for all their activities which is revised for specific projects, and this practice is also on the increase. An important point in the case of public works contracts is that, while insurance is not obligatory, damage resulting from natural calamity must be accepted as such by the authorities within a period of five days, otherwise the liability for making good remains with the contractor.

Contractors usually maintain insurance cover for the maintenance period which also covers the work of subcontractors. Insurance for the 10 year liability period is determined by agreement with the client and may either be held by the contractor or the client who may prefer to avoid the risk of a contractor going into liquidation.

Development control and standards

The planning laws are a matter for the regions and there are considerable differences from one region to another. Administration is further devolved to the *Provincia* and the *Comune* and the latter is the most important for land use planning and building control. There is a whole series of plans drawn up from broad strategies to detailed plans for specific types of development. Application for planning approval is complex and the information required is comprehensive, covering standards of construction as well as use. The application must be signed by the owner or title holder and designer. When permission is given work must be started within one year and completed within three years after which, if work is still not completed, a new application has to be made. The building must conform exactly to that which is approved.

It is difficult to assess exactly how all the regulations are enforced and this will vary by locality and over time. It is known that a substantial amount of building work including new housing is undertaken without proper permission, particularly in the South.

There are requirements for testing structural materials in accordance with legislation of 1972, which must be carried out in laboratories recognized by the *Ministero dei Lavori Pubblici*, and fire tests which must be carried out in laboratories recognized by the *Ministero dell'Interno* on materials for use in public places. In addition, there are national standards issued by the *Ente Nazionale per l'Unificazione nell'Industria* (UNI) which cover such aspects as safety associated with electrical and gas installations.

Italian specification practice is descriptive in character rather than prescriptive and rarely is reference made to performance criteria or specific standards. Generally, contractors enjoy much more freedom of interpretation with regard to the selection of materials, and consultants are prepared to concede a considerable level of autonomy to a trusted contractor. The consultant/contractor relationship tends to be less adversarial than in the UK - there is a level of mutual respect and pride in the quality of the end result.

The absence of a comprehensive, widely applied body of standards for building products has placed Italian manufacturers, many of whom are small family run concerns, at a disadvantage in the European market in view of the Construction Products Directive. Steps are underway to remedy this with new arrangements for the accreditation of testing laboratories and certification bodies.

New techniques and materials are not subject to a compulsory approval procedure. However, the *tituto Centrale per l'Industrializione e la Technologia Edilizia* (ICITE, Central Institute for Building Industrialization and Technology) and the *Consiglio Nazionale delle Ricerche* (National Council of Research) can issue a technical approval after laboratory tests.

Further reading

J L Meikle and P M Hillebrandt, The Italian Construction Industry : A Guide for UK Professionals, CIRIA, 1990.

RIBA, Architectural Practice in Europe : Italy, RIBA, 1991.

Association of Consulting Engineers et al, Engineering Consultancy in the European Community : Italy, London: Telford, 1990.

CONSTRUCTION COST DATA

Cost of labour

The figures below are typical of labour costs in Rome as at January 1991. The "wage rate" is the basis of an employee's income, while the "cost of labour" indicates the cost to a contractor of employing that employee. The difference between the two covers a variety of mandatory and voluntary contributions - a list of items which could be included is given in section 4.

	Wage rate (per hour) Lire	*Cost of labour (per hour) Lire*	*Number of hours worked per year*
Site operatives			
Mason/bricklayer	15 100	28 800	1880
Carpenter	15 100	28 800	1880
Plumber	15 100	28 800	1880
Electrician	15 100	28 800	1880
Structural steel erector	12 170	23 300	1880
Semi-skilled worker	12 170	27 600	1880
Unskilled labourer	13 580	26 000	1880
Equipment operator	15 100	28 800	1880
Watchman/security	-	-	1880
Site supervision			
General foreman	-	-	1880
Trades foreman	15 890	30 300	1880
Clerk of works	-	-	1880

	Wage rate (per hour) Lire	*Cost of labour (per hour) Lire*	*Number of hours worked per year*
Consultants' personnel			
Senior architect	-	75 000	1880
Senior engineer	-	55 000	1880
Senior surveyor	-	55 000	1880
Qualified architect	-	55 000	1880
Qualified engineer	-	55 000	1880
Qualified surveyor	-	55 000	1880

Cost of materials

The figures that follow are the costs of main construction materials, delivered to site in the Rome area, as incurred by contractors in January 1991. These assume that the materials would be in quantities as required for a medium sized construction project and that the location of the works would be neither constrained nor remote.

All the costs in this section exclude value added tax (VAT - see below).

	Units	*Cost Lire*
Cement and aggregate		
Ordinary portland cement in 50kg bags	tonne	110 000
Coarse aggregates for concrete	m^3	20 350
Fine aggregates for concrete	m^3	20 350
Ready mixed concrete (250kg Cement/m^3)	m^3	53 640
Ready mixed concrete (150kg Cement/m^3)	m^3	43 520
Steel		
Mild steel reinforcement	tonne	610 000
High tensile steel reinforcement	tonne	617 000
Structural steel sections	tonne	800 000
Bricks and blocks		
Common bricks (250 x 120 x120mm)	1000	350 000
Good quality facing bricks (240 x 120 x 60mm)	1000	500 000
Hollow concrete blocks (450 x 200 x 200mm)	each	1040
Solid concrete blocks (500 x 200 x 150mm)	each	1630
Precast concrete cladding units with exposed aggregate finish	m^2	120 000
Timber and insulation		
Softwood sections for carpentry	m^3	230 000
Softwood for joinery	m^3	700 000
Hardwood for joinery	m^3	1 300 000

	Units	Cost Lire
Exterior quality plywood (6mm)	m^2	15 000
Plywood for interior joinery (6mm)	m^2	7000
Softwood strip flooring (15mm)	m^2	18 500
Chipboard sheet flooring (15mm)	m^2	16 000
100 mm thick quilt insulation (55kg/m^3 density)	m^2	8150
100 mm thick polystyrene insulation self extinguishing (25kg/m^3)	m^2	14 900
Softwood internal door complete with frames and ironmongery	nr	200 000
Glass and ceramics		
Sheet/float glass (5 mm)	m^2	19 800
Sealed double glazing units (6-6-6)	m^2	77 000
Good quality ceramic wall tiles	m^2	30 000
Plaster and paint		
Plaster in 100 kg bags	tonne	100 000
Plasterboard (13mm thick)	m^2	5500
Emulsion paint in 5 litre tins	litre	5200
Gloss oil paint in 5 litre tins	litre	13 300
Tiles and paviors		
Clay floor tiles (100 x 150 x 7mm)	m^2	10 800
Vinyl floor tiles (250 x 250 x 1.5mm)	m^2	13 900
Precast concrete paving slabs (300 x 300 x 30mm)	m^2	8400
Clay roof tiles	1000	350 000
Precast concrete roof tiles	1000	650 000
Drainage		
WC suite complete	nr	200 000
Lavatory basin complete	nr	330 000
100mm diameter clay drain pipes	m	7600
150mm diameter stainless steel drain pipes	kg	8300

Unit rates

The descriptions below are generally shortened versions of standard descriptions listed in five languages (English, French, Italian, German and Spanish) in section 6. Where an item has a two digit reference number (e.g. 05 or 33), this relates to the full description against that number in section 6. Where an item has an alphabetic suffix (e.g. 12A or 34B) this indicates that the standard description has been modified. Where a modification is major the complete modified description is included here and the standard description should be ignored; where a modification is minor (e.g. the insertion of a named hardwood) the shortened description has been modified here but, in general, the full description in section 6 prevails.

The unit rates below are for main work items in a typical construction project in the Rome area in January 1991. The rates include all necessary labour, materials, equipment and, where appropriate, allowances for contractors overheads and profit, preliminary and general items and contractors profit and attendance on specialist trades. All the costs in this section exclude value added tax (VAT - see below).

		Unit	*Rate Lire*
Excavation			
01	Mechanical excavation of foundation trenches	m^3	9110
02	Hardcore filling making up levels	m^2	3320
03	Earthwork support	m^2	6960
Concrete work			
04	Plain insitu concrete in strip foundations in trenches	m^3	150 000
05	Reinforced insitu concrete in beds	m^3	122 000
06	Reinforced insitu concrete in walls	m^3	151 000
07	Reinforced insitu concrete in suspended floor or roof slabs	m^3	127 000
08	Reinforced insitu concrete in columns	m^3	140 000
09	Reinforced insitu concrete in isolated beams	m^3	181 000
10	Precast concrete slab	m^2	235 000
Formwork			
11A	Metal formwork to concrete walls	m^2	43 700
12A	Softwood formwork to concrete columns	m^2	63 200
13A	Metal formwork to horizontal soffits of slabs	m^2	42 300
Reinforcement			
14	Reinforcement in concrete walls	tonne	2 350 000
15	Reinforcement in suspended concrete slabs	tonne	2 160 000
16	Fabric reinforcement in concrete beds	m^2	12 500
Steelwork			
17	Fabricate, supply and erect steel framed structure	tonne	3 610 000
Brickwork and blockwork			
18A	Precast lightweight aggregate hollow concrete block walls	m^2	70 000
19	Solid (perforated) concrete blocks	m^2	47 700
20A	Solid (preforated) clay common bricks	m^2	83 500
21	Facing bricks	m^2	101 000
Roofing			
22	Concrete interlocking roof tiles 430 x 380mm	m^2	64 500
23	Plain clay roof tiles 260 x 160mm	m^2	72 500
24	Fibre cement roof slates 600 x 300mm	m^2	53 600
25	Sawn softwood roof boarding	m^2	59 300
26	Particle board roof coverings	m^2	63 300

		Unit	Rate Lire
27	3 layers glass-fibre based bitumen felt roof covering	m^2	46 800
28	Bitumen based mastic asphalt roof covering	m^2	46 800
29	Glass-fibre mat roof insulation 160mm thick	m^2	45 000
30	Rigid sheet loadbearing roof insulation 75mm thick	m^2	56 800
31	Troughed galvanized steel roof cladding	m^2	70 000
Woodwork and metalwork			
32	Preservative treated sawn softwood 50 x 100mm	m	10 200
33	Preservative treated sawn softwood 50 x 150mm	m	19 700
34A	Single glazed casement window in Douglas hardwood, size 650 x 900mm	each	285 000
35A	Two panel glazed door in Douglas hardwood, size 850 x 2000mm	each	924 000
36	Solid core half hour fire resisting hardwood internal flush doors, size 800 x 2000mm	each	891 000
37A	Aluminium double glazed window, size 1200 x 1200mm	each	886 000
38A	Aluminium double glazed door, size 850 x 2100mm	each	964 000
39A	Hardwood skirtings	m	15 100
40	Framed structural steelwork in universal joist sections	tonne	4 310 000
41	Structural steelwork lattice roof trusses	tonne	4 960 000
Plumbing			
42	UPVC half round eaves gutter	m	19 000
43	UPVC rainwater pipes	m	15 300
44	Light gauge copper cold water tubing	m	13 900
45	High pressure plastic for cold water supply	m	8410
46	Low pressure plastic pipes for cold water distribution	m	8220
47	UPVC soil and vent pipes	m	16 700
48	White vitreous china WC suite	each	474 000
49	White vitreous china lavatory basin	each	539 000
50	Glazed fireclay shower tray	each	134 000
51	Stainless steel single bowl sink and double drainer	each	291 000
Electrical work			
52	PVC insulated and copper sheathed cable	m	886
53	13 amp unswitched socket outlet	each	19 000
54	Flush mounted 20 amp, 1 way light switch	each	15 200
Finishings			
55	2 coats gypsum based plaster on brick walls	m^2	31 600
56A	White glazed tiles on plaster walls	m^2	89 800
57	Red clay quarry tiles on concrete floor	m^2	60 500
58	Cement and sand screed to concrete floors	m^2	20 500
59	Thermoplastic floor tiles on screed	m^2	47 800
60A	Mineral fibre tiles on concealed suspension system	m^2	51 000

		Unit	Rate Lire
Glazing			
61	Glazing to wood	m^2	39 300
Painting			
62	Emulsion on plaster walls	m^2	6580
63	Oil paint on timber	m^2	24 700

Approximate estimating

The building costs per unit area given below are averages incurred by building clients for typical buildings in the Rome area as at January 1991. They are based upon the total floor area of all storeys, measured between external walls and without deduction for internal walls.

Approximate estimating costs generally include for mechanical and electrical installations but exclude furniture, loose or special equipment, and external works; they also exclude fees for professional services. The costs shown are for specifications and standards appropriate to Italy and this should be borne in mind when attempting comparisons with similarly described building types in other countries. A discussion of this issue is included in section 4. Comparative data for countries in this publication, including construction cost data are presented in Part Three.

Approximate estimating costs must be treated with reserve; they cannot provide more than a rough guide to the probable cost of building.

All the rates in this section exclude value added tax (VAT - see below).

	Cost m^2 Lire	Cost ft^2 Lire
Industrial buildings		
Factories for letting	710 000	66 000
Factories for owner occupation (light industrial use)	850 000	79 000
Factories for owner occupation (heavy industrial use)	935 000	87 000
Factory/office (high-tech) for letting (shell and core only)	730 000	68 000
Factory/office (high-tech) for letting (ground floor shell, first floor offices)	800 000	74 000
Factory/office (high tech) for owner occupation (controlled environment, fully finished)	1 000 000	93 000
High tech laboratory (air conditioned)	1 500 000	139 000
Warehouses, low bay (6 to 8m high) for letting (no heating)	470 000	44 000
Warehouses, low bay for owner occupation (including heating)	550 000	51 000
Warehouses, high bay for owner occupation (including heating)	650 000	60 000

	Cost m^2 Lire	Cost ft^2 Lire
Administrative and commercial buildings		
Civic offices, non air conditioned	1 600 000	149 000
Civic offices, fully air conditioned	2 000 000	186 000
Offices for letting, 5 to 10 storeys, non air conditioned	1 500 000	139 000
Offices for letting, 5 to 10 storeys, air conditioned	1 750 000	163 000
Offices for letting, high rise, air conditioned	1 950 000	186 000
Offices for owner occupation 5 to 10 storeys, non air conditioned	2 100 000	163 000
Offices for owner occupation 5 to 10 storeys, air conditioned	2 300 000	181 000
Offices for owner occupation, high rise, air conditioned	2 300 000	195 000
Prestige/headquarters office, 5 to 10 storeys, air conditioned	2 500 000	214 000
Prestige/headquarters office, high rise, air conditioned	2 625 000	232 000
Health and education buildings		
General hospitals (100 beds)	1 850 000	172 000
Teaching hospitals (100 beds)	1 950 000	181 000
Private hospitals (100 beds)	2 200 000	204 000
Health centres	1 700 000	158 000
Nursery schools	1 800 000	167 000
Primary/junior schools	1 550 000	144 000
Secondary/middle schools	1 500 000	139 000
University (arts) buildings	1 550 000	144 000
University (science) buildings	1 750 000	162 000
Management training centres	1 700 000	158 000
Recreation and arts buildings		
Theatres (over 500 seats) including seating and stage equipment	2 950 000	274 000
Theatres (less than 500 seats) including seating and stage equipment	2 300 000	214 000
Concert halls including seating and stage equipment	4 500 000	418 000
Sports halls including changing and social facilities	1 400 000	130 000
Swimming pools (international standard) including changing and social facilities	2 400 000	223 000
Swimming pools (schools standard) including changing facilities	1 750 000	163 000
National museums including full air conditioning and standby generator	4 300 000	400 000
Local museums including air conditioning	2 000 000	186 000
City centre/central libraries	2 000 000	186 000
Branch/local libraries	1 650 000	153 000
Residential buildings		
Purpose designed single family housing 2 storey detached (single unit)	1 200 000	111 000
Social/economic apartment housing, low rise (no lifts)	1 000 000	93 000
Social/economic apartment housing, high rise (with lifts)	1 050 000	98 000

	Cost m^2 Lire	Cost ft^2 Lire
Private sector apartment building (standard specification)	1 200 000	111 000
Private sector apartment buildings (luxury)	2 000 000	186 000
Student/nurses halls of residence	1 400 000	130 000
Homes for the elderly (shared accommodation)	1 300 000	121 000
Homes for the elderly (self contained with shared communal facilities)	1 200 000	111 000
Hotel, 5 star, city centre	3 200 000	297 000
Hotel, 3 star,city/provincial	2 600 000	242 000
Motel	2 200 000	204 000

Regional variations

The approximate estimating costs are based on projects in Rome. Adjust these costs by the following factors for regional variations:

Milan : +10%
Bologna : +5%
Sicily : +10%

Value added tax (VAT)

The standard rate of value added tax (VAT) is currently 19%, chargeable on general building work.

Exchange rates and inflation

The combined effect of exchange rates and inflation on prices within a country and price comparisons between countries is discussed in section 4.

Exchange rates

The graph below plots the movement of the Italian Lira against sterling and the US dollar since 1980. The figures used for the graph are quarterly and the method of calculating these, and other related issues are discussed in section 4. Italy joined the European Monetary System (EMS) at its inception in 1979. The exchange rate at January 1991 was L 2198 to the pound sterling and L 1137 to the US dollar.

The Italian Lira against Sterling and the US Dollar

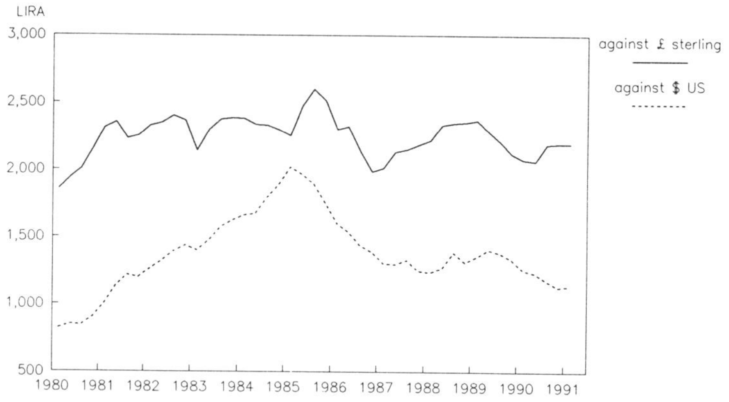

Inflation

The table below presents a number of indices showing the trends in inflation, prices and costs. All the indices have been rebased to 1980=100. Notes on inflation and the distinction between costs and prices are included in section 4.

Cost and price indices

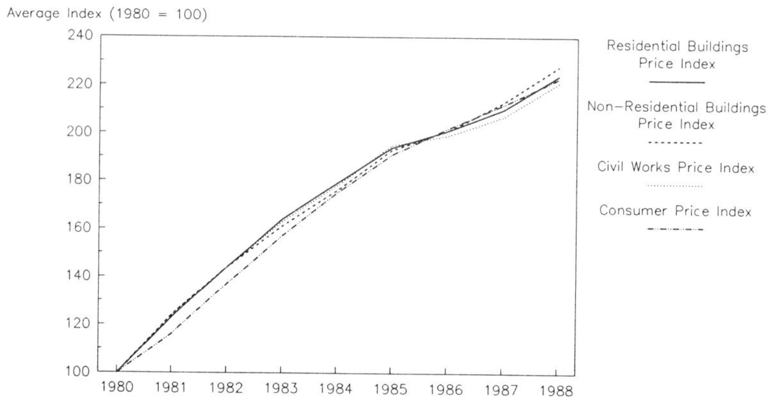

Consumer price inflation

The basis of the figures is the official consumer price index converted to quarterly averages. Inflation peaked in 1982 and has since fallen to an average annual rate of between 4% and 5%.

Consumer prices

1980 = 100

	1981	1982	1983	1984	1985	1986	1987	1988	1989	1990
average index	116	137	157	175	191	202	212	223	233	244
annual change (%)	16.0	18.1	14.6	11.5	9.1	5.8	5.0	5.2	4.5	4.7

Building cost indices

The Contractor's Association (ANCE) compiles a series of cost indices for residential, non-residential and civil works, based on data from the central statistics office, ISTAT (*Instituto Centrale di Statistica*). These are shown below.

Residential buildings

1980 = 100

	1981	1982	1983	1984	1985	1986	1987	1988
average index	123	144	164	179	194	201	210	224
annual change (%)	23.0	17.1	13.9	9.1	8.4	3.6	4.5	6.7

Non-residential buildings

1980 = 100

	1981	1982	1983	1984	1985	1986	1987	1988
average index	124	144	161	176	193	201	213	228
annual change (%)	24.0	16.1	11.8	9.3	9.7	4.1	6.0	7.0

Civil works

1980 = 100

	1981	1982	1983	1984	1985	1986	1987	1988
average index	124	144	163	178	195	199	207	221
annual change (%)	24.0	16.1	13.2	9.2	9.6	2.1	4.0	6.8

USEFUL ADDRESSES

Associazione Nazionale Costruttori Edili, (ANCE),
Via Guattani 16
00161 Roma
Tel: 06 84881. Fax: 06 84444364

Consiglio Nazionale delle Ricerche
Ufficio Relazioni Internazionali, Piazzale delle Science 7
00100 Roma
Tel: (6) 095 9345

Associazione delle Organizzazioni di Ingegneria e di
Consulenza Tecnico-Economica, (OICE) Via Adda 55
00198 Roma
Tel: 06 851685/8840547/868797
Fax: 06 851685

Associazione Ingegneri Consulenti Italiani, (AICI)
Via Nazionale 43
39018 Terlano
Tel: 0471 57248

Associazione Nazionale Ingegneri e Architetti Italiani, (ANIAI)
Via Sallustio 24
00187 Roma
Tel: 06 4744397

Sindicato Nazionale degli Ingegneri Liberi Professionisti, (SNILPI)
Via Salaria 292
00199 Roma
Tel: 06 859796

Ente Nationale Italiano di Unoificazione
Piazza Amando Diaz 2
1 - 201123 Milano
Tel: (02) 876914

Associazione Nationale Fra Produttori Di Manufatti in Cemento
Via Petitti 16
20149 Milano

Instituto Centrale Di Statistica (ISTAT)
Via Cesare Balbo 16
00184 Roma
Tel: (6) 4673

Associazione Ingegneri Consulente Italiani, AICI
Via Salaria 292
00199 Roma
Tel: (6) 861 871

Associazione Nazionale Construttori Edili
Via Guatani 16
00161 Roma

Fabbrica Italiana Tubi
Ferrotubi Spa, Via Lanzone 4
20123 Milano

Japan

KEY DATA

Population

Population (1989)	122.6m
Urban population	77%
Population under 15	18.5%
Population over 65	11.7%
Average annual growth rate (1983 to 1988)	0.5%

Geography

Land area	378 000 km²
Agricultural area	53 000 km²
Capital city	Tokyo (population 11 718 720)

Economy

Monetary unit	Yen (¥)
Exchange rate (January 1991) to	
the ECU	¥181
the pound sterling	¥ 259
the US dollar	¥134
Average annual inflation (1984 to 1989)	1.1%
Gross Domestic Product (GDP)	¥366 500bn
GDP per capita (1989)	¥2.99m
GDP growth (real) (1980 to 1988)	4.1%
Private consumption (% of total)	55.4%
Public consumption (% of total)	16.4%
Investment (% of total)	27.9%
Central government expenditure as percentage of GDP	27.2%

Construction

Total value of construction output (1989)	¥70 100bn
Construction output per capita	¥571 780
Construction output as a percentage of GDP	19.1%

THE CONSTRUCTION INDUSTRY

Construction output

During the 1980s domestic construction activity has increased steadily and in 1989 gross output of the Japanese construction industry was Yen 70.1 trillion or about 19% of GDP. Residential work accounts for some 33% of total output, non-residential 28% and civil engineering 39%. These figures include extension, reconstruction and remodelling as well as routine repair and maintenance work. Recently the construction industry has enjoyed a boom for new buildings so that the proportion of reconstruction and remodelling is currently low but is expected to grow. The proportion for routine repair and maintenance is also low, perhaps less than 10% of total output.

Over 50% of construction activity is concentrated in the Kanto and Kinlu regions which contain Tokyo and Osaka, Japan's two biggest urban areas. This emphasis may change in line with government policy to foster regional development.

Characteristics and structure of the industry

Construction companies in Japan usually carry out both design and construction, especially for private projects. Design departments of contractors may have as many as 1000 professionals.

Most civil engineering work is public sector sponsored and the public sector offices either have their own design sections or hire specialized consultant firms to design their projects. It is quite usual for firms of consultant engineers to employ over 500 people. The largest architectural consultancy firm has over 1000 employees and many have several hundred. There is co-operation as well as competition between consultants and contractors; contractors may be invited to participate in consultants' design work or vice versa.

In 1988 there were about 510 000 construction contractors of whom 8700 were licenced by the Ministry of Construction. The numbers are declining year by year. Most contractors are small but there are six large companies: Shimizu Corporation, Kajima Corporation, Taisei Corporation, Takenaka Corporation, Ohbayashi Corporation and Kumagai Gumi Company. They each have annual turnover of over 1000 billion yen. They all provide a comprehensive range of construction services in building, civil and heavy engineering; they can find construction sites for clients, help finance and then design, construct and maintain high quality buildings and engineering products. They all have head offices in Tokyo.

Major Japanese contractors

Major contractors	*Place in ENR's "Top 250 International Contractors"*	*Exports (% of turnover)*	*Work breakdown*
Kumagai Gumi Company Ltd	37	10	Building 58% Civil eng 35% Real estate 7%
Shimizu Corporation	31	5-10	Building 80% Civil eng 15% Real estate 5%
Kajima Corporation	39	5	Building 72% Civil eng 24% Real estate 4%
Ohbayashi Corporation	30	10	Building 74% Civil eng 23% Real estate 4%
Takenaka Corporation	38	5	n.a.
Taisei Corporation	70	less than 5	Building 74% Civil eng 23% Real estate 3%

Source:Engineering News Record

The similarities between the top contractors are quite remarkable. Moreover, it is understood that they all wish to increase their real estate and development share.

Japanese contractors have always had an expanding domestic market because Japan lags behind western developed countries in the provision of infrastructure, most notably roads, sewerage, housing and the city environments. However, the economy was affected by the oil crisis of the mid 1970s and growth has not been continuous.

The Contract Construction Business Law requires prospective contractors to obtain a licence to start a construction business. Nearly all site work in Japan is undertaken by trade contractors who maintain a special relationship with a general contractor, known as *Zenecons*. Under this relationship the general contractor will endeavour to provide continuous employment for his subcontractors, in return for which each subcontractor will allow the general contractor to stipulate a contract price, plus monitor both his financial and project performance. The very large companies do not have a permanent workforce, but a family of subcontractors who are loosely connected to them. Major Japanese construction firms are developing in a number of directions: internationally, diversifying into other businesses in some way linked to construction; strengthening the total engineering competence by research and development; and providing construction related finance. The Japanese have both the expertise and experience to compete with Western European and American contractors. During the 1980s Japanese international contractors have increasingly directed their efforts to the industrialized regions of the world. According to the Engineering News Record, 34 Japanese contractors in 1989 had 7.6% of the total construction export market. There were 15 Japanese contractors in the top 100 of Engineering News Record's "Top 250 International Contractors". Some of these are very specialized, for example in petrochemicals, but most have a broad range of operations.

One of the features of contracting organizations in Japan is that they undertake a considerable amount of research and development work. The range of research is very wide; from soil testing to air supported domes. Earthquake engineering is important and the Japanese are generally regarded as world leaders in research on the use of robots in construction and the development of intelligent buildings. Direct expenditure on research and development of the large construction firms is about 1% of turnover, but they also fund a considerable amount of outside research.

Clients and finance

In 1988, 64% of all new construction work was private, including over 95% of housing, over 80% of non-housing and about 24% of civil engineering.

Contractors (*Zenecons*) are increasingly moving into development projects, both in Japan and elsewhere.

Contractual arrangements

In the public sector, construction companies of the appropriate licence category and experience are invited to bid. In evaluating bids, central and local governments rank construction firms according to past orders obtained, sales, financial status and technological capabilities. Price is often less important than capability.

In the private sector the client may appoint a specific contractor or invite selected contractors to bid - the latter is the more common system.

The Japanese contractual system is based on trust and mutual understanding. It is very important for both parties to maintain a good and long term relationship. The Japanese rarely bring a lawyer into negotiations - that implies mistrust - and litigation is only undertaken as a last resort. Clients tend to work regularly with a contracting firm, and will often have in-house staff with knowledge of building design and construction who will have prepared outline drawings of the proposed works. The contractor co-ordinates the production of working drawings, which are prepared by specialist trade contractors.

The two contract forms in most common use are the Standard Form of Agreement and General Conditions of Government Contract for Works of Building and Civil Engineering, prepared and recommended by the Construction Industry Council of Japan, and the General Conditions of Construction Contracts (GCCC) approved by a number of architects' and contractors' associations. Contract documents, which are relatively short, normally consist of the written contract, general conditions, the design drawings and the specification. There is no bill of quantities but the contractor submits an itemized list of prices. Liquidated damages are payable if the project is delayed, and there is a guarantee period of two years for brick or concrete buildings and one year for timber structures. The employer is given express rights to vary the work and negotiations take place on dates and costs. Claims are rare.

Liability

The Registration Organization for Warrantied Houses, administered by the Construction Ministry, provides a warranty scheme. This gives a 10 year guarantee on the durability of structural components, including foundations, floors, walls and roofs; plus a five year warranty on the weather resistance of roofs. This scheme is available to single unit housebuilders using traditional Japanese housebuilding techniques. Prefabricated house builders, who compete with the single unit home builders, also provide a 10-year protection on structural components. Some condominium builders have recently started a similar 10-year guarantee. In response, the Housing and Urban Development Corporation, the government-managed house supplier, has, since 1983, developed a long-term warranty programme for some condominiums with warranties of 10 years for structural elements, including the roof, and 5 to 10 years for other elements.

Before this long-term warranty of houses and buildings can apply to all builders, a number of problems must be solved regarding such issues as design responsibilities, insurance systems, business profitability and so on. The principle behind long-term warranty is not to guarantee free repair services for 10 years, but to build structures in which defects will not occur for at least 10 years. Since this puts greater importance on quality, a long-term warranty is a necessity for all construction companies and an inevitable outcome in today's quality-conscious market. Those companies unable to offer such a warranty will eventually lose out in a competitive market.

Materials and construction methods

Although the basic materials and methods of construction used in Japan are similar to those of other developed countries, there are considerable differences in detail.

Generally, structures are heavy reinforced concrete or steel frame with spray applied fireproofing for large buildings, whereas for housing, timber frame is predominant. Although some prefabrication of components for smaller elements is used, a considerably greater amount of skilled trade work, for example cutting, fitting and welding of steelwork, cutting and threading of steel pipes, and the preparation of natural stone cladding, takes place on site. The use of tower cranes is unusual; instead materials are either distributed by hoists at the perimeter of buildings, or by small cranes mounted on the top construction level through holes left in the floors.

Further reading

Fumio Hasegawa and the Shimizu Group, Built by Japan, Wiley, 1988.

The Overseas Construction Association of Japan Inc Japan's Construction Today, 1990.

John Bennett, Roger Flanagan and George Norman, Capital and Counties Report: Japanese Construction Industy, Centre for Strategic Studies in Construction, University of Reading, May 1987.

Tokunosuke Hasegawa, "Public Investment in Japan", in Euroconstruct, Baukonjunktur 1990/91 in Westeuropa, IFO Institut für Wirtschaftsforschung, München, Dezember 1990.

CONSTRUCTION COST DATA

Cost of labour

The figures below are typical of labour costs in Tokyo as at January 1991. The "wage rate" is the basis of an employee's income, while the "cost of labour" is the cost to a contractor of employing that employee. The difference between the two covers a variety of mandatory and voluntary contributions - a list of items which could be included is given in section 4.

	Wage rate (per day) Yen	*Cost of labour (per hour) Yen*	*Number of hours worked per year*
Site operatives			
Mason/bricklayer	19 804	-	-
Carpenter	21 752	-	-
Plumber	18 824	-	-
Electrician	17 837	-	-
Structural steel erector	20 461	-	-
HVAC installer	13 162	-	-
Semi-skilled worker	12 374	-	-
Unskilled labourer	10 068	-	-
Equipment operator	17 400	-	-
Watchman/security	10 400	-	-

	Wage rate (per day) Yen	*Cost of labour (per hour) Yen*	*Number of hours worked per year*
Site supervision			
General Foreman	25 000	-	-
Trades foreman	23 000	-	-
Contractors' personnel			
Site manager	38 000	-	-
Resident engineer	30 000	-	-
Resident surveyor	30 000	-	-
Junior engineer	12 000	-	-
Junior surveyor	12 000	-	-
Planner	12 000	-	-
Consultants' personnel			
Senior architect	45 000	-	-
Senior engineer	45 000	-	-
Senior surveyor	45 000	-	-
Qualified architect	38 000	-	-
Qualified engineer	38 000	-	-
Qualified surveyor	38 000	-	-

Cost of materials

The figures that follow are the costs of main construction materials, delivered to site in the Tokyo area, as incurred by contractors in January 1991. These assume that the materials would be in quantities as required for a medium sized construction project and that the location of the works would be neither constrained nor remote.

	Units	*Cost Yen*
Cement and aggregate		
Ordinary portland cement in 50kg bags	tonne	17 500
Coarse aggregates for concrete	m^3	3200
Fine aggregates for concrete	m^3	3900
Ready mixed concrete (210kg Cement/cm^2)	m^3	11 800
Steel		
Mild steel reinforcement	tonne	59 500
High tensile steel reinforcement	tonne	69 500
Precompressing tandon	tonne	72 000
Structural steel sections	tonne	70 000

	Units	Cost Yen
Bricks and blocks		
Common bricks (210 x 100 x 60mm)	1000	88 000
Good quality facing bricks (210 x 100 x 60mm)	each	250
Hollow concrete blocks (190 x 190 x 390mm)	each	315
Solid concrete blocks (190 x 190 x 200mm)	each	1630
Precast concrete cladding units with exposed aggregate finish	m^2	21 000
Timber and insulation		
Softwood sections for carpentry	m^3	59 000
Softwood for joinery	m^3	128 000
Hardwood for joinery	m^3	170 000
Exterior quality plywood (12mm)	m^2	1000
Plywood for interior joinery (5mm)	m^2	720
Softwood strip flooring (15mm)	m^2	2700
Chipboard sheet flooring (15mm)	m^2	1200
Softwood internal door complete with frames and ironmongery	nr	60 000
Glass and ceramics		
Float glass (5 mm)	m^2	4490
Sealed double glazing units (FL3+A6+FL3) 12mm thick	m^2	9180
Good quality ceramic wall tiles	m^2	3280
Plaster and paint		
Plaster in 25 kg bags	tonne	36 000
Plasterboard (9mm thick)	m^2	395
Emulsion paint in 5 litre tins	kg	360
Gloss oil paint in 5 litre tins	kg	345
Tiles and paviors		
Clay floor tiles (200 x 200mm)	m^2	4860
Vinyl floor tiles (2 x 300 x 300mm)	m^2	18 480
Precast concrete paving slabs (300 x 300 x 60mm)	m^2	4110
Clay roof tiles	1000	115 000
Precast concrete roof tiles	1000	95 000
Drainage		
WC suite complete	nr	42 700
Lavatory basin complete	nr	50 300
100mm diameter clay drain pipes	m	2290
150mm diameter stainless steel drain pipes	m	7160

Unit rates

The descriptions below are generally shortened versions of standard descriptions listed in five languages (English, French, Italian, German and Spanish) in section 6. Where an item has a two digit reference number (e.g. 05 or 33), this relates to the full description against that number in section 6. Where an item has an alphabetic suffix (e.g. 12A or 34B) this indicates that the standard description has been modified. Where a modification is major the complete modified description is included here and the standard description should be ignored; where a modification is minor (e.g. the insertion of a named hardwood) the shortened description has been modified here but, in general, the full description in section 6 prevails.

The unit rates below are for main work items in a typical construction project in the Tokyo area in January 1991. The rates include all necessary labour, materials, equipment and, where appropriate, allowances for contractors overheads and profit, preliminary and general items and contractors profit and attendance on specialist trades.

		Unit	*Rate Yen*
Excavation			
01	Mechanical excavation of foundation trenches	m^3	1230
02	Hardcore filling making up levels	m^2	936
03	Earthwork support	m^2	14 000
Concrete work			
04	Plain insitu concrete in strip foundations in trenches	m^3	20 400
05	Reinforced insitu concrete in beds	m^3	20 400
06	Reinforced insitu concrete in walls	m^3	16 100
07	Reinforced insitu concrete in suspended floor or roof slabs	m^3	16 100
08	Reinforced insitu concrete in columns	m^3	16 100
09	Reinforced insitu concrete in isolated beams	m^3	16 100
10	Precast concrete slab	m^2	24 600
Formwork			
11A	Softwood formwork to concrete walls	m^2	7840
12A	Softwood formwork to concrete columns	m^2	7840
13	Softwood or metal formwork to horizontal soffits of slabs	m^2	7840
Reinforcement			
14	Reinforcement in concrete walls	tonne	157 000
15	Reinforcement in suspended concrete slabs	tonne	157 000
Steelwork			
17	Fabricate, supply and erect steel framed structure	tonne	301 000

		Unit	*Rate Yen*
Brickwork and blockwork			
18	Precast lightweight aggregate hollow concrete block walls	m^2	5500
19A	Solid (perforated) concrete blocks	m^2	20 500
Roofing			
22	Concrete interlocking roof tiles 430 x 380mm	m^2	5850
23	Plain clay roof tiles 260 x 160mm	m^2	4330
24	Fibre cement roof slates 600 x 300mm	m^2	2340
25	Sawn softwood roof boarding	m^2	6320
26	Particle board roof coverings	m^2	5270
27	3 layers glass-fibre based bitumen felt roof covering	m^2	7610
28	Bitumen based mastic asphalt roof covering	m^2	6440
29	Glass-fibre mat roof insulation 160mm thick	m^2	2110
31	Troughed galvanized steel roof cladding	m^2	4970
Woodwork and metalwork			
32	Preservative treated sawn softwood 50 x 100mm	m	2930
35	Two panel glazed door in hardwood size 850 x 2000mm	each	211 000
36	Solid core half hour fire resisting hardwood internal flush doors, size 800 x 2000mm	each	160 000
37	Aluminium double glazed window, size 1200 x 1200mm	each	178 000
38	Aluminium double glazed door, size 850 x 2100mm	each	187 000
39	Hardwood skirtings	m	1760
40	Framed structural steelwork in universal joist sections	tonne	170 000
41	Structural steelwork lattice roof trusses	tonne	190 000
Plumbing			
42	UPVC half round eaves gutter	m	1640
43	UPVC rainwater pipes	m	4100
44	Light gauge copper cold water tubing	m	1520
45	High pressure plastic for cold water supply	m	1050
46	Low pressure plastic pipes for cold water distribution	m	1280
47	UPVC soil and vent pipes	m	4550
48	White vitreous china WC suite	each	90 100
49	White vitreous china lavatory basin	each	81 900
51	Stainless steel single bowl sink and double drainer	each	64 000
Electrical Work			
52	PVC insulated and copper sheathed cable	m	328
53	13 amp unswitched socket outlet	each	5150
54	Flush mounted 20 amp 1 way light switch	each	5200

	Unit	Rate Yen
Finishings		
55A 2 coats gypsum based plaster on concrete walls 20mm thick	m^2	6330
56 White glazed tiles on plaster walls	m^2	15 800
57 Red clay quarry tiles on concrete floor	m^2	17 000
58A Cement and sand screed to concrete floors 30mm thick	m^2	1990
59 Thermoplastic floor tiles on screed	m^2	5030
60 Mineral fibre tiles on concealed suspension system	m^2	3240
Glazing		
61 Glazing to wood	m^2	7020
Painting		
62 Emulsion on plaster walls	m^2	1580
63 Oil paint on timber	m^2	1460

Approximate estimating

The building costs per unit area given below are averages incurred by building clients for typical buildings in the Tokyo area as at January 1991. They are based upon the total floor area of all storeys, measured between external walls and without deduction for internal walls.

Approximate estimating costs generally include for mechanical and electrical installations but exclude furniture, loose or special equipment, and external works; they also exclude fees for professional services. The costs shown are for specifications and standards appropriate to Japan and this should be borne in mind when attempting comparisons with similarly described building types in other countries. A discussion of this issue is included in section 4. Comparative data for countries covered in this publication, including construction cost data are presented in Part Three.

Approximate estimating costs must be treated with reserve; they cannot provide more than a rough guide to the probable cost of building.

	Cost m^2 Yen	Cost ft^2 Yen
Industrial buildings		
Factories for letting	270 000	25 000
Warehouses, low bay (6 to 8m high) for letting (no heating)	200 000	18 500
Administrative and commercial buildings		
Civic offices, fully air conditioned	360 000	33 500
Offices for letting, high rise, air conditioned	405 000	37 500
Prestige/headquarters office, high rise, air conditioned	450 000	42 000

	Cost m^2 Yen	Cost ft^2 Yen
Health and Education Buildings		
General hospitals (300 beds)	550 000	51 000
Secondary/middle schools	420 000	39 000
Recreation and arts buildings		
Theatres (over 500 seats) including seating and stage equipment	610 000	56 500
Residential buildings		
Social/economic apartment housing, high rise (with lifts)	300 000	28 000
Private sector apartment buildings (luxury)	390 000	36 000
Hotel, 5 star, city centre	500 000	46 000
Hotel, 3 star, city/provincial	400 000	37 000

Regional Variations

The approximate estimating costs are based on projects in Tokyo. Adjust these costs by the following factors to take account of regional variations:

Nagoya	:	-10%
Osaka	:	-6%
Hiroshima	:	-9%
Fukuoka	:	-9%
Sapporo	:	-5%

EXCHANGE RATES AND INFLATION

The combined effect of exchange rates and inflation on prices within a country and price comparisons between countries is discussed in section 4.

Exchange rates

The graph below plots the movement of the Japanese yen against sterling and the US dollar since 1980. The figures used for the graph are quarterly and the method of calculating these, and other related issues are discussed in section 4. The exchange rate at January 1991 was Yen 259 to the pound sterling and Yen 134 to the US dollar.

The Japanese Yen against Sterling and the US Dollar

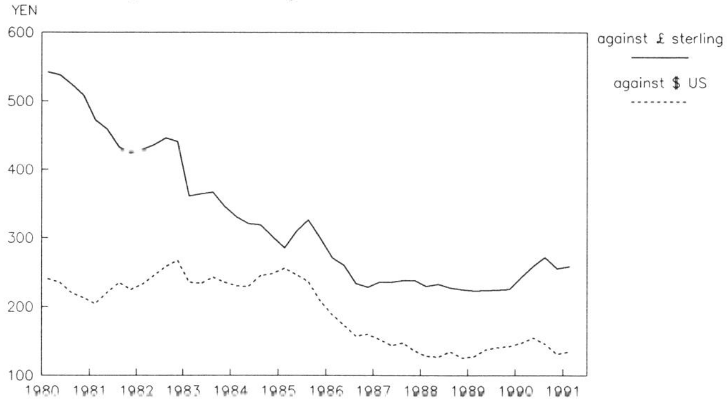

Inflation

The graph below presents a number of indices showing the trends in inflation, prices and costs. All the indices have been rebased to 1980=100. Notes on inflation and the distinction between costs and prices are included in section 4.

Cost and price indices

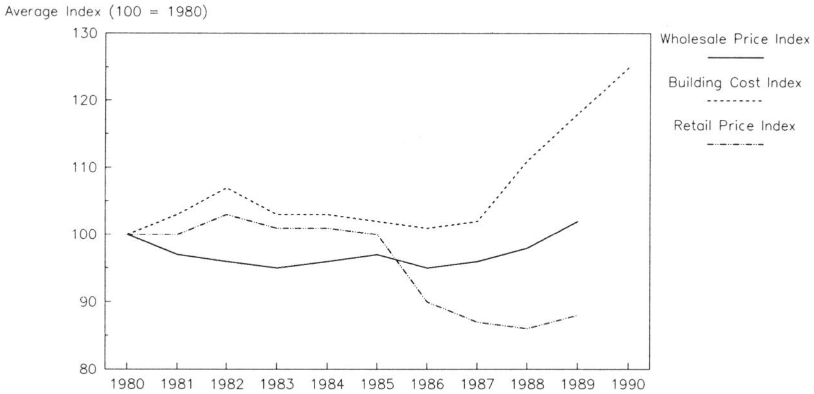

Consumer price inflation

The table below presents general inflation in Japan since 1980. The basis of the figures is the official consumer price index converted to quarterly averages. The annual change is the percentage change between the last quarters of consecutive years. Inflation peaked in 1982; progress since then in reducing inflation has been ahead of the OECD area as a whole.

Consumer price inflation 1980 = 100

	1981	1982	1983	1984	1985	1986	1987	1988	1989	1990
average index	100	103	101	101	100	90	87	86	88	n.a.
annual change (%)	0.0	3.0	-1.9	0.0	-1.0	-10.0	-3.3	-1.1	2.3	n.a.

Building cost inflation

The tables below present cost and price indices for general building work. These indicate that the rapid increase in building costs at the end of the decade has not been accompanied by a corresponding increase in wholesale prices.

Building cost 1980 = 100

	1981	1982	1983	1984	1985	1986	1987	1988	1989	1990
average index	103	104	103	103	102	101	102	111	118	125
annual change (%)	3.0	1.0	-1.0	0.0	-1.0	-1.0	1.0	8.8	6.3	5.9

Wholesale price 1980 = 100

	1981	1982	1983	1984	1985	1986	1987	1988	1989	1990
average index	97	96	95	96	97	95	96	98	102	n.a.
annual change (%)	-3.0	-1.0	-1.0	1.1	1.0	-2.1	1.1	2.1	4.1	n.a.

USEFUL ADDRESSES

Japan Federation of Construction Contractors, Inc.
Tokyo Kensetsu Bldg., 5 - 1 Hacchobori
2 - chome, Chuo-ku
Tokyo 104
Tel: 553 0701
Fax: 552 2360

Management Research Society for Construction Industry
Tokyo Kensetsu Bldg., 5 - 1 Hacchobori
2 - chome, Chuo-ku
Tokyo 104
Tel: 553 4832

Japan Construction Consultants Association
Shin Kudan Bldg., 2 - 4 Kudan Minami
2 - chome, Chiyoda-ku
Tokyo 102
Tel: 239 7992

Japan Civil Engineering Constractors' Association Inc
Tokyo Kensetsu Bldg., 5 - 1 Hacchobori
2 - chome, Chuo-ku
Tokyo 104
Tel: 553 3201
Fax: 552 3206

National Land Agency, 2 - 1 Kasumigaseki
1 - chome, Chiyoda-ku
Tokyo 100
Tel: 3593 3311

Ministry of Construction, 1 - 3 Kasumigaseki
2 - chome, Chiyoda-ku
Tokyo 100
Tel: 3580 4311

Luxembourg

KEY DATA

Population

Population (1989)	375 000
Urban population	2%
Population under 15	17.1%
Population over 65	13.4%
Average annual growth rate (1983 to 1988)	0.2%

Geography

Land area	2586 km^2
Agricultural area	n.a.
Capital city	Luxembourg (population 76 000)

Economy

Monetary unit	Lux Franc (LFr)
Exchange rate (January 1991) to	
the ECU	LFr 42.2
the pound sterling	LFr 60.3
the US dollar	LFr 31.2
Average annual inflation (1984 to 1989)	4.2%
Gross Domestic Product (GDP)	LFr 260bn
GDP per capita (1989)	LFr 693 333
GDP growth (real) (1980 to 1988)	n.a.
Private consumption (% of total)	57.1%
Public consumption (% of total)	17.0%
Investment (% of total)	24.29%
Central government expenditure as percentage of GDP	46.2%

Construction

Total value of construction output (1989)	LFr 30bn
Construction output per capita	LFr 80 000
Construction output as a percentage of GDP	11.5%

THE CONSTRUCTION INDUSTRY

Construction output

In 1989 the value of the output of new building and civil engineering work in Luxembourg was about LFr 30 billion, equivalent to 690 million ECUs or 11.5% of GDP. Adding an allowance for repair and maintenance might increase this to around 13% of GDP. Output has grown steadily during the 1980s. In 1989, housing contributed about 34% to the total output of new work. New non-residential building is also a little over a third and civil engineering is less than a third.

In 1987 there were over 1000 contracting firms of which only 5 employed over 200 persons.

There are about 220 architects and 30 independent suppliers of engineering consultancy services in Luxembourg, employing about 350 persons. Most of the turnover (about 95%) of these companies arises inside Luxembourg and the remainder mostly outside Europe. Two large engineering companies are affiliated to German engineering consultancy companies. In addition, the government has in-house engineering departments which cover most of the civil engineering infrastructure work.

Public tendering is regulated by government rules. Private tendering methods are variable. Informal contacts are very often used to obtain work.

There is a ten year major defect liability period. Professional indemnity insurance is not obligatory.

Further reading

Association of Consulting Engineers, et al, Engineering Consultancy in the European Community: Belgium, Luxembourg. London: Telford 1990.

CONSTRUCTION COST DATA

Cost of labour

The figures below are typical of labour costs in Luxembourg as at January 1991. The "wage rate" is the basis of an employee's income, while the "cost of labour" indicates the cost to a contractor of employing that employee. The difference between the two covers a variety of mandatory and voluntary contributions - a list of items which could be included is given in section 4.

Spon's Construction Cost and Price Indices Handbook

M C Fleming and B A Tysoe

This comprehensive handbook explains, clearly and with many examples how construction cost and price indices are compiled and how they can be used by the estimator in tender analysis, pricing and price adjustment, cost planning and forecasting.

The book contains over 100 tables and graphs and all the main indices are covered individually with full details of their movements, including some long-running series going back over 100 years.

It is important to have an understanding of how indices are compiled and the authors have considered the problems involved in measuring construction costs and price trends and the different methodological approaches to measurement that may be made.

Contents: List of abbreviations and acronyms. Introduction. **Part A: Construction indices: uses and methodology.** Uses of construction indices. Problems and methods of measurements. **Part B: Currently compiled construction indices.** Introduction. Output price indices. Tender price indices. DOE public sector building, 1968. DOE QSSD Index of building tender prices. BCIS tender price index, 1974. DOE road construction tender price index, 1970. DOE price index for public sector house building (PIPSH, 1964. SLD housing tender price index (HTPI), 1970. D B & E tender price index, 1966. Cost indices. BCIS general building cost index, 1971. Spon's cost indices. Building cost 1965. Electrical services cost 1965. Civil engineering cost 1970. Landscaping cost index 1976. APSAB cost index 1970. Building housing costs index 1973. SDD housing costs index 1970. BIA/BCIS house rebuilding costs index 1978. Association of Cost Engineers erected process plant indices 1958. BMCIS maintenance cost 1970. Summary comparison of indices and commentary. **Part C: Historical construction indices.** Introduction. Historical Cost and Price Indices. Maiwald's indices 1845-1938. Jones/Saville index 1845-1956. Venning index 1914. MOW/DOE 'CNC' indices 1939, 1946-1980 Q1. BRS measured work index 1939-1969 Q2. Summary comparison of indices and commentary. Appendix. General indices of prices. Index of total home cost. The retail price index. Index of capital goods cost. Glossary of Relevant Terms. Subject index.

February 1991: 234x156: 382pp Hardback: 0-419-15330-6: £25.00

2-6 Boundary Row, London SE1 8HN

	Wage rate (per hour) LFr	*Cost of labour (per hour) LFr*	*Number of hours worked per year*
Site operatives			
Mason/bricklayer	-	645	1800
Carpenter	-	645	1800
Plumber	-	720	1800
Electrician	-	780	1800
Structural steel erector	-	640	1800
HVAC installer	-	720	1800
Semi-skilled worker	-	580	1800
Unskilled labourer	-	540	1800
Watchman/security	-	550	1800
Site supervision			
General foreman	-	975	1800
Clerk of works	-	700	1800
Contractors' personnel			
Site manager	-	2300	1800
Resident engineer	-	2556	1800
Resident surveyor	-	2556	1800
Junior engineer	-	1600	1800
Junior surveyor	-	1600	1800
Planner	-	1600	1800
Consultants' personnel			
Senior architect	-	2556	1800
Senior engineer	-	2556	1800
Senior surveyor	-	2556	1800
Qualified architect	-	1986	1800
Qualified engineer	-	1986	1800
Qualified surveyor	-	1986	1800

Cost of materials

The figures that follow are the costs of main construction materials, delivered to site in the Luxembourg city area, as incurred by contractors in January 1991. These assume that the materials would be in quantities as required for a medium sized construction project and that the location of the works would be neither constrained nor remote.

All the costs in this section exclude value added tax (VAT - see below).

	Units	*Cost LFr*
Cement and aggregate		
Ordinary portland cement in 50kg bags	tonne	5700
Coarse aggregates for concrete	m^3	875
Fine aggregates for concrete (river sand)	m^3	890
Ready mixed concrete (B15/C2)	m^3	2050
Ready mixed concrete (B25/C3)	m^3	2350
Steel		
Mild steel reinforcement	tonne	22 000
High tensile steel reinforcement	tonne	22 000
Structural steel sections	tonne	28 000
Bricks and blocks		
Common bricks (2DF)	1000	12 000
Good quality facing bricks (2DF)	1000	15 000
Hollow concrete blocks (490 x 175 x 175mm)	1000	46 000
Solid concrete blocks (490 x 175 x 175mm)	1000	42 000
Precast concrete cladding units with exposed aggregate finish	m^2	780
Timber and insulation		
Softwood sections for carpentry	m^3	11 500
Softwood for joinery	m^3	30 000
Hardwood for joinery	m^3	48 000
Exterior quality plywood (10mm)	m^2	386
Plywood for interior joinery (10mm)	m^2	312
Softwood strip flooring (22mm)	m^2	1695
Chipboard sheet flooring (10mm)	m^2	200
100mm thick quilt insulation (4DM 25)	m^2	635
100mm thick rigid slab insulation	m^2	185
Softwood internal door complete with frames and ironmongery	nr	6000
Glass and ceramics		
Float glass (12mm)	m^2	220
Sealed double glazing units	m^2	1500
Plaster and paint		
Good quality ceramic wall tiles	m^2	1000
Plaster in 50 kg bags	tonne	4000
Plasterboard (20mm thick)	m^2	685
Emulsion paint in 5 litre tins	litre	125
Gloss oil paint in 5 litre tins	litre	550

	Units	Cost LFr
Tiles and paviors		
Clay floor tiles	m^2	900
Vinyl floor tiles	m^2	700
Precast concrete paving slabs	m^2	775
Clay roof tiles	m^2	750
Precast concrete roof tiles	m^2	500
Drainage		
WC suite complete	nr	20 000
Lavatory basin complete	nr	15 000
100mm diameter clay drain pipes	m	490
150mm diameter cast iron drain pipes	m	585

Unit rates

The descriptions below are generally shortened versions of standard descriptions listed in five languages (English, French, Italian, German and Spanish) in section 6. Where an item has a two digit reference number (e.g. 05 or 33), this relates to the full description against that number in section 6. Where an item has an alphabetic suffix (e.g. 12A or 34B) this indicates that the standard description has been modified. Where a modification is major the complete modified description is included here and the standard description should be ignored; where a modification is minor (e.g. the insertion of a named hardwood) the shortened description has been modified here but, in general, the full description in section 6 prevails.

The unit rates below are for main work items on a typical construction project in the Luxembourg city area in January 1991. The rates include all necessary labour, materials, equipment and, where appropriate, allowances for contractors overheads and profit, preliminary and general items and contractors profit and attendance on specialist trades. All the rates in this section exclude value added tax (VAT - see below).

		Unit	Rate LFr
Excavation			
01	Mechanical excavation of foundation trenches	m^3	465
02	Hardcore filling making up levels	m^2	985
03	Earthwork support	m^2	2000
Concrete work			
04	Plain insitu concrete in strip foundations in trenches	m^3	3300
05	Reinforced insitu concrete in beds	m^3	3600
06	Reinforced insitu concrete in walls	m^3	3600
07	Reinforced insitu concrete in suspended floor or roof slabs	m^3	3600
08	Reinforced insitu concrete in columns	m^3	4500

		Unit	Rate LFr
09	Reinforced insitu concrete in isolated beams	m^3	3650
10	Precast concrete slab	m^2	1500
Formwork			
11	Softwood formwork to concrete walls	m^2	1250
12	Softwood or metal formwork to concrete columns	m^2	1585
13	Softwood or metal formwork to horizontal soffits of slabs	m^2	685
Reinforcement			
14	Reinforcement in concrete walls	tonne	36 000
15	Reinforcement in suspended concrete slabs	tonne	38 000
16	Fabric reinforcement in concrete beds	m^2	34
Steelwork			
17	Fabricate, supply and erect steel framed structure	tonne	80 000
Brickwork and blockwork			
18	Precast lightweight aggregate hollow concrete block walls	m^2	900
19	Solid (perforated) concrete blocks	m^2	1100
21	Facing bricks	m^2	1800
Roofing			
22	Concrete interlocking roof tiles 430 x 380mm	m^2	900
23	Plain clay roof tiles 260 x 160mm	m^2	1200
24	Fibre cement roof slates 600 x 300mm	m^2	1500
25	Sawn softwood roof boarding	m^2	500
28	Bitumen based mastic asphalt roof covering	m^2	750
29	Glass-fibre mat roof insulation 160mm thick	m^2	850
30	Rigid sheet loadbearing roof insulation 75mm thick	m^2	650
31	Troughed galvanized steel roof cladding	m^2	600
Woodwork and metalwork			
32A	Preservative treated sawn softwood	m^3	23 000
35	Two panel glazed door in hardwood, size 850 x 2000mm	each	30 000
36	Solid core half hour fire resisting hardwood internal flush doors, size 800 x 2000mm	each	18 000
37	Aluminium double glazed window, size 1200 x 1200mm	each	35 000
38	Aluminium double glazed door, size 850 x 2100mm	each	42 000
40	Framed structural steelwork in universal joist sections	tonne	120 000
41	Structural steelwork lattice roof trusses	tonne	170 000

		Unit	Rate LFr
Plumbing			
42A	Zinc half round eaves gutter	m	750
43A	Zinc rainwater pipes	m	980
44	Light gauge copper cold water tubing	m	852
45	High pressure plastic for cold water supply	m	523
46A	Galvanised steel pipes for cold water distribution	m	523
47A	Cast iron soil and vent pipes	m	1100
48	White vitreous china WC suite	each	12 000
49	White vitreous china lavatory basin	each	10 000
50	Glazed fireclay shower tray	each	14 000
Electrical work			
52	PVC insulated and copper sheathed cable	m	200
53A	16 amp socket outlet	each	1300
Finishings			
55	2 coats gypsum based plaster on brick walls	m^2	260
56	White glazed tiles on plaster walls	m^2	1800
57	Red clay quarry tiles on concrete floor	m^2	2500
58	Cement and sand screed to concrete floors	m^2	550
60	Mineral fibre tiles on concealed suspension system	m^2	1000
Glazing			
61	Glazing to wood	m^2	600
Painting			
62	Emulsion on plaster walls	m^2	250
63	Oil paint on timber	m^2	400

Approximate estimating

The building costs per unit area given below are averages incurred by building clients for typical buildings in the Luxembourg city area as at January 1991. They are based upon the total floor area of all storeys, measured between external walls and without deduction for internal walls.

Approximate estimating costs generally include for mechanical and electrical installations but exclude furniture, loose or special equipment, and external works; they also exclude fees for professional services. The costs shown are for specifications and standards appropriate to Luxembourg and this should be borne in mind when attempting comparisons with similarly described building types in other countries. A discussion of this issue is included in section 4. Comparative data for countries covered in this publication, including construction cost data are presented in Part Three.

Approximate estimating costs must be treated with reserve; they cannot provide more than a rough guide to the probable cost of building.

All the rates in this section exclude value added tax (VAT - see below).

	Cost m²LFr	*Cost ft²LFr*
Industrial buildings		
Warrehouses, factories and high tech units	10 000 to 40 000	930 to 3700
Administrative and commercial buildings		
Offices	50 000 to 70 000	4650 to 6500
Health and education buildings		
Hospitals and schools	60 000	5600

	Cost m³LFr	*Cost ft²LFr*
Recreation and arts buildings		
Theatres, concert halls, swimming pools, sports halls, museums	15 000	-

	Cost m²LFr	*Cost ft²LFr*
Residential buildings		
Social/economic single family housing (multiple units)	35 000	3250
Private/mass market single family housing 2 storey detached/semidetached (multiple units)	35 000	3250
Purpose designed single family housing 2 storey detached (single unit)	50 000	4650
Social/economic apartment housing, low rise (no lifts)	35 000	3250
Social/economic apartment housing, high rise (with lifts)	40 000	3700
Private sector apartment building (standard specification)	40 000	3700
Private sector apartment buildings (luxury)	55 000	5100
Student/nurses halls of residence	50 000	4650
Homes for the elderly (shared accommodation)	50 000	4650
Homes for the elderly (self contained with shared communal facilities)	70 000	6500
Hotel, 5 star, city centre	80 000	7450
Hotel, 3 star, city/provincial	60 000	5600
Motel	55 000	5100

Regional variations

The approximate estimating costs are based on projects in and around the city of Luxembourg. For other parts of the country it is not necessary to adjust these costs for regional variations.

Value added tax (VAT)

The standard rate of value added tax (VAT) is currently 12%, chargeable on general building work.

EXCHANGE RATES AND INFLATION

The combined effect of exchange rates and inflation on prices within a country and price comparisons between countries is discussed in section 4.

Exchange rates

The graph below plots the movement of the Luxembourg Franc against sterling and the US dollar since 1980. The figures used for the graph are quarterly and the method of calculating these, and other related issues are discussed in section 4. Luxembourg joined the European Monetary System (EMS) at its inception in 1979. The exchange rate at January 1991 was LFr 60.3 to the pound sterling and LFr 31.2 to the US dollar.

The Luxembourg Franc against Sterling and the US Dollar

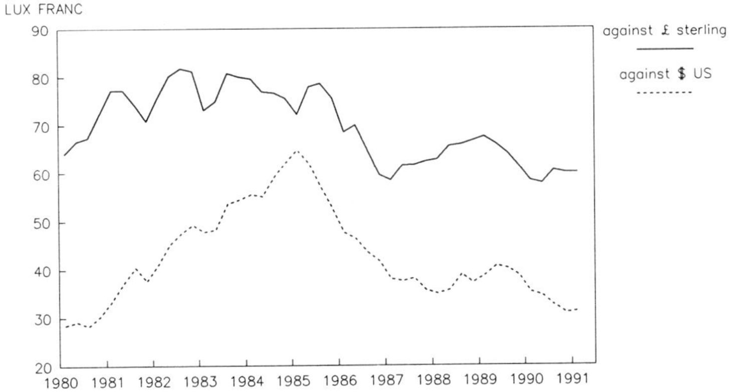

Inflation

The graph below presents a number of indices showing the trends in inflation, prices and costs. All the indices have been rebased to 1980=100. Notes on inflation and the distinction between costs and prices are included in section 4.

Cost and price indices

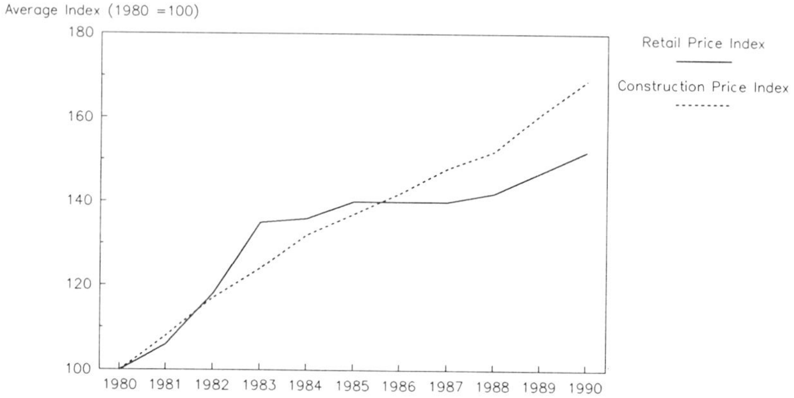

Consumer price inflation

The basis of the figures is the official consumer price index converted to quarterly averages. Inflation peaked in 1983. Since then the inflation rate has been low.

The table below presents general inflation in Luxembourg since 1980. The annual change is the percentage change between the last quarters of consecutive years.

Retail price index 1980 = 100

	1981	1982	1983	1984	1985	1986	1987	1988	1989	1990
average index	106	118	135	136	140	140	140	142	147	152
annual change (%)	8.0	11.3	14.4	0.7	2.9	0.0	0.0	1.4	3.5	3.4

Building cost inflation

The table below presents a general building cost index prepared by the National Institute of statistics in Luxembourg.

Cost of construction 1980 = 100

	1981	1982	1983	1984	1985	1986	1987	1988	1989	1990
average index	108	117	124	132	137	142	148	152	161	169
annual change (%)	8.0	8.2	6.0	6.5	3.8	3.6	4.2	2.7	5.9	5.0

USEFUL ADDRESSES

Chambre des Ingénierus-Conseils du Grand-Duche de Luxembourg
CICL
Rue Wilson, 36
L-2732 Luxembourg

Malta

KEY DATA

Population

Population (1989)	349 000
Urban population	n.a.
Population under 15	n.a.
Population over 65	10.2%
Average annual growth rate (1983 to 1988)	n.a.

Geography

Land area	316 km^2
Agricultural area	n.a.
Capital city	Valletta (population 14 000)

Economy

Monetary unit	Maltese Lira ML
Exchange rate (January 1991) to	
the ECU	ML 0.404
the pound sterling	ML 0.577
the US dollar	ML 0.30
Average annual inflation (1984 to 1989)	1.9%
Gross Domestic Product (GDP)	ML 0.65bn
GDP per capita (1989)	ML 1862
GDP growth (real) (1980 to 1988)	n.a.
Private consumption (% of total)	n.a.
Public consumption (% of total)	n.a.
Investment (% of total)	n.a.
Central government expenditure as percentage of GDP	n.a.

Construction

Total value of construction output (1989)	ML 55m
Construction output per capita	ML 158
Construction output as a percentage of GDP	8.5%

THE CONSTRUCTION INDUSTRY

Construction output

In 1989, the value of gross domestic fixed capital formation on construction (which may be taken as a measure of output of the construction indsutry on new work) was ML55 million; equivalent to 8.5% of GDP. This figure is, however, rather different from the official figure for gross output of the construction industry of ML23 million. The difference may well be accounted for by the public utilities of electricity and water.

There were 379 construction establishments, apart from the public utilities, employing 3474 persons in 1988.

Government policy

The policy of the government is to diversify the economy and to reduce reliance on tourism and the public sector. In the late 1980s Chase Manhattan Bank was asked to advise the government on developing facilities to encourage the establishment of offshore companies. This led to the three key components of Malta's business plan: the Industrial Development Act of 1988 to attract foreign companies to the country; the International Business Activities Act; and the Malta Free Port which creates a free zone for transhipment and warehousing.

Malta is attempting, with considerable success, to attract industry to the island. The Malta Development Corporation (MDC) promotes and supports industrial development and administers assistance under the incentives package. Incentives include ten year tax holidays for new export-oriented companies; fiscal incentives including investment and Accelerated Depreciation Allowances for existing industry; soft loans; training grants; ready built factories at subsidized rents; duty exemptions on plant, machinery and all materials; and a liberal policy on work permits.

In addition, the Industrial Development Act of 1988 created the Malta International Business Authority (MIBA). A wide range of benefits for all kinds of financial services is available.

At the same time the infrastructure is being improved and tourism is being developed, the latter development directed, in particular, at business travellers and conferences.

CONSTRUCTION COST DATA

Cost of labour

The figures below are typical of labour costs in Malta as at January 1991. The "wage rate" is the basis of an employee's income, while the "cost of labour" indicates the cost to a contractor of employing that employee. The difference between the two covers a variety of mandatory and voluntary contributions - a list of items which could be included is given on in section 4.

	Wage rate (per week) ML	*Cost of labour (per week) ML*	*Number of hours worked per year*
Site operatives			
Mason/bricklayer	60	77	1760
Carpenter	60	84	1760
Plumber	53	84	1760
Electrician	53	74	1760
HVAC installer	53	74	1760
Semi-skilled worker	50	70	1760
Unskilled labourer	45	63	1760
Equipment operator	55	77	1760
Watchman/security	40	56	1760
Site supervision			
General foreman	70	98	1760
Trades foreman	70	98	1760
Contractors' personnel			
Site manager	70	98	1760
Resident engineer	110	154	1760
Resident surveyor	90	126	1760
Junior engineer	50	70	1760
Junior surveyor	50	70	1760
Consultants' personnel			
Senior architect	125	175	1400
Senior engineer	125	175	1400
Senior surveyor	100	140	1400

Cost of materials

The figures that follow are the costs of main construction materials, delivered to site in Valletta, as incurred by contractors in January 1991. These assume that the materials would be in quantities as required for a medium sized construction project and that the location of the works would be neither constrained nor remote.

	Units	*Cost ML*
Cement and aggregate		
Ordinary portland cement in 50kg bags	tonne	32.00
Coarse aggregates for concrete	m^3	3.83
Fine aggregates for concrete	m^3	3.32
Ready mixed concrete (Grade 20)	m^3	18.00
Ready mixed concrete (Grade 30)	m^3	21.50
Steel		
Mild steel reinforcement	tonne	185.00
High tensile steel reinforcement	tonne	185.00
Structural steel sections	tonne	200.00
Bricks and blocks		
Common stone blocks	1000	300.00
Hollow concrete blocks (18" x 6" x 10") internal quality	1000	250.00
Hollow concrete blocks (18" x 9" x 10") internal quality	1000	310.00
Hollow concrete blocks (18" x 9" x 10") external quality	1000	310.00
Precast concrete cladding units with exposed aggregate finish	m^2	10.50
Timber and insulation		
Softwood sections for carpentry	ft^3	2.95
Softwood for joinery	ft^3	3.00
Hardwood for joinery	ft^3	10.76
Exterior quality plywood (12mm)	m^2	4.00
Plywood for interior joinery (15mm)	m^2	3.35
Chipboard (12mm)	m^2	2.90
100 mm thick rigid slab insulation	m^2	3.00
Softwood internal door complete with frames and ironmongery	nr	50.00
Glass and ceramics		
Float glass	m^2	2.00
Sealed double glazing units	m^2	25.00

	Units	Cost ML
Plaster and paint		
Good quality ceramic wall tiles	m^2	5.50
Plaster in 50 kg bags	tonne	60.00
Emulsion paint in 5 litre tins	litre	4.50
Gloss oil paint in 5 litre tins	litre	6.00
Tiles and paviors		
Precast concrete paving slabs (60 x 500 x 50mm)	m^2	3.00
Precast concrete roof tiles	m^2	1.70
Drainage		
WC suite complete	nr	40.00
Lavatory basin complete	nr	45.00
100mm asbestos cement pipes	m	1.40
100mm PVC pipes	m	1.70

Unit rates

The descriptions below are generally shortened versions of standard descriptions listed in five languages (English, French, Italian, German and Spanish) in section 6. Where an item has a two digit reference number (e.g. 05 or 33), this relates to the full description against that number in section 6. Where an item has an alphabetic suffix (e.g. 12A or 34B) this indicates that the standard description has been modified. Where a modification is major the complete modified description is included here and the standard description should be ignored; where a modification is minor (e.g. the insertion of a named hardwood) the shortened description has been modified here but, in general, the full description in section 6 prevails. The suffix XX refers to items not covered in the standard descriptions.

The unit rates below are for main work items on a typical construction project in Valletta in January 1991. The rates include all necessary labour, materials, equipment and, where appropriate, allowances for contractors overheads and profit, preliminary and general items and contractors profit and attendance on specialist trades.

		Unit	Rate ML
Excavation			
01	Mechanical excavation of foundation trenches	m^3	2.90
02	Hardcore filling making up levels	m^2	1.75
Concrete work			
04	Plain insitu concrete in strip foundations in trenches	m^3	24.00
05	Reinforced insitu concrete in beds	m^3	27.00
06	Reinforced insitu concrete in walls	m^3	27.00
07	Reinforced insitu concrete in suspended floor or roof slabs	m^3	28.00
08	Reinforced insitu concrete in columns	m^3	30.00

		Unit	Rate ML
09	Reinforced insitu concrete in isolated beams	m^3	30.00
10	Precast concrete slab	m^2	11.80
Formwork			
11	Softwood formwork to concrete walls	m^2	2.00
12	Softwood or metal formwork to concrete columns	m^2	4.00
13	Softwood or metal formwork to horizontal soffits of slabs	m^2	2.00
Reinforcement			
14	Reinforcement in concrete walls	tonne	380.00
15	Reinforcement in suspended concrete slabs	tonne	350.00
16A	Fabric reinforcement in concrete beds (Type C503)	m^2	1.58
Steelwork			
17	Fabricate, supply and erect steel framed structure	tonne	450.00
Brickwork and blockwork			
18A	Precast lightweight aggregate hollow concrete block walls	m^2	5.50
XX	External hollow wall of 2 skins x 230mm thick walls, fair faced both sides	m^2	12.90
Roofing			
31	Troughed galvanised steel roof cladding	m^2	6.00
XX	4mm polyester reinforced bitumen membrane	m^2	3.70
Woodwork and metalwork			
34A	Single glazed casement window in Mahogany hardwood, size 650 x 900mm	each	75.00
35A	Two panel glazed door in Mahogany hardwood, size 850 x 2000mm	each	175.00
36	Solid core half hour fire resisting hardwood internal flush doors, size 800 x 2000mm	each	90.00
37A	Aluminium double glazed window, size 1200 x 1200mm	each	106.00
38A	Aluminium double glazed door, size 850 x 2100mm	each	132.00
39	Hardwood skirtings	m	1.75
XX	Aluminium single leaf door 800 x 1925mm including 4mm glass and all ironmongery	each	80.00
Plumbing			
43	UPVC rainwater pipes	m	5.80
44	Light gauge copper cold water tubing	m	2.00
45	High pressure plastic for cold water supply	m	3.00
46	Low pressure plastic pipes for cold water distribution	m	3.50
47	UPVC soil and vent pipes	m	5.80

		Unit	Rate ML
48	White vitreous china WC suite	each	49.00
49	White vitreous china lavatory basin	each	54.00
50	Glazed fireclay shower tray	each	48.00
51	Stainless steel single bowl sink and double drainer	each	45.00
Electrical work			
52	PVC insulated and copper sheathed cable	m	1.80
53	13 amp unswitched socket outlet	each	12.00
54	Flush mounted 20 amp, 1 way light switch	each	10.00
Finishings			
55	2 coats gypsum based plaster on brick walls	m^2	2.60
XX	Rendering in two coats: 3mm thick cement sand (1:3) and 6mm thick cement line sand (1:1½:3) to walls and ceilings	m^2	1.05
56A	150 x 150mm white glazed wall tiles on and including cement sand mortar bed	m^2	7.50
58	Cement and sand screed to concrete floors	m^2	2.50
60	Mineral fibre tiles on concealed suspension system	m^2	6.50
Glazing			
61	Glazing to wood	m^2	5.00
Painting			
62	Emulsion on plaster walls	m^2	0.60
63	Oil paint on timber	m^2	1.50

Approximate estimating

The building costs per unit area given below are averages incurred by building clients for typical buildings in Valletta as at January 1991. They are based upon the total floor area of all storeys, measured between external walls and without deduction for internal walls.

Approximate estimating costs generally include for mechanical and electrical installations but exclude furniture, loose or special equipment, and external works; they also exclude fees for professional services. The costs shown are for specifications and standards appropriate to Malta and this should be borne in mind when attempting comparisons with similarly described building types in other countries. A discussion of this issue is included in section 4. Comparative data for countries covered in this publication, including construction cost data are presented in Part Three.

Approximate estimating costs must be treated with reserve; they cannot provide more than a rough guide to the probable cost of building.

	Cost m^2 ML	Cost ft^2 ML
Industrial buildings		
Factories for letting	100	9.30
Administrative and commercial buildings		
Offices for letting, 5 to 10 storeys, air conditioned	283	26.00
Health and education buildings		
Private hospitals (80 beds)	461	43.00
Residential buildings		
Hotel tourist class provincial	222	21.00

Regional variations

The approximate estimating costs are based on projects in Valletta. No adjustment is necessary for other parts of the country.

EXCHANGE RATES AND INFLATION

The combined effect of exchange rates and inflation on prices within a country and price comparisons between countries is discussed in section 4.

Exchange rates

The graph below plots the movement of the Maltese Lira against sterling and the US dollar since 1980. The figures used for the graph are quarterly and the method of calculating these, and other related issues are discussed in section 4. The exchange rate at January 1991 was ML 0.58 to the pound sterling and ML 0.30 to the US dollar.

The Maltese Lira against Sterling and the US Dollar

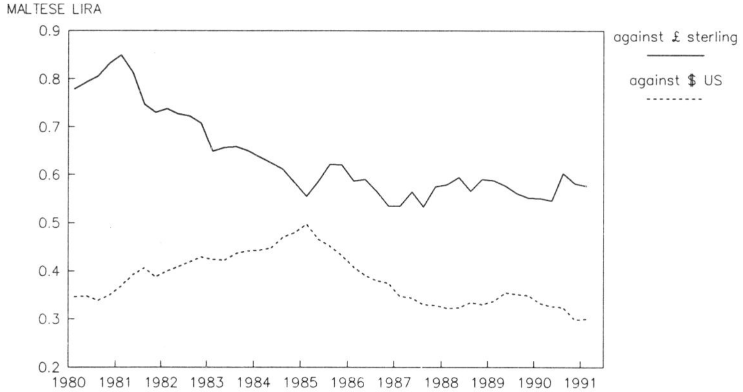

Inflation

The graph and the table below present the consumer price index showing the trend in inflation since 1980. The index has been rebased to 1980=100. Notes on inflation and the distinction between costs and prices are included in section 4.

Consumer price inflation

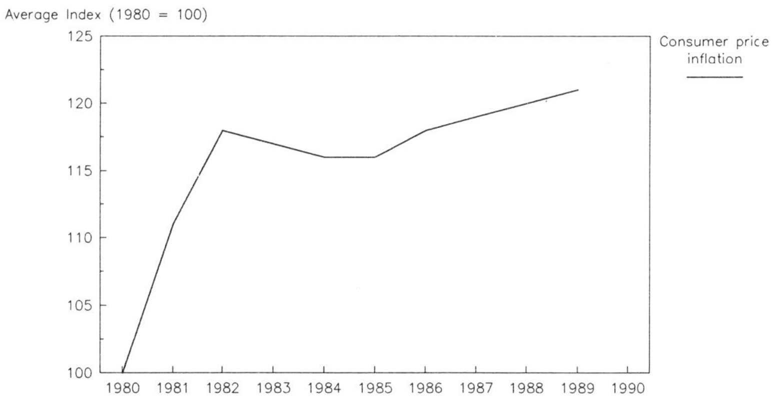

The basis of the figures is the official consumer price index converted to quarterly averages. Inflation peaked in 1981; since then progress in reducing inflation has reduced the annual rate to below 1% since 1987.

Consumer price inflation 1980 = 100

	1981	1982	1983	1984	1985	1986	1987	1988	1989	1990
average index	112	118	117	117	116	119	119	120	121	n.a.
annual change (%)	11.5	5.8	-0.8	-0.5	-0.2	2.0	0.5	0.9	0.9	n.a.

USEFUL ADDRESSES

Malta Development Corporation
House of Catalunya
Marsamxetto Road
Valletta
Tel: (+356) 221431, 223688, 221523
Telex: 1275 DEVCOR MW
Fax: (+356) 606407

Netherlands

KEY DATA

Population

Population (1989)	14.8m
Urban population	88%
Population under 15	17.8%
Population over 65	12.9%
Average annual growth rate (1983 to 1988)	0.7%

Geography

Land area	37 000 km²
Agricultural area	20 000 km²
Capital city	Amsterdam (population 1 031 000)

Economy

Monetary unit	Guilder (G)
Exchange rate (January 1991) to	
the ECU	G 2.31
the pound sterling	G 3.30
the US dollar	G 1.71
Average annual inflation (1984 to 1989)	0.7%
Gross Domestic Product (GDP)	G 451bn
GDP per capita (1989)	G 30 473
GDP growth (real) (1980 to 1988)	1.4%
Private consumption (% of total)	60.1%
Public consumption (% of total)	15.6%
Investment (% of total)	20.8%
Central government expenditure as percentage of GDP	53.3%

Construction

Total value of construction output (1989)	G 53.7bn
Construction output per capita	G 3630
Construction output as a percentage of GDP	11.9%

THE CONSTRUCTION INDUSTRY

Construction output

In 1989 the value of the gross output of the Dutch construction industry was G 53.7 billion, or 23 billion ECUs. This represents 11.9% of GDP. Overall, the Dutch construction industry has grown steadily since 1983 but a slowing down in growth occurred in 1990 and a fall is expected in 1991. The breakdown by type of work is shown below.

Construction output 1989

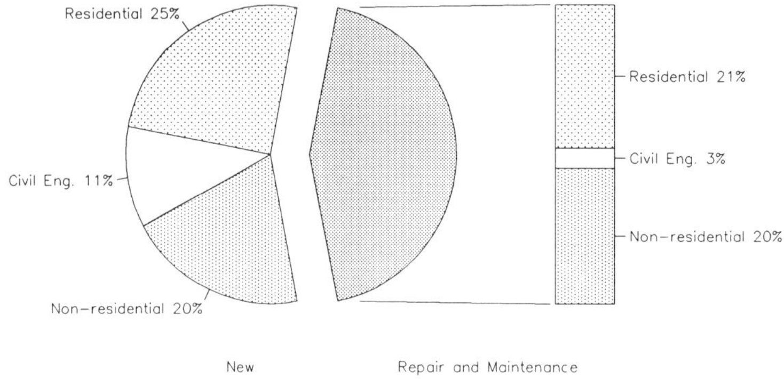

Repair and maintenance account for a very high proportion of total output while the proportion devoted to civil engineering is low. A breakdown by value of building projects (in preparation only) by province compared with population is given in the table below.

Building projects in preparation by province and population percentages (1989)

Province	*% Population*	*% Building projects*
Groningen	3.8	3
Friesland	4.0	2
Drenthe	3.0	2
Overijssel	6.9	5
Flevoland	1.3	2
Gelderland	12.1	10
Utrecht	6.8	8
Noord-Holland	16.0	24
Zuid-Holland	21.6	27
Zeeland	2.4	2
Noord-Brabant	14.7	11
Limburg	7.4	5
Total	100	100

Source: Centraal bureau voor de statistiek, maandstatistiek bouwjmijverheid

Characteristics and structure of the industry

Most of the technical ministries have well established in-house facilities including design departments. It has been estimated that about 60% of the public sector market for consultancy services is taken by in-house offices. The largest office is in the Ministry of Public Works which used to do all its own design and supervision of major projects but now lets about 10% to external consultants. The Ministry of Housing will increase the work it lets to external designers from 50% to 75-80%. Other Ministries are also increasing their external work, and Provincial departments are being slimmed down, giving more work to private firms. In addition, several Dutch contractors have in-house engineering departments for turnkey projects and many suppliers also have design offices.

Private sector engineering firms have a relatively small but increasing market. Many of them are expanding into areas quite separate from engineering and are becoming risk-taking business enterprises. About four firms in this sector employ over 500 staff but at the other extreme there are a large number of very small firms.

There are about 2000 architects registered in the Netherlands. The title of architect is not protected by law and there is no requirement to employ an architect on building projects, though in practice architects are used for all but the very smallest projects. Most architects are in private practice.

In 1986 there were some 18 000 contracting companies in the Netherlands employing 260 000 workers; around 25 companies have more than 500 employees. It is normal practice to employ a main contractor rather than the separate trades system.

Dutch contractors have traditionally worked overseas on building and civil engineering and, particularly, on water related projects. The Netherlands has long been renowned for port and harbour engineering, dredging, land reclamation, coastal defences, waterway engineering and offshore pipeline engineering.

In 1989 there were 22 Dutch contractors in Building magazine's "Top 300 European Contractors" and eight in the top 100. Seven were in Engineering News Record's "Top 250 International Contractors".

Major Dutch contractors

Major contractors	*Place in Building's "Top 300 European Contractors"*	*Place in ENR's "Top 250 International Contractors"*	*Exports (% of turnover)*	*Main work types*
Hollandse Beton Groep (HBG)	23	29	50	Building, power, transport.
Koninklijke Volker Stevin	37			
NBM Amstelland	44			
Ballast Nedam	51	28	70	Building, industrial, transport.
TBI Holding	55			
Wilma International	63			
BAM Holding	67			
Heymens Bedryven	97			

Source: Building; Engineering News Record

Clients and finance

The government finances about half of all new construction work in the Netherlands, directly or indirectly .

Housing falls into three categories:

- housing built under the Housing Act by municipalities or, more usually, housing associations with a high level of state contribution to the cost of development (about 35% of new housing in the period 1971 to 1984)
- housing built for sale or rent by housing associations, financial institutions, pension funds and the private sector with a moderate level of state subsidy (43% of new housing in the period 1971 to 1984)
- free sector housing built for sale or rent by the private sector with no subsidy for the cost of construction though still eligible for mortgage tax relief (19% of new housing in the period 1971 to 1984).

Mortgages for house purchase are available from banks but they do not usually lend more than 60% to 70% of the value of the house without additional security. However the local authority may guarantee the loan and in that case loans of up to 100% of the value of the property are possible.

There are about 850 housing associations throughout the Netherlands usually operating locally and with a small number of developments under their individual control.

Private non-residential building is let either by the users themselves or by property developers (*projectontwikkelaarsz*). The latter may be investment arms of financial institutions such as banks, insurance companies, pensions funds, property companies, subsidiaries of major industrial concerns or estate agents. Public sector utilities contracts are placed by departments of Ministries or local governments.

About 75% of civil engineering - often known as GWW (ground, water and roadbuilding) - contracts are let by the public sector: central, provincial municipal or local government organizations.

Selection of design consultants

Dutch law does not prescribe any specific tender or selection procedures for consultants.

If an external design consultant is employed, he may be appointed after direct negotiation, after a call for proposals from two or more consultants or pre-selection from a restricted number of candidates followed by a call for proposals. Negotiation with one consultant is common though other methods of appointment are being increasingly used.

Contractual arrangements

There are three main methods of obtaining tenders:

- by public advertisement
- by selective tendering, that is, by invitation after public announcement of intention to let a contract
- by negotiation with a single contractor.

Public advertisement is used for many Government and local authority contracts, but selective tendering (even with the extra cost of tendering due to the practice of reimbursing tendering costs) is steadily gaining favour. Normally, selective and negotiated tendering methods are used in private development. In both open and selective tendering the contract is awarded after consideration of the tender sum; only in exceptional circumstances will the contract not be awarded to the lowest tenderer.

When tenders are negotiated the contractor is brought into the design team during the planning process, at the end of which he is asked to state his price for the work. If this cannot be negotiated successfully the contractor is paid for the preliminary work which he has done and further tenders are sought.

For housing and general building, tender documents usually consist of a specification and drawings. From these the contractor is expected to prepare his tender and to do this he often has to measure quantities. For civil engineering works, bills of quantities are usually provided. Once a contract is underway either the architect or the contractor prepares detailed working drawings. A standard form of contract, *Uniforme Administrative Voorwaarden voor de Uitvoering van Werken* (the UAV 68) is widely used.

When compiling a tender, the contractor prepares his own internal schedule of quantities, in operational order, from the specification and drawings that he prices, keeping labour and materials separate. However, this document is confidential to the tenderer and does not form any part of the contract unless he is successful when he will normally make it available. Sometimes the prices (but not the quantities) will be made part of the contract for the purpose of valuing variations.

It is usual to employ a general contractor who is responsible for all work except that carried out on a direct contract with the client. Examples of direct contracts include piling works and mechanical and electrical installations. Nominated sub-contracting occurs to a limited extent.

Interim payments on account may be made to the contractor and the frequency of these is set out in the contract documents. A common formula is to pay 5% of the contract sum once 15% of the work is complete, rising by increments of 5% on both scales thereafter. On completion of the contract a further 7.5% is paid, followed by the final payment of 2.5% at the end of a three months maintenance period.

Clauses may be included in the contract documents to cover the reimbursement of variations in the costs of labour and materials. Where fluctuations are allowed, they are usually paid according to an official index (*Risicoregeling*). Authorized variations to the contract works are permitted and the prices of these are negotiated. Variations are normally agreed before the works are carried out as required by the UAV 68. The final account therefore takes the form of a summary of the previously agreed variations.

Liability and insurance

Liability for contractors and designers lasts normally for 10 years but may be for 30 years. However, costs awarded against architects may amount to only half the designers' fees and it is possible to opt out of liability as part of the contract engagement. Although insurance is not obligatory, in practice approximately 80% of new projects are covered.

Development control and standards

The Physical Planning Act of 1962 (amended 1985) created a three tier planning structure. Central government takes the key decisions; the provinces produce regional plans and the municipalities produce both structure and legally binding plans. The municipalities have a duty to adopt their own building regulations and to issue building permits for all construction, as well as some other activities such as tree planting and demolition. The planning system and the building regulations system operate as one. Planning permission can be refused only when a proposal is not in compliance with the approved plan, building regulations or permission under the Monuments and Historic Buildings Act. Any refusal must have specific stated reasons and an appeal is possible. The normal time limit for a decision is two months, though this can be extended to four months.

Building Regulations are administered with planning control. Under the Housing Act the Minister of Housing and Physical Planning can issue specific regulations and the act is also the basis of municipal building bye-laws. The Society of Netherlands Municipalities drafts and keeps up to date model bye-laws, though within limits these may be amended by individual municipalities.

Standards are prepared and published according to the regulations of the Netherlands Standards Institution (NNI), a private organization. Standards have no legal status though they may be referred to in model bye-laws and other documents.

A system of testing and approval exists for building products and building components on a private basis. Several private organisations are involved in this work and they issue certificates of approval. The use of certified products is promoted nationally but the scheme is nevertheless purely voluntary.

Further reading

Association of Consulting Engineers et al, Engineering Consultancy in the Europen Community : The Netherlands London: Telford, 1990.

H W E Davies "The Netherlands" in Department of the Environment, Planning Control in Western Europe London: HMSO, 1989, pp 337-407.

CONSTRUCTION COST DATA

Cost of labour

The figures below are typical of labour costs in Utrecht as at January 1991. The "wage rate" is the basis of an employee's income, while the "cost of labour" indicates the cost to a contractor of employing that employee. The difference between the two covers a variety of mandatory and voluntary contributions - a list of items which could be included is given in section 4.

	Wage rate (per hour) G	*Cost of labour (per hour) G*	*Number of hours worked per year*
Site operatives			
Skilled tradesman	50	-	-
Electrician	48	-	-
HVAC installer	48	-	-

Cost of materials

The figures that follow are the costs of main construction materials, delivered to site in the Utrecht area, as incurred by contractors in January 1991. These assume that the materials would be in quantities as required for a medium sized construction project and that the location of the works would be neither constrained nor remote.

All the costs in this section exclude value added tax (VAT - see below).

	Units	*Cost G*
Cement and aggregate		
Ordinary portland cement in 50kg bags	tonne	180
Coarse aggregates for concrete	m^3	50
Fine aggregates for concrete	m^3	33
Ready mixed concrete (B.300)	m^3	120
Ready mixed concrete (B.375)	m^3	125
Steel		
Mild steel reinforcement	tonne	1500
High tensile steel reinforcement	tonne	1550
Structural steel sections	tonne	1300
Bricks and blocks		
Common bricks (215 x 100 x 55mm)	1000	250
Good quality facing bricks (50 x 100 x 210mm)	1000	350
Hollow concrete blocks (290 x 140 x 190mm)	1000	1100
Solid concrete blocks (210 x 150 x 85mm)	1000	550
Precast concrete cladding units with exposed aggregate finish	m^2	350
Timber and insulation		
Softwood sections for carpentry	m^3	750
Softwood for joinery	m^3	850
Hardwood for joinery	m^3	1100
Exterior quality plywood (10mm)	m^2	75
Plywood for interior joinery (10mm)	m^2	25
Softwood strip flooring (22mm)	m^2	20
Chipboard sheet flooring (22mm)	m^2	8

	Units	Cost G
100mm thick quilt insulation	m^2	6
100mm thick rigid slab insulation	m^2	7
Softwood internal door complete with frames and ironmongery	nr	375
Glass and ceramics		
Float glass (5mm)	m^2	35
Sealed double glazing units (4+5mm)	m^2	75
Good quality ceramic wall tiles (150 x 150mm)	m^2	27
Plaster and paint		
Plaster in 50 kg bags (coarse)	m^3	75
Plaster in 50 kg bags (fine grained)	m^3	225
Plasterboard (12.5mm thick)	m^2	5
Emulsion paint in 5 litre tins	litre	5
Gloss oil paint in 5 litre tins	litre	6
Tiles and paviors		
Clay floor tiles (150 x 150 x 10mm)	m^2	32
Vinyl floor tiles	m^2	22
Precast concrete paving slabs (300 x 300 x 45mm)	m^2	13
Clay roof tiles	1000	1350
Precast concrete roof tiles	1000	1075
Drainage		
WC suite complete	nr	450
Lavatory basin complete	nr	175
110mm diameter hard pvc pipes	m	13
150mm diameter cast iron drain pipes	m	45

Unit rates

The descriptions below are generally shortened versions of standard descriptions listed in five languages (English, French, Italian, German and Spanish) in section 6. Where an item has a two digit reference number (e.g. 05 or 33), this relates to the full description against that number in section 6. Where an item has an alphabetic suffix (e.g. 12A or 34B) this indicates that the standard description has been modified. Where a modification is major the complete modified description is included here and the standard description should be ignored; where a modification is minor (e.g. the insertion of a named hardwood) the shortened description has been modified here but, in general, the full description in section 6 prevails.

The unit rates below are for main work items on a typical construction project in the Utrecht area in January 1991. The rates include all necessary labour, materials, equipment and, where appropriate, allowances for contractors overheads and profit, preliminary and general items and contractors profit and attendance on specialist trades. All the rates in this section exclude value added tax (VAT - see below).

		Unit	*Rate G*
Excavation			
01	Mechanical excavation of foundation trenches	m^3	8
02	Hardcore filling making up levels	m^2	6
03	Earthwork support	m^2	6
Concrete work			
04	Plain insitu concrete in strip foundations in trenches	m^3	206
05	Reinforced insitu concrete in beds	m^3	188
06	Reinforced insitu concrete in walls	m^3	238
07	Reinforced insitu concrete in suspended floor or roof slabs	m^3	194
08	Reinforced insitu concrete in columns	m^3	306
09	Reinforced insitu concrete in isolated beams	m^3	200
10	Precast concrete slab	m^2	163
Formwork			
11	Softwood or metal formwork to concrete walls	m^2	175
12	Softwood or metal formwork to concrete columns	m^2	188
13	Softwood or metal formwork to horizontal soffits of slabs	m^2	156
Reinforcement			
14	Reinforcement in concrete walls	tonne	6000
15	Reinforcement in suspended concrete slabs	tonne	5500
16	Fabric reinforcement in concrete beds	m^2	13
Steelwork			
17	Fabricate, supply and erect steel framed structure	tonne	4750
Brickwork and blockwork			
18	Precast lightweight aggregate hollow concrete block walls	m^2	100
19	Solid (perforated) concrete blocks	m^2	125
20	Solid (perforated) sand lime bricks	m^2	113
21	Facing bricks	m^2	144
Roofing			
22	Concrete interlocking roof tiles 430 x 380mm	m^2	44
23	Plain clay roof tiles 260 x 160mm	m^2	50
26	Particle board roof coverings	m^2	56
27	3 layers glass-fibre based bitumen felt roof covering	m^2	63

		Unit	Rate G
28	Bitumen based mastic asphalt roof covering	m^2	50
29	Glass-fibre mat roof insulation 160mm thick	m^2	38
31	Troughed galvanized steel roof cladding	m^2	100
	Woodwork and metalwork		
34A	Single glazed casement window in hardwood, size 650 x 900mm	m^2	813
36	Solid core half hour fire resisting hardwood internal flush doors, size 800 x 2000mm	each	438
37	Aluminium double glazed window, size 1200 x 1200mm	m^2	1000
40	Framed structural steelwork in universal joist sections	tonne	6100
	Plumbing		
48	White vitreous china WC suite	each	813
49	White vitreous china lavatory basin	each	313
51	Stainless steel single bowl sink and double drainer	each	219
	Finishings		
55	2 coats gypsum based plaster on brick walls	m^2	25
56	White glazed tiles on plaster walls	m^2	81
57	Red clay quarry tiles on concrete floor	m^2	100
58	Cement and sand screed to concrete floors	m^2	25
59	Thermoplastic floor tiles on screed	m^2	44
60	Mineral fibre tiles on concealed suspension system	m^2	56
	Glazing		
61	Glazing to wood	m^2	125
	Painting		
62	Emulsion on plaster walls	m^2	19
63	Oil paint on timber	m^2	31

Approximate estimating

The building costs per unit area given below are averages incurred by building clients for typical buildings in the Utrecht area as at January 1991. They are based upon the total floor area of all storeys, measured between external walls and without deduction for internal walls.

Approximate estimating costs generally include for mechanical and electrical installations but exclude furniture, loose or special equipment, and external works; they also exclude fees for professional services. The costs shown are for specifications and standards appropriate to Netherlands and this should be borne in mind when attempting comparisons with similarly described building types in other countries. A discussion of this issue is included in section 4. Comparative data for countries covered in this publication, including construction cost data are presented in Part Three.

Approximate estimating costs must be treated with reserve; they cannot provide more than a rough guide to the probable cost of building.

	Cost *m² G*	*Cost* *ft² G*
Industrial buildings		
Factories for letting	950	88
Factories for owner occupation (light industrial use)	1200	111
Factories for owner occupation (heavy industrial use)	1450	135
Factory/office (high-tech) for letting (shell and core only)	1800	167
Factory/office (high-tech) for letting (ground floor shell, first floor offices)	2500	232
Factory/office (high tech) for owner occupation (controlled environment, fully finished)	3200	297
High tech laboratory (air conditioned)	3750	350
Warehouses, low bay (6 to 8m high) for letting (no heating)	750	70
Warehouses, low bay for owner occupation (including heating)	1000	93
Warehouses, high bay for owner occupation (including heating)	1250	116
Administrative and commercial buildings		
Civic offices, non air conditioned	1500	139
Civic offices, fully air conditioned	2250	209
Offices for letting, 5 to 10 storeys, non air conditioned	1800	167
Offices for letting, 5 to 10 storeys, air conditioned	2400	220
Offices for letting, high rise, air conditioned	2800	260
Offices for owner occupation 5 to 10 storeys, non air conditioned	2000	186
Offices for owner occupation 5 to 10 storeys, air conditioned	2650	245
Offices for owner occupation high rise, air conditioned	3000	279
Prestige/headquarters office, 5 to 10 storeys, air conditioned	3250	302
Prestige/headquarters office, high rise, air conditioned	3750	350
Health and education buildings		
General hospitals (300 to 350 beds)	3950	367
Teaching hospitals (800+ beds)	4500	418
Health centres	2000	186
Nursery schools	1100	102
Primary/junior schools	1200	111
Secondary/middle schools	1250	116
University (arts) buildings	1800	167
University (science) buildings	2100	195
Management training centres	1800	167

	Cost m^2 G	Cost ft^2 G
Recreation and arts buildings		
Theatres (over 500 seats) including seating and stage equipment	2000	186
Theatres (less than 500 seats) including seating and stage equipment	1750	163
Sports halls including changing and social facilities	1250	116
Swimming pools (international standard) including changing and social facilities	1750	163
Swimming pools (schools standard) including changing facilities	1500	139
City centre/central libraries	1800	167
Branch/local libraries	1400	130

	Cost per dwelling G	
Residential buildings		
Social/economic single family housing (multiple units)	100 000	-
Private/mass market single family housing 2 storey detached/semidetached (multiple units)	150 000	-
Purpose designed single family housing 2 storey detached (single unit)	200 000	-
Social/economic apartment housing, low rise (no lifts)	75 000	-
Social/economic apartment housing, high rise (with lifts)	95 000	-
Private sector apartment building (standard specification)	125 000	-
Private sector apartment buildings (luxury)	375 000	-
Student/nurses halls of residence	50 000	-
Homes for the elderly (shared accommodation)	85 000	-
Homes for the elderly (self contained with shared communal facilities)	115 000	-

Regional variations

The approximate estimating costs are based on projects in Utrecht. Adjust these costs by the following factors to take account of regional variations:

Central Netherlands : 0%
Northern/southern provinces : -10%

Value added tax

The standard rate of value added tax (VAT) is currently 18.5%, chargeable on general building work.

EXCHANGE RATES AND INFLATION

The combined effect of exchange rates and inflation on prices within a country and price comparisons between countries is discussed in section 4.

Exchange rates

The graph below plots the movement of the Dutch Guilder against sterling and the US dollar since 1979. The figures used for the graph are quarterly and the method of calculating these, and other related issues are discussed in section 4. Netherlands joined the European Monetary System (EMS) at its inception in 1979. The exchange rate at January 1991 was G 3.30 to the pound sterling and G 1.71 to the US dollar.

The Dutch Guilder against Sterling and the US Dollar

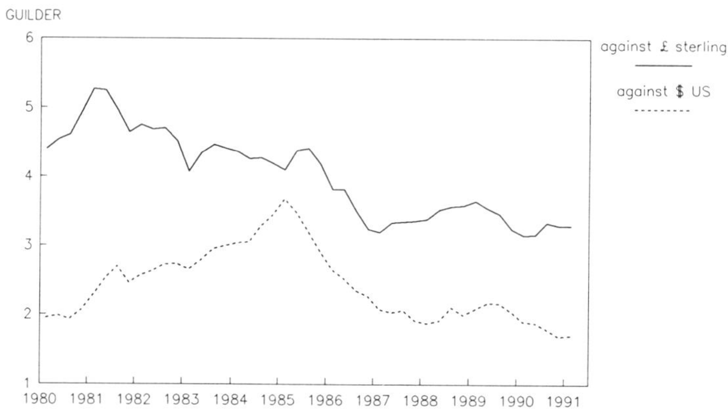

Inflation

The table below presents a number of indices showing the trends in inflation, prices and costs. All the indices have been rebased to 1980=100. Notes on inflation and the distinction between costs and prices are included in section 4.

Cost and price indices

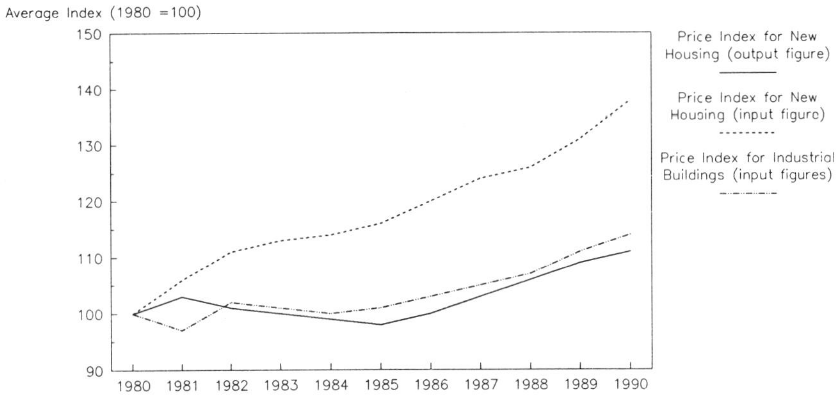

Construction cost and price inflation

The table below is an output price index for new housing compiled by the Central Office for Statistics.

Housing price index (output) 1980 = 100

	1981	1982	1983	1984	1985	1986	1987	1988	1989	1990
average index	103	101	100	99	98	100	103	106	109	111
annual change (%)	3.0	-1.9	-1.0	-1.0	-1.0	2.0	3.0	2.9	2.8	1.8

The table below is an input price index for new housing compiled by the Foundation Building Research.

Housing price index (input) 1980 = 100

	1981	1982	1983	1984	1985	1986	1987	1988	1989	1990
average index	106	111	113	114	116	120	124	126	131	138
annual change (%)	6.0	4.7	1.8	0.9	1.8	3.4	3.3	1.6	4.0	5.3

The table below is a privately produced input price index for new industrial buildng.

Industrial building price index (input) 1980 = 100

	1981	1982	1983	1984	1985	1986	1987	1988	1989	1990
average index	97	102	101	100	101	103	105	107	111	114
annual change (%)	-	5.2	-1.0	-1.0	1.0	2.0	1.9	1.9	3.7	2.7

USEFUL ADDRESSES

Ministre van Volkshuisvesting, Ruimtelijke Ordening en Milieu
(Ministry of Housing, Physical Planning and Environment)
Postbus 20951, 2500 EZ's-Gravenhage
Tel: (070) 32 64 201
Fax: (070) 32 45 478

Ministerie van Verkeer en Waterstaat
(Ministry of Transport and Public Works)
Postbus 20901
2500 EX's-Gravenhage.
Tel: (070) 37 45 745
Fax: (070) 37 44 335

Rijkswaterstaat
(Government Institution for Public Works)
Koningskade 4
2595 AA's-Gravenhage
Tel: (070) 37 45 745
Fax: (070) 37 44 335

Nederlands Verbond van Ondernemers in de Bouwnijverheid NVOB
(Netherlands Organization of Building Contractors)
Postbus 320, 3740 Ah Baarn
Tel: (02154) 27 911
Fax: (02154) 11 044

Orde van Nederlandse Raadgevende Ingenieurs (ONRI)
(Association of Consulting Engineers of The Netherlands)
Koninginnegracht 22
2514 AB's-Gravenhage
Tel: (070) 36 30 756
Fax: (070) 36 00 661

Bond Nederlandse Architecten, BNA
Royal Institute of Dutch Architects
Keizersgracht 321 1016 EE Amsterdam
Tel: (020) 22 81 11
Fax: (020) 23 26 79

International Federation of Consulting Engineers
205 Groot Hertoginnelaan, PO Box 17334
2052 CH
The Hague
Tel: 60 48 28

Nederlands Nomalisatie Instituit
Kalfjeslaan 2, PO Box 5059
2600 GB Delft

Nederaindse Technische Vereniging voor Verwarming en Luch Behandeling
Prinses Julianadlein lA
Amersfoort 3811 NM

Centraal Bureau voor de Statistiek
Prinses Beatrixlaan 428, Postbus 959
2270 AZ Voorburg
Tel: 070 694341

Boowcentrumgebouw
Weena 700, Postbus 299
300AG Rotterdam

Centraal Bureau voor de Statistiek
Kloosterweg 1, Postbus 4481
6401 C2 Heerlen
Tel: 045 736666

Norway

KEY DATA

Population

Population (1989)	4.2m
Urban population	74%
Population under 15	18.8%
Population over 65	16.4%
Average annual growth rate (1983 to 1988)	0.3%

Geography

Land area	324 000 km^2
Agricultural area	10 000 km^2
Capital city	Oslo (population 450 808)

Economy

Monetary unit	Norwegian Krone (Nkr)
Exchange rate (January 1991) to	
the ECU	Nkr 8.00
the pound sterling	Nkr 11.4
the US dollar	Nkr 5.91
Average annual inflation (1984 to 1989)	6.6%
Gross Domestic Product (GDP)	Nkr 594bn
GDP per capita (1989)	Nkr 141 429
GDP growth (real) (1980 to 1988)	2.7%
Private consumption (% of total)	51.9%
Public consumption (% of total)	20.6%
Investment (% of total)	28.1%
Central government expenditure as percentage of GDP	47.6%

Construction

Total value of construction output (1989)	Nkr 93.5bn
Construction output per capita	Nkr 22 260
Construction output as a percentage of GDP	15.7%

THE CONSTRUCTION INDUSTRY

Construction output

In 1989 the value of the gross output of the construction industry in Norway was Nkr 93.5 billion, equivalent to 12.3 billion ECUs. This represents about 15.7% of GDP. The breakdown by types of work is shown below.

Construction output 1989

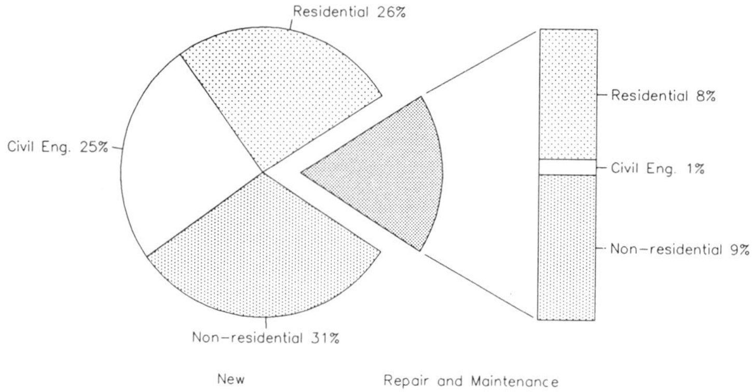

The proportion of total output accounted for by repair and maintenance is low while that for civil engineering is very high, though it had previously been even higher, but is now not expected to fall greatly.

In addition a further Nkr 15 billion or 2.0 billion ECUs is probably accounted for by other sectors including DIY.

Construction output decreased in 1989 and is expected to fall further. In particular, private non-residential building has been falling since 1987 due to over-supply in the previous boom.

Characteristics and structure of the industry

In 1987 there were some 9400 construction firms of which 8 employed more than 500 persons. There are five Norwegian contracting company in Building magazine's "Top 300 European Contractors" and only one in the top 100. One is in Engineering News Record's "Top 250 International Contractors": Block Watne International a.s. (No. 234), which is a building company located in Sandness with only about 2½% of its turnover in exports.

Major Norwegian contractors

Major contractors	*Place in Building's "Top 300 European Contractors"*	*Main work types*
Selmer-Sande Entreprenor a.s.	58	Building and civil engineering (with regional building offices).
Veidekke a.s.	111	Most building and civil engineering including offshore construction.
Aker Entreprenor(a)	149	All building and civil engineering including housing, oil drilling, etc.
EEG-Henriksen	152	All building and civil engineering.
Selvaagbygg a.s.	274	Building including housing and development.

Source: Building

When a general contractor is employed he is responsible for site co-ordination. He is usually allowed to sub-let work of a specialist nature which he has neither the staff nor the experience to do himself. Less frequently, and in certain regions only, the client enters into contracts with individual trade contractors. On a small job site, co-ordination is usually undertaken by the architect, but on a large project a special project supervisor is often engaged. Mechanical and electrical contracts are nearly always separate contracts entered into by the client.

Architects take the leading role in design. They do not have to be registered. The profession is two tiered: fully qualified architects and technicians. Architects undertake the design work and prepare all the working drawings but they do not always supervise the works. They are responsible for bills of quantities.

Engineers play a significant role in the design of building as well as being the main profession for civil works. There are three tiers: civil engineer, engineer (technical) and technician.

Contractual arrangements

The three main methods of obtaining tenders in Norway are by public advertisement, selective tendering and negotiation. Of these, public advertisement is by far the most common, even for private work. Cost reimbursement contracts are occasionally used but, in the main, are limited to industrial building.

Contract documents usually comprise the following:

- drawings
- specification, which includes a description of the works and Norwegian "standard" documents
- contract conditions (NS 3401) - a standard form of contract
- the deed of agreement between the employer and the contractor (also NS 3401)
- form of tender.

Unit quantities are often prepared, usually by the architect, to assist the contractor in submitting a lump sum tender. However, these bills of quantities are not usually regarded as one of the contract documents but, once the successful contractor has been chosen and often after further negotiation, he is asked to check the bill of quantities before signing the contract. The quantities are later used for valuing the work.

Most contracts are subject to fluctuations but variations in labour and material rates have to be officially recognized. The only accepted labour increases are those brought about by an agreement between the National Employers Federation and the Building Workers' Union.

Variations are permitted, providing the scope of the work is not changed, and existing unit rates are used as far as possible to value the change. If the work is so different that it bears no relationship to previous work, new rates are agreed but, in the event of no agreement, the work is charged as daywork. The contractor or contractors have to prepare detailed statements to support claims for payment throughout the progress of the works. These interim payments are usually subject to 10% retention. Either stage or interim valuation payments are checked by the architect or project supervisor. Once 95% of the contract is completed the retention percentage falls to 2% and this is held over until the end of the maintenance period, usually one year. Alternatively the contractor may be allowed to take out a guarantee to cover his contractual liability and this entitles him to be paid for all work done without deduction. The final account is prepared by the contractor and checked by the architect.

Development control and standards

A new Building Act was passed in 1965 covering rules concerning local building authorities and their field of work, regulations on the use of land, general technical requirements concerning buildings, rules concerning application for permission to build and building control by local building authorities. The system of control covers the whole country for buildings. Special regulations are issued by the various authorities responsible for civil works.

The Building Regulations comprise detailed rules concerning applications for building permission, technical requirements for specific types of building and for building elements including structural systems. These technical requirements are given as performance requirements and relate to both test and design methods given in Norwegian Standards. In general these regulations express what is required of the finished building, and have very few rules concerning how this should be brought about. The Ministry of Municipal Affairs and Labour issues advisory documents which give "deemed to satisfy" clauses that indicate various practical methods of complying with the requirements of the regulations. Although Norwegian standards are referred to by the Building Regulations, they are not mandatory, but are examples of technical solutions which meet the requirements.

The local building authorities have the duty and the right to inspect work on site to the extent they find necessary.

CONSTRUCTION COST DATA

Cost of labour

The figures below are typical of labour costs in Oslo as at January 1991. The "wage rate" is the basis of an employee's income, while the "cost of labour" indicates the cost to a contractor of employing that employee. The difference between the two covers a variety of mandatory and voluntary contributions - a list of items which could be included is given in section 4.

	Wage rate (per hour) Nkr	*Cost of labour (per hour) Nkr*
Skilled labour	88.30	185.60
Semi-skilled labour	87.10	178.00
Unskilled labour	75.80	160.00

Cost of materials

The figures that follow are the costs of main construction materials, delivered to site in the Oslo area, as incurred by contractors in January 1991. These assume that the materials would be in quantities as required for a medium sized construction project and that the location of the works would be neither constrained nor remote.

All the costs in this section exclude value added tax (VAT - see below).

	Units	*Cost Nkr*
Cement and aggregate		
Ordinary portland cement in 50kg bags	tonne	1200
Coarse aggregates for concrete	m^3	50
Fine aggregates for concrete	m^3	85
Ready mixed concrete (K25)	m^3	610
Ready mixed concrete (K30)	m^3	700
Steel		
Mild steel reinforcement	tonne	5500
High tensile steel reinforcement	tonne	6000
Structural steel sections	tonne	7000
Bricks and blocks		
Good quality facing bricks (225 x 115 x 75mm)	1000	2100
Hollow concrete blocks (450 x 225 x 225mm)	1000	6600
Solid concrete blocks (450 x 225 x 100mm)	1000	10 000
Precast concrete cladding units with exposed aggregate finish	m^2	120
Timber and insulation		
Softwood sections for carpentry	m^3	2100
Softwood for joinery	m^3	3100
Hardwood for joinery	m^3	32 500
Exterior quality plywood (12.5mm)	m^2	75
Plywood for interior joinery (7.5mm)	m^2	155
Softwood strip flooring (24mm)	m^2	245
Chipboard sheet flooring (3.2mm)	m^2	50
100mm thick quilt insulation	m^2	50
100mm thick rigid slab insulation	m^2	75
Softwood internal door complete with frames and ironmongery	nr	2000
Glass and ceramics		
Float glass	m^2	150
Sealed double glazing units	m^2	1675
Plaster and paint		
Good quality ceramic wall tiles (100 x 100mm)	m^2	60
Plaster in 50 kg bags	tonne	3650
Plasterboard (13mm thick)	m^2	35
Emulsion paint in 5 litre tins	litre	35
Gloss oil paint in 5 litre tins	litre	90

	Units	Cost Nkr
Tiles and paviors		
Clay floor tiles (150 x 150mm)	m^2	90
Vinyl floor tiles (200 x 200 x 2.5mm)	m^2	200
Precast concrete paving slabs (600 x 600 x 50mm)	m^2	120
Clay roof tiles	1000	5340
Precast concrete roof tiles	1000	7000
Drainage		
WC suite complete	nr	5500
Lavatory basin complete	nr	2000
100mm diameter clay drain pipes	m	65
150mm diameter cast iron drain pipes	m	110

Unit rates

The descriptions below are generally shortened versions of standard descriptions listed in five languages (English, French, Italian, German and Spanish) in section 6. Where an item has a two digit reference number (e.g. 05 or 33), this relates to the full description against that number in section 6. Where an item has an alphabetic suffix (e.g. 12A or 34B) this indicates that the standard description has been modified. Where a modification is major the complete modified description is included here and the standard description should be ignored; where a modification is minor (e.g. the insertion of a named hardwood) the shortened description has been modified here but, in general, the full description in section 6 prevails.

The unit rates below are for main work items on a typical construction project in the Oslo area in January 1991. The rates include all necessary labour, materials, equipment and, where appropriate, allowances for contractors overheads and profit, preliminary and general items and contractors profit and attendance on specialist trades. All the rates in this section exclude value added tax (VAT - see below).

		Unit	Rate Nkr
Excavation			
01	Mechanical excavation of foundation trenches	m^3	2200
02	Hardcore filling making up levels	m^2	85
03	Earthwork support	m^2	1200
Concrete work			
04	Plain insitu concrete in strip foundations in trenches	m^3	1740
05	Reinforced insitu concrete in beds	m^3	1920
06	Reinforced insitu concrete in walls	m^3	2010

		Unit	Rate Nkr
07	Reinforced insitu concrete in suspended floor or roof slabs	m^3	2225
08	Reinforced insitu concrete in columns	m^3	2350
09	Reinforced insitu concrete in isolated beams	m^3	2170
10	Precast concrete slab	m^2	430
Formwork			
11	Softwood formwork to concrete walls	m^2	415
12	Softwood or metal formwork to concrete columns	m^2	650
13	Softwood or metal formwork to horizontal soffits of slabs	m^2	1210
Reinforcement			
14	Reinforcement in concrete walls	tonne	10 900
15	Reinforcement in suspended concrete slabs	tonne	11 100
16	Fabric reinforcement in concrete beds	m^2	75
Steelwork			
17	Fabricate, supply and erect steel framed structure	tonne	8200
Brickwork and blockwork			
18	Precast lightweight aggregate hollow concrete block walls	m^2	375
19	Solid (perforated) concrete blocks	m^2	410
20	Sand lime (perforated) bricks	m^2	375
21	Facing bricks	m^2	400
Roofing			
22	Concrete interlocking roof tiles 430 x 380mm	m^2	500
23	Plain clay roof tiles 260 x 160mm	m^2	375
24	Fibre cement roof slates 600 x 300mm	m^2	460
25	Sawn softwood roof boarding	m^2	870
26	Particle board roof coverings	m^2	200
27	3 layers glass-fibre based bitumen felt roof covering	m^2	300
28	Bitumen based mastic asphalt roof covering	m^2	525
29	Glass-fibre mat roof insulation 160mm thick	m^2	500
30	Rigid sheet loadbearing roof insulation 75mm thick	m^2	625
31	Troughed galvanized steel roof cladding	m^2	365
Woodwork and metalwork			
32	Preservative treated sawn softwood 50mm x 100mm	m	35
33	Preservative treated sawn softwood 50 x 150mm	m	50
34A	Single glazed casement window in hardwood, size 650 x 900mm	each	950
35A	Two panel glazed door in hardwood, size 850 x 2000mm	each	2550

		Unit	Rate Nkr
36	Solid core half hour fire resisting hardwood internal flush doors, size 800 x 2000mm	each	1750
37A	Aluminium triple glazed window, size 1200 x 1200mm	each	3750
38	Aluminium double glazed door, size 850 x 2100mm	each	4450
40	Framed structural steelwork in universal joist sections	tonne	9400
41	Structural steelwork lattice roof trusses	tonne	10 500
Plumbing			
42	UPVC half round eaves gutter	m	60
43	UPVC rainwater pipes	m	75
44	Light gauge copper cold water tubing	m	50
45	High pressure plastic for cold water supply	m	45
46	Low pressure plastic pipes for cold water distribution	m	45
47	UPVC soil and vent pipes	m	90
Electrical work			
52	PVC insulated and copper sheathed cable	m	65
53	13 amp unswitched socket outlet	each	210
54	Flush mounted 20 amp, 1 way light switch	each	275
Finishings			
55	2 coats gypsum based plaster on brick walls	m^2	115
56	White glazed tiles on plaster walls	m^2	270
57	Red clay quarry tiles on concrete floor	m^2	320
58	Cement and sand screed to concrete floors	m^2	75
59	Thermoplastic floor tiles on screed	m^2	170
60	Mineral fibre tiles on concealed suspension system	m^2	300
Glazing			
61	Glazing to wood	m^2	280
Painting			
62	Emulsion on plaster walls	m^2	75
63	Oil paint on timber	m^2	120

Approximate estimating

The building costs per unit area given below are averages incurred by building clients for typical buildings in the Oslo area as at January 1991. They are based upon the total floor area of all storeys, measured between external walls and without deduction for internal walls.

Approximate estimating costs generally include for mechanical and electrical installations but exclude furniture, loose or special equipment, and external works; they also exclude fees for professional services. The costs shown are for specifications and standards appropriate to Norway and this should be borne in mind when attempting comparisons with similarly described building types in other countries. A discussion of this issue is included in section 4. Comparative data for countries covered in this publication, including construction cost data are presented in Part Three.

Approximate estimating costs must be treated with reserve; they cannot provide more than a rough guide to the probable cost of building.

All the rates in this section exclude value added tax (VAT - see below).

	Cost m^2 *Nkr*	*Cost* ft^2 *Nkr*
Industrial buildings		
Factories for letting	2100	195
Factories for owner occupation (light industrial use)	3850	360
Factory/office (high-tech) for letting (shell and core only)	3000	280
Factory/office (high tech) for owner occupation (controlled environment, fully finished)	4700	440
Warehouses, low bay (6 to 8m high) for letting (no heating)	1850	170
Administrative and commercial buildings		
Offices for letting, non air conditioned	5750	540
Prestige/headquarters office, air conditioned	8250	768
Health and education buildings		
General hospitals	9500	885
Long term hospitals	9500	885
Nursery schools	5000	465
Primary/secondary schools	5000	465
Recreation and arts buildings		
Sports halls including changing and social facilities	4250	400
Swimming pools (international standard) including changing and social facilities	5750	535
Residential buildings		
Private/mass market single family housing 2 storey detached/semidetached (multiple units)	5000	465
Private sector apartment building (standard specification)	7000	650

Value added tax (VAT)

The standard rate of value added tax (VAT) is currently 25%, chargeable on general building work.

EXCHANGE RATES AND INFLATION

The combined effect of exchange rates and inflation on prices within a country and price comparisons between countries is discussed in section 4.

Exchange rates

The graph below plots the movement of the Norwegian Krone against the American dollar and sterling since 1980. The figures used for the graph are quarterly and the method of calculating these, and other related issues are discussed in section 4. The exchange rate at January 1991 was Nkr 5.91 to the US$ and Nkr 11.4 to the pound sterling.

The Norwegian Krone against Sterling and the US Dollar

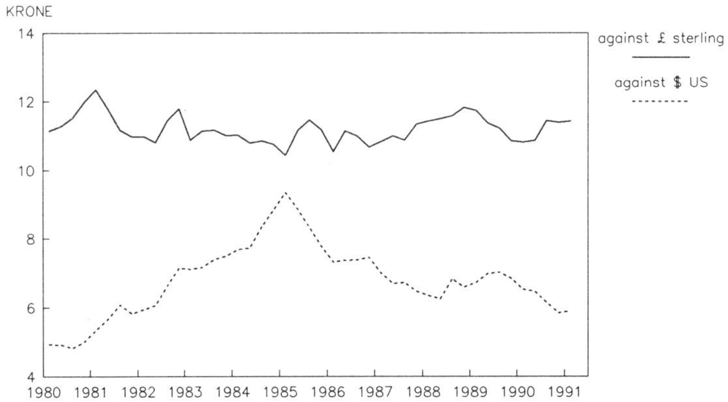

Inflation

The graph and the table below present the consumer price index showing the trend in inflation since 1981. The index has been rebased to 1981=100. Notes on inflation and the distinction between costs and prices are included in section 4.

Consumer price inflation

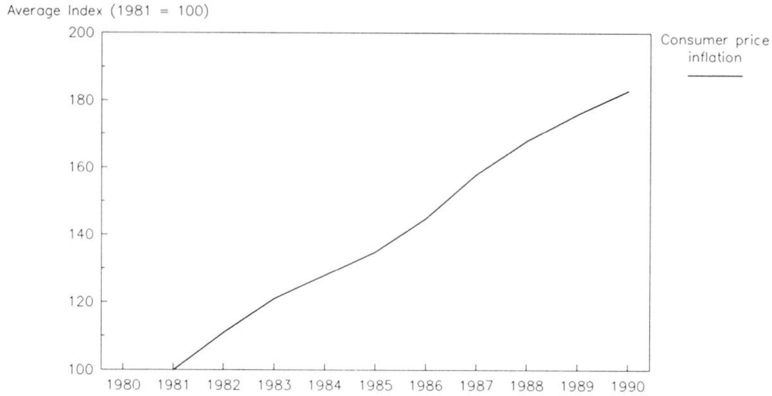

The basis of the figures is the official consumer price index converted to quarterly averages. Inflation peaked in 1982. In recent years inflation has fallen below 5% for the first time in almost a decade.

Notes on inflation are included on section 4.

Consumer price inflation 1981 = 100

	1982	1983	1984	1985	1986	1987	1988	1989	1990
average index	111	121	128	135	145	158	168	176	183
annual change (%)	11.4	8.4	6.2	5.6	7.2	8.7	6.7	4.5	4.1

Poland

KEY DATA

Population

Population (1989)	37.9m
Urban population	61%
Population under 15	25.2%
Population over 65	9.9%
Average annual growth rate (1983 to 1988)	0.8%

Geography

Land area	313 000 km^2
Agricultural area	188 760 (60%)
Capital city	Warsaw (population 1 649 000)

Economy

Monetary unit	Zloty (Zl)
Exchange rate (January 1991) to	
the ECU	Zl 12 900
the pound sterling	ZL 18 500
the US dollar	Zl 9573
Average annual inflation (1984 to 1989)	75.6%
Gross Domestic Product (GDP)	n.a.
GDP per capita (1989)	n.a.
GDP growth (real) (1980 to 1988)	n.a.
Private consumption (% of total)	n.a.
Public consumption (% of total)	n.a.
Investment (% of total)	n.a.
Central government expenditure as percentage of GDP	40.4%

Construction

Total value of construction output (1989)	Zl 14 668bn
Construction output per capita	Zl 387 000
Construction output as a percentage of GDP	n.a.

THE CONSTRUCTION INDUSTRY

Construction output

In 1989 the value of the gross output of the construction industry in Poland was Zl 14668 billion equivalent to at 3 July 1989 exchange rates to 16.5 billion ECUs. According to indirect indicators this may be nearer to 27 billion ECUs. Repair and maintenance in 1989 was 20% to 30% of output. About 26% of the public sector output (which accounts for 85% of the total) was housing; 32% civil engineering and 35% non-residential buildings. No breakdown is available for private work but it would most likely consist of building with a high proportion of housing.

Output of the industry has fallen in the last year or so, and an improvement is expected over the next few years.

The geographical distribution of output is shown in the table below.

Population and Construction Output 1989

Region	*Population %*	*Output %*
Southern	17.5	20.9
South-eastern	18.2	15.5
South-western	9.4	10.5
Central	8.3	8.0
Central-eastern	5.4	4.7
Central-western	13.4	13.2
Northern	9.8	11.3
North-eastern	5.9	5.6
Warsaw	12.1	10.3
TOTAL	100.0	100.0

In most countries the capital city has a disproproportionate share of construction output. In Poland, however, this is not the case.

In 1989, construction exports to the value of Zl 1152 billion (3.8% of domestic output) were made, 72% going to other Eastern European countries and the remainder mainly to Asia (especially Japan) and Africa (especially Libya).

The output of state and co-operative firms is wide-ranging in terms of end use and three quarters of their output is accounted for by housing, land engineering, industrial and commercial work.

Characteristics and structure of the industry

In 1989 only some 14% of the total output of the construction industry was undertaken by private firms, the remainder accounted for by "socialized" building firms and other organizations.

Breakdown of construction output by type of firm, 1989

	Zl bn	*%*
Socialized		
Building firms	7379	50
Other organizations	5173	35
Total	12552	86
Private companies	2116	14
Total	14668	100

There are no data on the size or type of the "other" organizations carrying out 35% of the work. However, data for the public and private companies undertaking 65% of the work are shown below.

	No. of firms	*Average output per firm Zl million*
Public Companies		
State	1 350	5 198
Co-operative	451	621
Private Companies	146 709	14
Total	148 510	63

The number of private companies has grown by over 50% since 1985. In 1989 there were 14 State construction and assembly enterprises employing over 3000 persons. However, over 75% of the work was done by those employing between 200 and 2000 persons. Polish state contractors are splitting up and preparing for privatization which is being organized jointly by the Ministry of Construction and the Ministry of Ownership Changes.

Polish contractors do not feature in Building magazine's "Top 300 European Contractors" but one is in Engineering News Record's "Top 250 International Contractors": Budimex of Warsaw (No. 212) which is a power specialist.

Many of the large construction firms are involved in mechanical and electrical work as well as building or civil construction.

There are some 57 000 designers of various specialities in state engineering and installation design offices. However, as with contractors, many of these will be electrical or mechanical engineers. It is expected that, increasingly, private design practices will be established.

Slow construction work is a problem in all sectors of the industry. There are believed to be 200 000 units under construction in multi-occupancy blocks and 350 000 single family homes. The number of dwelling units actually completed is about 150 000 a year.

Clients and finance

Investment, of which construction is part, is financed as follows: state manufacturing plants 50%; housing co-operatives 15%; local government enterprises 10%; government investments 5%; and private sector 20%.

About 60% of the finance comes from the resources of enterprises and individuals; 27% from bank credits; and 13% from budget subsidies.

Contract procedures

The technical, spatial and administrative aspects of building design and construction, as well as the maintenance and demolition of structures are regulated by the Building Law Act. There are also Regulations concerning contracts, execution and acceptance of works. It is expected that the Building Law and Regulations will soon be revised along free-market economy lines.

Development control and standards

Physical planning is regulated by the Physical Planning Act 1984. Planning policy is determined by Central Government and by the *Voiedvodship* Council. The Planning Authority draws up broad regional development plans and detailed proposals and plans are drawn up at local level. The public is able to participate in the preparation of plans.

Building construction is regulated by the Building Act 1974. There are obligatory Technical standards for construction which are established by the Polish Committee for Standard Measurements and Quality. There are three types of technical standards:

- state standards, number prefix PN/B
- branch standards, number prefix BN
- factory or local standards.

Presently, work is underway to align Polish standards with the European EUROCODE (EC).

Quality standards for construction works are described in the "Technical requirements for the execution and acceptance of construction and assembly works."

Research and development

The main organizations involved in construction industry research and development are:

- The Technical Building Institute (ITB)
- The Institute for Organization, Management and Economy in Building (ORGBUD)
- The National Research and design Centre for Industrial Housing and Public Buildings (COBPBO)
- The National Research and Design Centre for Industrial Buildings (BISTYP).

Both ITB and ORGBUD are research institutes only, while COBPBO and BISTYP also undertake design. Research is financed partly from the national budget and partly by contracts.

Acknowledgements

The authors are grateful to Dr Eng Tadeusz Krawczyk of the *Instytut Organizacji Zarzadzania i Ekonomiiki Przemyslu Budowlanego* (ORGBUD) for providing information for this report.

Further reading

Rolf Bollinger, "Chancen und Risken für die Westeuropäische Bauwirtschaft nach den Wende in Osteuropa" in Euroconstruct, Baukonjunktur 1990/91 in Westeuropa, IFO Institut für Wirtschaftsforschung, München, Dezember 1990.

CONSTRUCTION COST DATA

Cost of labour

The figures below are typical of labour costs in Warsaw as at January 1991. The "wage rate" is the basis of an employee's income, while the "cost of labour" indicates the cost to a contractor of employing that employee. The difference between the two covers a variety of mandatory and voluntary contributions - a list of items which could be included is given on in section 4.

	Wage rate (per hour) Zl	*Cost of labour (per hour)* Zl	*Number of hours worked per year*
Site operatives			
Mason/bricklayer	4660	7690	2112
Carpenter	4660	7690	2112
Plumber	4775	7880	2112
Electrician	4770	7870	2112
Structural steel erector	5940	9800	2112
	(per month)*	*(per month*)*	
Site supervision	Zl 000s	Zl 000s	
General foreman	2250	3710	2112
Trades foreman	2000	32[illegible]	2112
Clerk of works	1400	[illegible]10	2112
Contractors' personnel			
Site manager	4000	6600	2112
Resident engineer	3250	5360	2112
Resident surveyor	200[illegible]	3300	2112
Junior engineer	175[illegible]	2885	2112
Junior surveyor	1[illegible]50	2885	2112
Planner	1500	2475	2112
Consultants' personnel			
Senior architect	3500	5775	2112
Senior engineer	2500	4125	2112
Senior surveyor	2500	4125	2112
Qualified architect	2000	3300	2112
Qualified engineer	2000	3300	2112
Qualified surveyor	2000	3300	2112

* Monthly wage rates include a bonus of up to +35%.

Cost of materials

The figures that follow are the costs of main construction materials, delivered to site in the Warsaw area, as incurred by contractors in January 1991. These assume that the materials would be in quantities as required for a medium sized construction project and that the location of the works would be neither constrained nor remote.

	Units	*Cost Zl*
Cement and aggregate		
Ordinary portland cement in 50kg bags	tonne	343 200
Coarse aggregates for concrete	m^3	50 100
Fine aggregates for concrete	m^3	26 750
Ready mixed concrete (B-10)	m^3	262 100
Ready mixed concrete (B-20)	m^3	332 000
Steel		
Mild steel reinforcement	tonne	3 600 000
High tensile steel reinforcement	tonne	4 180 000
Structural steel sections	tonne	13 000 000
Bricks and blocks		
Common bricks (250 x 120 x 65mm)	1000	870 000
Good quality facing bricks (250 x 120 x 65mm)	1000	1 650 000
Hollow concrete blocks (490 x 240 x 102mm)	each	3050
Solid concrete blocks (250 x 250 x 140mm)	each	2950
Timber and insulation		
Softwood for joinery	m^3	1 182 000
Hardwood for joinery	m^3	2 542 000
Exterior quality plywood (4mm)	m^2	31 600
Plywood for interior joinery (4mm)	m^2	29 100
Softwood strip flooring (32mm)	m^2	53 000
Chipboard sheet flooring (8mm)	m^2	54 900
100mm thick quilt insulation	m^2	31 000
100mm thick rigid slab insulation	m^2	34 600
Softwood internal door complete with frames and ironmongery	nr	330 000
Glass and ceramics		
Float glass (3mm)	m^2	28 100
Sealed double glazing units	m^2	52 000

	Units	Cost Zl
Plaster and paint		
Good quality ceramic wall tiles (100 x 100mm)	m^2	61 100
Plaster in 50 kg bags	tonne	740 000
Emulsion paint in 5 litre tins	litre	13 000
Gloss oil paint in 5 litre tins	litre	28 250
Tiles and paviors		
Clay floor tiles (150 x 150mm)	m^2	57 300
Vinyl floor tiles (200 x 200 x 2.5mm)	m^2	25 450
Precast concrete paving slabs (600 x 600 x 50mm)	m^2	36 000
Clay roof tiles	1000	2 200 000
Precast concrete roof tiles	1000	1 320 000
Drainage		
WC suite complete	nr	384 000
Lavatory basin complete	nr	205 000
100mm diameter clay drain pipes	m	21 650
150mm diameter cast iron drain pipes	m	92 900

Unit rates

The descriptions below are generally shortened versions of standard descriptions listed in five languages (English, French, Italian, German and Spanish) in section 6. Where an item has a two digit reference number (e.g. 05 or 33), this relates to the full description against that number in section 6. Where an item has an alphabetic suffix (e.g. 12A or 34B) this indicates that the standard description has been modified. Where a modification is major the complete modified description is included here and the standard description should be ignored; where a modification is minor (e.g. the insertion of a named hardwood) the shortened description has been modified here but, in general, the full description in section 6 prevails.

The unit rates below are for main work items on a typical construction project in the Warsaw area in January 1991. The rates include all necessary labour, materials, equipment and, where appropriate, allowances for contractors overheads and profit, preliminary and general items and contractors profit and attendance on specialist trades.

		Unit	Rate Zl
Excavation			
01	Mechanical excavation of foundation trenches	m^3	112 700
02	Hardcore filling making up levels	m^2	18 900
03	Earthwork support	m^2	5000

		Unit	Rate Zl
Concrete work			
04	Plain insitu concrete in strip foundations in trenches	m^3	456 500
05	Reinforced insitu concrete in beds	m^3	435 700
06	Reinforced insitu concrete in walls	m^3	526 400
07	Reinforced insitu concrete in suspended floor or roof slabs	m^3	505 500
08	Reinforced insitu concrete in columns	m^3	531 800
09	Reinforced insitu concrete in isolated beams	m^3	542 500
10	Precast concrete slab	m^2	158 800
Formwork			
11	Softwood formwork to concrete walls	m^2	60 400
12	Softwood or metal formwork to concrete columns	m^2	63 000
13	Softwood or metal formwork to horizontal soffits of slabs	m^2	40 000
Reinforcement			
14	Reinforcement in concrete walls	tonne	5 000 000
15	Reinforcement in suspended concrete slabs	tonne	5 400 000
16	Fabric reinforcement in concrete beds	m^2	118 800
Steelwork			
17	Fabricate, supply and erect steel framed structure	tonne	15 000 000
Brickwork and blockwork			
18	Precast lightweight aggregate hollow concrete block walls	m^2	55 300
19	Solid (perforated) concrete blocks	m^2	99 100
20	Solid (perforated) sand lime blocks	m^2	70 700
21	Facing bricks	m^2	198 000
Roofing			
22	Concrete interlocking roof tiles 430 x 380mm	m^2	69 500
23	Plain clay roof tiles 260 x 160mm	m^2	164 900
24	Fibre cement roof slates 600 x 300mm	m^2	38 300
25	Sawn softwood roof boarding	m^2	66 900
26	Particle board roof coverings	m^2	58 800
27	3 layers glass-fibre based bitumen felt roof covering	m^2	61 400
28	Bitumen based mastic asphalt roof covering	m^2	24 100
29	Glass-fibre mat roof insulation 160mm thick	m^2	92 600
30	Rigid sheet loadbearing roof insulation 75mm thick	m^2	68 100
31	Troughed galvanized steel roof cladding	m^2	65 300
Woodwork and metalwork			
32	Preservative treated sawn softwood 50mm x 100mm	m	5000
33	Preservative treated sawn softwood 50 x 150mm	m	6600
34A	Single glazed casement window in hardwood, size 650 x 900mm	each	456 000

		Unit	Rate Zl
35A	Two panel glazed door in hardwood, size 850 x 2000mm	each	1 911 000
36	Solid core half hour fire resisting hardwood internal flush doors, size 800 x 2000mm	each	778 700
37	Aluminium double glazed window, size 1200 x 1200mm	each	2 377 800
38	Aluminium double glazed door, size 850 x 2100mm	each	2 224 800
39	Hardwood skirtings	m	8000
40	Framed structural steelwork in universal joist sections	tonne	18 440 000
41	Structural steelwork lattice roof trusses	tonne	24 440 000

Plumbing

42	UPVC half round eaves gutter	m	58 300
43	UPVC rainwater pipes	m	58 700
44	Light gauge copper cold water tubing	m	76 900
45	High pressure plastic for cold water supply	m	62 700
46	Low pressure plastic pipes for cold water distribution	m	19 700
47	UPVC soil and vent pipes	m	61 900
48	White vitreous china WC suite	each	428 100
49	White vitreous china lavatory basin	each	188 300
50	Glazed fireclay shower tray	each	146 900
51	Stainless steel single bowl sink and double drainer	each	1 066 700

Electrical work

52	PVC insulated and copper sheathed cable	m	28 700
53	13 amp unswitched socket outlet	each	48 200
54	Flush mounted 20 amp, 1 way light switch	each	40 100

Finishings

55	2 coats gypsum based plaster on brick walls	m^2	14 800
56	White glazed tiles on plaster walls	m^2	146 700
57	Red clay quarry tiles on concrete floor	m^2	120 200
58	Cement and sand screed to concrete floors	m^2	46 600
59	Thermoplastic floor tiles on screed	m^2	51 700
60	Mineral fibre tiles on concealed suspension system	m^2	173 600

Glazing

61	Glazing to wood	m^2	99 700

Painting

62	Emulsion on plaster walls	m^2	8400
63	Oil paint on timber	m^2	32 300

Approximate estimating

The building costs per unit area given below are averages incurred by building clients for typical buildings in the Warsaw area as at January 1991. They are based upon the total floor area of all storeys, measured between external walls and without deduction for internal walls.

Approximate estimating costs generally include for mechanical and electrical installations but exclude furniture, loose or special equipment, and external works; they also exclude fees for professional services. The costs shown are for specifications and standards appropriate to Poland and this should be borne in mind when attempting comparisons with similarly described building types in other countries. A discussion of this issue is included in section 4. Comparative data for countries covered in this publication, including construction cost data are presented in Part Three.

Approximate estimating costs must be treated with caution; they cannot provide more than a rough guide to the probable cost of building.

	Cost m^2 *Zl*	*Cost* ft^2 *Zl*
Industrial buildings		
Factories for owner occupation (heavy industrial use)	1 770 000	164 500
Administrative and commercial buildings		
Civic offices, non air conditioned	2 500 000	233 000
Health and education buildings		
Health centres	2 468 000	229 000
Nursery schools	2 970 000	276 000
Residential buildings		
Social/economic single family housing (multiple units)	4 250 000	395 000
Social/economic apartment housing, low rise (no lifts)	2 567 000	238 500
Social/economic apartment housing, high rise (with lifts)	2 705 000	251 500

EXCHANGE RATES AND INFLATION

The combined effect of exchange rates and inflation on prices within a country and price comparisons between countries is discussed in section 4.

Exchange rates

The graph below plots the movement of the Polish Zloty against sterling and the US dollar since 1980. The figures used for the graph are quarterly and the method of calculating these, and other related issues are discussed in section 4. The exchange rate at January 1991 was Zl 18500 to the pound sterling and Zl 9573 to the US dollar.

The Polish Zloty against Sterling and the US Dollar

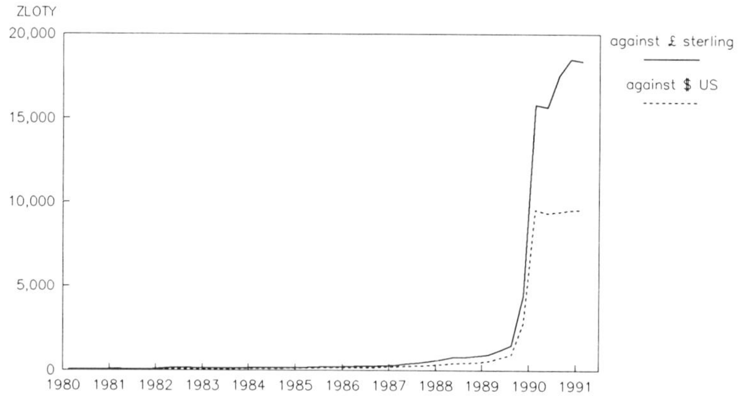

Inflation

The graph below presents a number of indices showing the trends in inflation, prices and costs. All the indices have been rebased to 1980=100. Notes on inflation and the distinction between costs and prices are included in section 4.

Cost and price indices

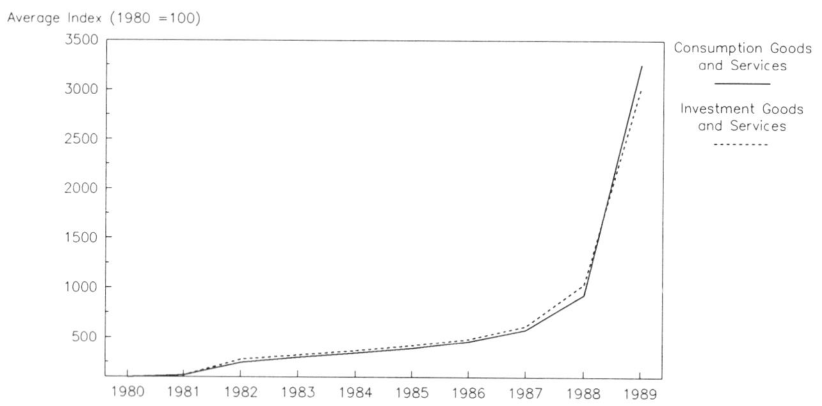

Consumer price inflation

The following tables present indices of price movements in consumer goods and in investment goods, indicating that trends in both have been broadly in line since 1980. Inflation rose dramatically in 1988 and, in 1989, averaged some 250% for consumer goods and just under 200% for investment goods.

Consumer goods and services 1980 = 100

	1981	1982	1983	1984	1985	1986	1987	1988	1989
average index	121	243	297	342	393	463	580	929	3261
annual change (%)	21	101	22	15	15	18	25	60	251

Investment goods and services 1980 = 100

	1981	1982	1983	1984	1985	1986	1987	1988	1989
average index	121	276	323	366	419	487	614	1037	3071
annual change (%)	21	128	17	13	14	16	26	69	196

USEFUL ADDRESSES

Ministerstwo Gospodarki Przestrzennej i Budownictwa
(Ministry of Spatial Planning and Building)
ul. Wspólna 2, 00-505 WARSZAWA
Tel: (022) 21-03-51

Glówny Urzad Statystyczny
(Central Office of Statistics)
A1. Niepodleglosci 208
00-925 WARSZAWA
Tel: (022) 25-32-41

Useful addresses

Centralny Osrodek Informacji Budownictwa, ul. Senatorska 27
(Central Office of Building Information (collecting and disseminating information about building companies, producers of building materials, technologies etc.)
00-099 WARSZAWA
Tel: (022) 26-82-71

Polska Izba Handlu Zagranicznego
(Polish Chamber of International Trade)
ul. Trebacka 4
00-074 WARSZAWA
Tel: (022) 26-02-21

Instytut Organizacji Zarzadzania i Economiki
(Institute of Management Organization and Economics of the Building Industry, ORGUD)
Przemyslu Budowlanego ORGUD
ul. Filtrowa 1
00-925 WARSZAWA
Tel: (022) 25-52-81

Stala Wystawa Budownictwa
(Permanent Exhibition of Construction)
ul. Bartycka 26
00-716 WARSZAWA
Tel: (022) 41-59-44

Instytut Techniki Budowlanej
(Institute of Construction Technology)
ul. Filtrowa 1
00-925 WARSZAWA
Tel: (022) 25-04-70

Portugal

KEY DATA

Population

Population (1989)	10.4m
Urban population	32%
Population under 15	21.2%
Population over 65	12.9%
Average annual growth rate (1983 to 1988)	0.8%

Geography

Land area	92 000 km²
Agricultural area	33 000 km²
Capital city	Lisbon (population 2 128 000)

Economy

Monetary unit	Escudo (Esc)
Exchange rate (January 1991) to	
the ECU	Esc 182
the pound sterling	Esc 261
the US dollar	Esc 135
Average annual inflation (1984 to 1989)	12.5%
Gross Domestic Product (GDP)	Esc 6008bn
GDP per capita (1989)	Esc 0.578m
GDP growth (real) (1980 to 1988)	2.2%
Private consumption (% of total)	64.8%
Public consumption (% of total)	15.2%
Investment (% of total)	26.5%
Central government expenditure as percentage of GDP	40.4%

Construction

Total value of construction output (1989)	Esc 1000bn
Construction output per capita	Esc 96 150
Construction output as a percentage of GDP	16.6%

THE CONSTRUCTION INDUSTRY

Construction output

The value of gross output of the Portuguese construction industry in 1989 was probably about ESC 1000 billion or about 5.8 billion ECUs. This represents around 17% of GDP. The estimated breakdown by type of work is shown below.

Construction output 1989

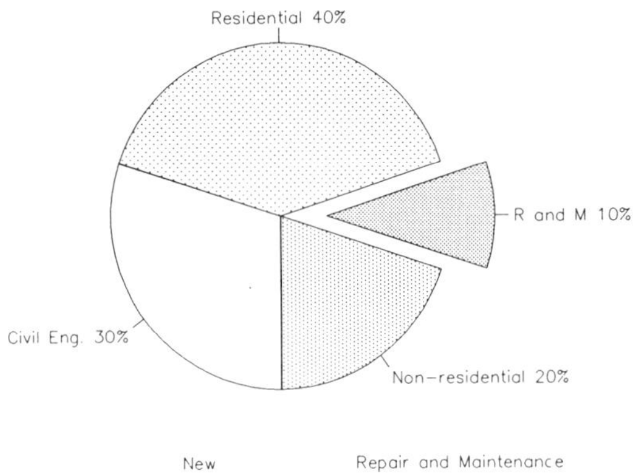

The proportion of both residential building and civil engineering in total output are very high but repair and maintenance represents a relatively low proportion of the total. It may be that the real importance of residential construction is greater than the diagram indicates. It has been estimated that 40% of all houses constructed in the mid 1980s were built outside the official formal economy by very small builders.

The level of Portuguese construction activity started to rise in 1986 when the effects of European Community funds began to take effect. Continued buoyancy of the civil engineering sector is anticipated in response to government determination to improve the national infrastructure backed by European Community Structural Funds. The surge in construction activity has been concentrated in the coastal regions. Recent figures published by the Portuguese Public Works Association indicate that 75% of activity is in the main coastal towns and cities, with Lisbon accounting for 25% and Oporto 18%.

The share of construction output in each region compared with the percentage of total population is shown in the table below.

Population and construction output 1987 by region

Region	*% Population*	*% Construction output*
Aveiro	6.5	4.2
Beja	1.7	1.9
Braga	7.5	6.3
Braganca	1.8	0.9
Castelo Branco	2.2	1.4
Coimbra	4.3	4.4
Evora	1.7	0.8
Faro	3.3	8.0
Guarda	1.9	1.4
Leiria	4.2	2.7
Lisboá	20.7	25.2
Portalegre	1.3	0.7
Porto	16.3	17.6
Santarem	4.5	2.8
Setúbal	7.6	8.1
Viana do Castelo	2.6	2.0
Vila Real	2.6	2.3
Viseu	4.1	2.8
Azores	2.5	3.5
Madeira	2.6	3.2
Total	100	100

Source: Instituto Nacional Estatistica, Lisboa

Characteristics and structure of the industry

Both the main contractor system and the separate trades systems operate in Portugal. The contractor usually undertakes detailed design.

Engineers are dominant in the design of construction projects and there are about 15 engineers for every one architect. Engineers' education is more broadly based then in the UK and engineers hold many of the key posts in government and in private industry. There are two types of engineers: those with a five year university degree and those with a three year technical college qualification, usually referred to as technical engineers. Engineers are involved not only in design, but also in site and project management and administration.

Architects have improved their status and influence in recent years. The Association of Architects is an organization to which, by law, all Portuguese architects working in Portugal and all resident European Community architects must belong. Non-resident

European Community architects have to register with the Association to do work in Portugal but need not be members. Most architects practice as individuals and there are few large practices. A law of 1988 redefines the lines of professional competence between the architect and engineer but, since the allocation is not mandatory, the architects have no monopoly over the design of buildings. The architect has no role in project supervision.

Estimates of the number of contractors vary from 20 000 to 28 000 firms. The industry is however, made up of a large number of small and medium sized enterprises. There are very few large firms but they do a large proportion of the work, especially in the civil engineering sector. There are four Portuguese companies in Building magazine's "Top 300 European Contractors" but there are none in the top 100. They do not feature in Engineering News Record's "Top 250 International Contractors".

Major Portuguese contractors

Major contractors	*Place in Building's "Top 300 European Contractors"*	*Exports (% of turnover)*	*Main work types*
Engil	239	n.a.	
Construçoes Tecnicas	266	about 50%	
Mague	278	small	
Soares do Costa	(largest in Portugal)	some	Building, civil engineering, power, etc.
Mota and Companhia	291	about 30%	Roads
Teixeira Duarte	(2nd largest in Portugal)	some	Civil engineering and building.

Source: Building

Both Soares do Costa and Teixeira Duarte should, on 1988 performance, be in the "Building" list and are larger than the other companies which are included.

Most work abroad is undertaken in former Portuguese colonies.

All contractors need a licence to operate and the category of work for which the licence is given depends on various factors, including the financial and technical capacity of the company. All work over ESC 4 million which requires local authority approval must be carried out by a licenced contractor.

Clients and finance

Sources of finance for public sector projects are dominated by funds from the European Community. Structural funds from the European Regional Development Fund, the European Social Fund and the European Agricultural Guidance and Guarantee Fund in 1989 were expected to total ESC 167 billion (0.985 billion ECUs). In addition, other European Community funds of ESC 51 billion (0.30 billion ECUs) were expected, bringing the total to ESC 218 billion (1.285 billion ECUs). This represents about 4% of GDP. Not all these funds are intended for construction but they would provide a large part of the funds for civil engineering works and industrial buildings.

Over 95% of housing is built by private sector clients - about 60% private individuals, 7% co-operatives and the rest other private bodies. A housing fund, operated through the banking system, has been established for the purchase and construction of residential property.

New property investment companies (*Sociedade gestora de investimento imobiliario* - SGILs) have been established to invest in construction and these receive major tax benefits. The three institutions which specialize in property financing are the General Savings Bank (*Caixa Geral dos Depositos* - CGD) for civil construction and owner occupied housing, the Real Estate Credit Bank (*Credito Predial Portugues* - CPP) for mortgage finance and the National Development Bank (*Banco Nacional de Fomento* - BNF) for development projects.

Contractual arrangements

A law of 1986 regulates public procurement procedures for construction by what is known as the *régime juridico*. The *régime* is a document of 236 articles which constitute a legal framework for contracts and of which certain measures will become embedded in the contract unless specifically excluded. It also determines procedures and payment methods.

The normal contractor selection method is "open" so long as firms have the right category of licence to undertake the work. Choice is then usually based on price.

There is special legislation for the construction of large and/or complex projects, for which special prequalification is required. Requirements include technical and financial capabilities, track record on projects and holdings of equipment. About five contractors or consortia are chosen and given final drawings and documentation for the tender process. Proposals must follow the rules in decree 348-A/86 and in *Portarais*

6QS-A-B-C/86. Each stage of the process is separately considered and the contractor must submit documents in separate envelopes as follows:

- qualifications, financial capacity and proof of payment of taxes
- specifications of materials and equipment in accordance with tender documents
- cost and price data including unit prices of materials, catalogues and specifications of equipment or systems to be used
- documents setting out how the contract will be carried out including for example, the programme, contract period, payment revision procedures, bonuses and penalties and defects liability period.

The lowest price is not the sole criterion as all the other factors are included in the final analysis.

The three most popular contractual methods for both evaluating the cost of work and making payment to contractors are set out below.

- Total lump sum (fixed fee) contracts: Payment is in accordance with a schedule and programme of works presented with the proposal

- Unit rates and quantities contracts: The evaluation of quantities and progress of works is the responsibility of the client or his team known as *fiscals*. Payment is made monthly based on unit prices applied to the quantity of work done

- Cost plus: The quantities of works may be defined by the client at the time of signing the contract or gradually with the evolution of the project. An agreed management margin is added to the actual cost of the work by the contractor. This type of contract needs highly qualified cost-control personnel on both sides to operate efficiently.

Many private contracts follow the public system using either a main contractor or separate trades. However, in Portugal there is great flexibility in the way contracts are organized and it is difficult to say which is the prevalent method.

Liability and insurance

The Civil Code established in 1867 and revised for construction in 1966 and 1977 deals with civil liability for construction for all parties to the process. Specific articles are 1207 and 1230. In theory a builder can be held responsible for breaches under the building contract and the Civil Code, though the courts have yet to exercise such powers outside of the contractual terms.

Post-construction liability for private works under the Civil Code extends to five years. Post-construction liability for public works is regulated by the *régime juridico* and lasts for two years though this is usually reduced by the contract to one year.

All post-construction liability rests with the contractor who normally undertakes detailed design. There are no specific laws dealing with the engineer's or architect's liability. If the fault is a major design fault, it is up to the courts to determine liability. However, the Portuguese legal system is very slow; it may well take over 10 years for a case to go through the system. The danger of an architect, engineer, or indeed a contractor, being taken to court is slight.

Contractors usually insure with an all-risks policy and a latent defects policy. Some professionals take out indemnity insurance for all their work. However, as their risks are small, this is not very common.

Development control and standards

The relevant basic legal text on planning for private work is decree 166/70 of 15 April 1970. Public works need no formal permission from local authorities. It is recommended that plans should be produced by local authorities, ranging from general strategic development plans to detailed plans for a certain area. However, under the Portuguese constitution, municipalities are autonomous organizations and if they have drawn no plans, permission is given, or not, on an ad hoc basis. Before applying for permission, any prospective developer can ask for a description of the type of work which may be undertaken on a specific piece of land.

There are different procedures for small projects, large projects and tourism projects. The small and large project applications are submitted to municipalities, but the tourism applications are submitted to the Ministry of Tourism. For large projects in particular, the procedure involves submission or consultation with up to eighteen government or sub-government departments. Obtaining planning permission is a lengthy, bureaucratic and uncertain process.

The main objectives of technical control legislation in Portugal are the protection of health and safety. The building regulations are set out in ministerial decree 718/87 of 21 August 1987. They require little control of building design, choice of materials or fire resistance except in Lisbon and Oporto and on very special projects.

While the Ministry of Public Works has overall responsibility for building standards and norms, it is the Commission for Revision and Establishment of Technical Regulations, together with the Superior Council for Public Works which produce the regulations. The *Laboratorio Nacional de Engenharia Civil* (LNEC) is responsible for approval, classification of building materials and components plus quality control. However, few products are tested and approved, partly because of the long duration of the process. Moreover, LNEC is mainly interested in civil engineering research, not in approval of building materials. Standards set are usually European Community standards adopted by Portugal but they are not always observed and some materials are of poor quality.

Regulations are, in theory, enforced by the municipalities but in practice the law requires only that the designer declares that he has conformed to the construction norms, observed any relevant regulations and any conditions laid down by the municipality. The site is inspected before foundations are laid and after the foundations are built. On completion, an occupation licence is granted if all regulations have been observed.

Further reading

S D Reynolds and S Sheppard, The Iberian Construction Industry : A Guide for UK Professionals, London: CIRIA, 1989.

RIBA, Architectural Practice in Europe : Portugal, London : RIBA, 1991.

Association of Consulting Engineers et al, Engineering Consultancy in the European Community : Portugal, London : Telford, 1990.

CONSTRUCTION COST DATA

Cost of labour

The figures below are typical of labour costs in Lisbon as at January 1991. The "wage rate" is the basis of an employee's income, while the "cost of labour" indicates the cost to a contractor of employing that employee. The difference between the two covers a variety of mandatory and voluntary contributions - a list of items which could be included is given in section 4.

	Wage rate (per hour) Esc	*Cost of labour (per hour) Esc*
Skilled labour	320	645
Semi-skilled labour	285	580
Unskilled labour	260	515

Cost of materials

The figures that follow are the costs of main construction materials, delivered to site in the Lisbon area, as incurred by contractors in January 1991. These assume that the materials would be in quantities as required for a medium sized construction project and that the location of the works would be neither constrained nor remote.

All the costs in this section exclude value added tax (VAT - see below).

	Units	*Cost Esc*
Cement and aggregate		
Ordinary portland cement in 50kg bags	tonne	12 480
Coarse aggregates for concrete	m^3	2240
Fine aggregates for concrete	m^3	2240
Ready mixed concrete (K25)	m^3	8500
Steel		
Mild steel reinforcement	tonne	78 000
High tensile steel reinforcement	tonne	92 000
Bricks and blocks		
Good quality facing bricks (225 x 115 x 75mm)	1000	55 000
Hollow concrete blocks (450 x 225 x 225mm)	1000	105 000
Precast concrete cladding units with exposed aggregate finish	m^2	3200
Timber and insulation		
Softwood sections for carpentry	m^3	81 390
Softwood for joinery	m^3	95 000
Hardwood for joinery	m^3	185 000
Exterior quality plywood (12.5mm)	m^2	550
Plywood for interior joinery (7.5mm)	m^2	350
Softwood strip flooring (24mm)	m^2	6500
Chipboard sheet flooring (3.2mm)	m^2	1000
100mm thick quilt insulation	m^2	500
Glass and ceramics		
Float glass	m^2	850
Sealed double glazing units	m^2	8500
Plaster and paint		
Plaster in 50 kg bags	tonne	5750
Plasterboard (13mm thick)	m^2	675
Emulsion paint in 5 litre tins	litre	410
Gloss oil paint in 5 litre tins	litre	650
Tiles and paviors		
Clay floor tiles (150 x 150mm)	m^2	2110
Vinyl floor tiles (200 x 200 x 2.5mm)	m^2	2475
Precast concrete paving slabs (600 x 600 x 50mm)	m^2	2700
Clay roof tiles	1000	41 700
Precast concrete roof tiles	1000	65 500

Drainage

	Units	Cost Esc
WC suite complete	nr	16 000
Lavatory basin complete	nr	12 500

Unit rates

The descriptions below are generally shortened versions of standard descriptions listed in five languages (English, French, Italian, German and Spanish) in section 6. Where an item has a two digit reference number (e.g. 05 or 33), this relates to the full description against that number in section 6. Where an item has an alphabetic suffix (e.g. 12A or 34B) this indicates that the standard description has been modified. Where a modification is major the complete modified description is included here and the standard description should be ignored; where a modification is minor (e.g. the insertion of a named hardwood) the shortened description has been modified here but, in general, the full description in section 6 prevails.

The unit rates below are for main work items on a typical construction project in the Lisbon area in January 1991. The rates include all necessary labour, materials, equipment and, where appropriate, allowances for contractors overheads and profit, preliminary and general items and contractors profit and attendance on specialist trades. All the rates in this section exclude value added tax (VAT - see below).

		Unit	Rate Esc
Excavation			
01	Mechanical excavation of foundation trenches	m^3	1344
02	Hardcore filling making up levels	m^2	672
03	Earthwork support	m^2	874
Concrete work			
04	Plain insitu concrete in strip foundations in trenches	m^3	10 368
05	Reinforced insitu concrete in beds	m^3	11 508
06	Reinforced insitu concrete in walls	m^3	11 520
07	Reinforced insitu concrete in suspended floor or roof slabs	m^3	12 036
08	Reinforced insitu concrete in columns	m^3	12 168
09	Reinforced insitu concrete in isolated beams	m^3	12 048
10	Precast concrete slab	m^2	3624
Formwork			
11	Softwood formwork to concrete walls	m^2	2460
12	Softwood or metal formwork to concrete columns	m^2	1296
13	Softwood or metal formwork to horizontal soffits of slabs	m^2	2976

		Unit	Rate Esc
Reinforcement			
14	Reinforcement in concrete walls	tonne	125 760
15	Reinforcement in suspended concrete slabs	tonne	121 920
16	Fabric reinforcement in concrete beds	m^2	672
Brickwork and blockwork			
18	Precast lightweight aggregate hollow concrete block walls	m^2	2500
19	Solid (perforated) concrete blocks	m^2	2400
20	Sand lime (perforated) bricks	m^2	4500
21	Facing bricks	m^2	5500
Roofing			
22	Concrete interlocking roof tiles 430 x 380mm	m^2	2688
23	Plain clay roof tiles 260 x 160mm	m^2	6384
24	Fibre cement roof slates 600 x 300mm	m^2	6048
25	Sawn softwood roof boarding	m^2	1044
26	Particle board roof coverings	m^2	2928
27	3 layers glass-fibre based bitumen felt roof covering	m^2	3960
28	Bitumen based mastic asphalt roof covering	m^2	2112
29	Glass-fibre mat roof insulation 160mm thick	m^2	1332
30	Rigid sheet loadbearing roof insulation 75mm thick	m^2	1998
31	Troughed galvanized steel roof cladding	m^2	3732
Woodwork and metalwork			
32	Preservative treated sawn softwood 50mm x 100mm	m	1734
33	Preservative treated sawn softwood 50 x 150mm	m	269
34	Single glazed casement window in hardwood, size 650 x 900mm	each	23 976
35A	Two panel glazed door in hardwood, size 850 x 2000mm	each	51 072
36	Solid core half hour fire resisting hardwood internal flush doors, size 800 x 2000mm	each	53 520
37	Aluminium double glazed window, size 1200 x 1200mm	each	42 432
38	Aluminium double glazed door, size 850 x 2100mm	each	53 040
39A	Hardwood skirtings 12 x 44mm	m	459
40	Framed structural steelwork in universal joist sections	tonne	118 260
41	Structural steelwork lattice roof trusses	tonne	239 112
Plumbing			
42	UPVC half round eaves gutter	m	1196
43	UPVC rainwater pipes	m	1327
44	Light gauge copper cold water tubing	m	1074
45	High pressure plastic for cold water supply	m	798
46	Low pressure plastic pipes for cold water distribution	m	659

		Unit	Rate Esc
47	UPVC soil and vent pipes	m	1404
48	White vitreous china WC suite	each	26 256
49	White vitreous china lavatory basin	each	18 384
50	Glazed fireclay shower tray	each	19 692
51	Stainless steel single bowl sink and double drainer	each	13 104
Electrical work			
52	PVC insulated and copper sheathed cable	m	484
53	13 amp unswitched socket outlet	each	2618
54	Flush mounted 20 amp, 1 way light switch	each	3803
Finishings			
55	2 coats gypsum based plaster on brick walls	m^2	790
56	White glazed tiles on plaster walls	m^2	5309
57	Red clay quarry tiles on concrete floor	m^2	3920
58	Cement and sand screed to concrete floors	m^2	883
59	Thermoplastic floor tiles on screed	m^2	2374
60	Mineral fibre tiles on concealed suspension system	m^2	3971
Glazing			
61	Glazing to wood	m^2	1618
Painting			
62	Emulsion on plaster walls	m^2	426
63	Oil paint on timber	m^2	974

Approximate estimating

The building costs per unit area given below are averages incurred by building clients for typical buildings in the Lisbon area as at January 1991. They are based upon the total floor area of all storeys, measured between external walls and without deduction for internal walls.

Approximate estimating costs generally include for mechanical and electrical installations but exclude furniture, loose or special equipment, and external works; they also exclude fees for professional services. The costs shown are for specifications and standards appropriate to Portugal and this should be borne in mind when attempting comparisons with similarly described building types in other countries. A discussion of this issue is included in section 4. Comparative data for countries covered in this publication, including construction cost data are presented in Part Three.

Approximate estimating costs must be treated with caution; they cannot provide more than a rough guide to the probable cost of building.

All the rates in this section exclude value added tax (VAT - see below).

	Cost m^2 *Esc*	*Cost* ft^2 *Esc*
Industrial buildings		
Factories for letting	44 300	4100
Factories for owner occupation (light industrial use)	65 500	6090
Factory/office (high tech) for owner occupation (controlled environment, fully finished)	75 000	6970
Warehouses, low bay (6 to 8m high) for letting (no heating)	44 300	4100
Administrative and commercial buildings		
Offices for letting, 5 to 10 storeys, non air conditioned	87 600	8140
Prestige/headquarters office, 5 to 10 storeys, air conditioned	140 000	13 000
Health and education buildings		
General hospitals	125 000	11 610
Nursery schools	96 000	8920
Primary/secondary schools	90 000	8360
Recreation and arts buildings		
Sports halls including changing and social facilities	94 000	8730
Swimming pools (international standard) including changing and social facilities	105 000	9750
Residential buildings		
Private/mass market single family housing 2 storey detached/semidetached (multiple units)	68 000	6300
Private sector apartment building (standard specification)	85 000	7890

Value added tax (VAT)

The standard rate of value added tax (VAT) is currently 25%, chargeable on general building work.

EXCHANGE RATES AND INFLATION

The combined effect of exchange rates and inflation on prices within a country and price comparisons between countries is discussed in section 4.

Exchange rates

The graph below plots the movement of the Portuguese Escudo against the American dollar and sterling since 1980. The figures used for the graph are quarterly and the method of calculating these, and other related issues are discussed in section 4. The exchange rate at January 1991 was Esc 135 to the US$ and Esc 261 to the pound sterling.

The Portuguese Escudo against the Sterling and the US Dollar

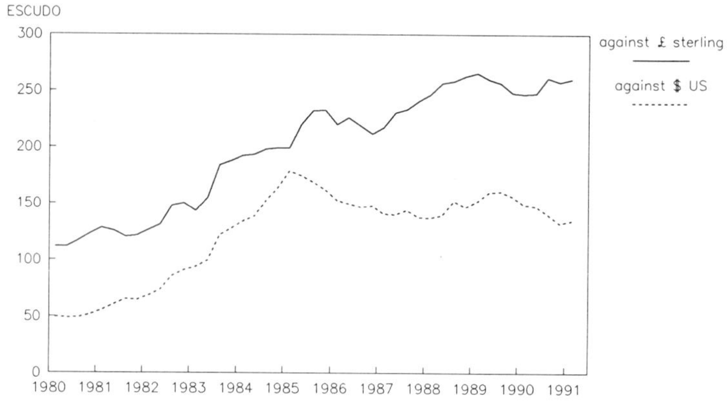

Inflation

The graph and the table below present the consumer price index showing the trend in inflation since 1980. The index has been rebased to 1980=100. Notes on inflation and the distinction between costs and prices are included in section 4.

Consumer price inflation

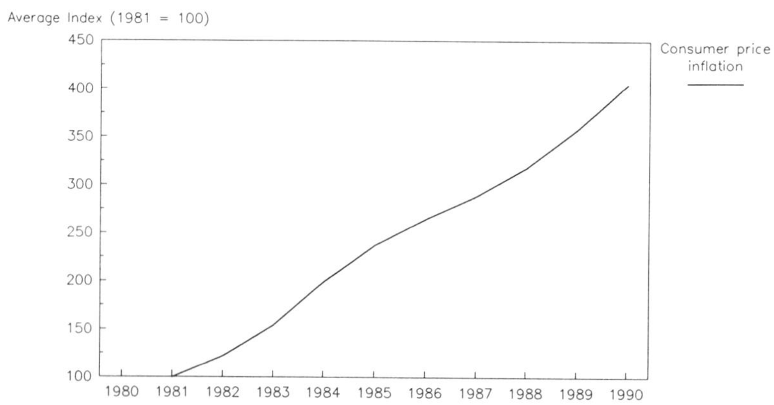

Consumer price inflation

The basis of the figures is the official consumer price index converted to quarterly averages. Inflation peaked in 1984; since then some progress has been made in reducing inflation but by 1990 the annual rate was still one of the highest in the EC at just over 13%.

Notes on inflation are included on section 4.

Consumer price inflation 1981 = 100

	1982	1983	1984	1985	1986	1987	1988	1989	1990
average index	122	154	199	237	265	290	318	358	406
annual change (%)	22.4	25.5	29.3	19.3	11.7	9.4	9.7	12.6	13.4

USEFUL ADDRESSES

Ministério do Planeamento
(Planning Ministry)
Praça do Comércio
1110 Lisboa
Tel: 010 351 369394

Ministério das Obras Públicas
(Public Works Ministry)
Praça do Comércio
1110 Lisboa
Tel: 010 351 369394

Laboratório Nacional de Engenharia Civil (LNEC)
(National Civil Engineering Laboratory)
Avenue do Brasil 101
1700 Lisboa
Tel: 010 351 882131
Fax: 897660

Associaçao de Empresas de Construçao e Obras Públicas do Sul (AECOPS)
(Contractor's Association - South)
Rua António Enes
1000 Lisboa
Tel: 010 351 533193
Fax: 562816

Associaçao dos Arquitectos Portuguese
(Association of Portuguese Architects)
Av. 24 de Julho, 52-1° Esq
1200 Lisboa
Tel: 010 351 543512
Fax: 543667

Ordem dos Engenheiros
(The "Order" of Engineers)
Av. António Augusta Aguiar 3-D
1000 Lisboa
Tel: 010 351 562438

Instituto Portugues para a Qualidade
(Portuguese Quality Assurance Association)
Rua José Estevao, 83-1-1/3D
1100 Lisboa
Tel: 010 351 523978
Fax: 530033

Instituto Nacional de Estatistica
(National Statistical Institute)
Avenida António José Almeida
1000 Lisboa
Tel: 010 351 8470050
Fax: 8489480

Associaçao Portuguesa de Projectistas e Consultores, (APPC)
(Portuguese Consultants Association)
Av. António Augusta Aguiar 126,7
1000 Lisboa

Associaçao dos Industriais da Construçao Civil e Obras Públicas do Norte (AICCOPN)
(Constractor's Assocation - North)
Rua Alvares Cabral 306
Porto

Sindicato Nacional Dos Arquitectos
(National Syndicate of Architects)
Rua Barata Salgueiro 36
Lisboa 2

Grupo Portugues De Betao Manufacturado
Rua Da Filipa De Vihena 9-2D
1400 Lisboa

Spain

KEY DATA

Population

Population (1989)	39.1m
Urban population	77%
Population under 15	20.4%
Population over 65	13.0%
Average annual growth rate (1983 to 1988)	0.5%

Geography

Land area	505 000 km^2
Agricultural area	308 000 km^2
Capital city	Madrid (population 4 731 224)

Economy

Monetary unit	Pesetas (Pta)
Exchange rate (January 1991) to	
the ECU	Pta 129
the pound sterling	Pta 184
the US dollar	Pta 95
Average annual inflation (1984 to 1989)	6.9%
Gross Domestic Product (GDP)	Pta 39 440bn
GDP per capita (1989)	Pta 1.009m
GDP growth (real) (1980 to 1988)	2.6%
Private consumption (% of total)	63.0%
Public consumption (% of total)	14.3%
Investment (% of total)	23.9%
Central government expenditure as percentage of GDP	36.1%

Construction

Total value of construction output (1989)	Pta 5537bn
Construction output per capita	Pta 141 970
Construction output as a percentage of GDP	14%

THE CONSTRUCTION INDUSTRY

Construction output

In 1989 the value of the gross output of the Spanish construction industry, including new work, renovation and modernization, was about Pta 5537 billion, equivalent to about 42 billion ECUs. A further Pta 800 billion of construction work may be produced informally by other sectors. The industry's gross output represents about 14% of GDP. Output has been growing strongly since 1985 prior to which the industry had been in a state of depression; it is expected to continue its rapid growth. Of the total output, 76% is new work. The broad divisions for new work and renovation and modernization are shown below.

Construction output 1989

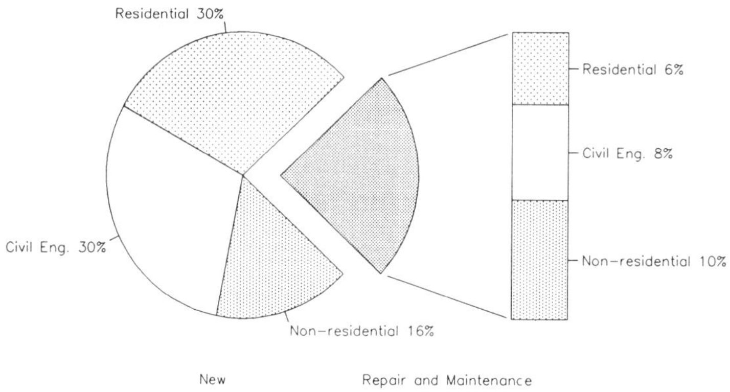

The outstanding feature of the distribution by type of work is the very high level of new civil engineering work - one of the highest in Europe. This is due to the recent rapid development of Spain and the emphasis on the improvement of infrastructure. The share of non-residential building in total output, on the other hand, is low with over 75% in industrial, office and commercial buildings.

Characteristics and structure of the industry

The professions, particularly architects, have a highly protected position in Spain. Every profession has its own *colegio* (college) to which all professionals must belong and the law defines their duties, practices and fee scales. It is obligatory to employ an architect on all building projects; he is responsible for design and supervision, and he employs the other professionals. In addition, there are technical architects who are formally responsible for site organization, quality control and safety, but also undertake many of the functions of a quantity surveyor. Private sector architectural practices are nearly all very small.

There are 17 *Colegios Oficiales de Arquitectura* (COAs), one in each of the autonomous communities. The COAs are represented by the *Consejo Superior de los Colegios de Arquitectura de España* (CSCA) which is a very influential body with offices in Madrid. It represents the interests of COAs and hence architects at a national and international level. The COAs are substantial organizations whose functions, in addition to those relating to the conduct, rights and interests of architects, include the control and approval (*visado*) of all project documents and the channelling of all payments to architects. This means that they can ensure that clients actually pay architects and that bad debtors cannot engage an architect. It also means that they are an integral part of the building control process and, by their certification that documents are prepared in compliance with the rules, ensure that architects conform to planning and building regulations.

Engineers do not have an equivalent statutory role. Some 40% to 50% of engineers are employed in the public sector, working in client organizations. The remainder are in the private sector. Some of the engineering firms are quite substantial and indeed most are stock companies, often with participation from banks, contractors or suppliers. Engineering consultancy firms grew in the 1960s and 1970s. In 1988, Tecniberia, the main engineering association, had 74 corporate members. These were mainly engineering companies but also included some management consultants. Together they employed 13 000 people of whom some 4000 were engineers.

A new law on the building process - *Ley de Ordenación de la Edificación* (LOE) is in draft and is intended to resolve the controversy over the various roles of the professions.

All contractors operating in Spain must be legally registered and hold a licence. To qualify for public contracts, contractors must also be classified by the Treasury Ministry according to types of work, size of contract and maximum value of contracts at any one time. Contractors from other European Community states need not be so qualified, but must present the documents to show their status in their own countries, agree to submit to Spanish law, and furnish proof of financial and technical capability.

The contracting industry is dominated by a large number of strong and successful contractors. The top ten firms contribute 20% of construction activity and the top 20 over a quarter. They have some export business but only two feature in Engineering News Record's "Top 250 International Contractors": Dragados and Cubiertas. They normally operate in both the building and civil engineering fields and as developers. The larger contractors are organizing themselves to undertake projects as managing contractors and their share of total construction employment is falling as result. Subcontracting is increasing.

Major Spanish contractors

Major contractors	*Head office*	*Place in Building's "Top 300 European Contractors"*	*Exports (% of turnover)*
Dragados y Construcciónes	Madrid	25	approx 10%
Cubiertas y Mzov SA	Barcelona	39	approx 15%
Ferrovial	Madrid	41	
Agroman Empresa Constructora SA	Madrid	43	some
Obras y Construcciónes Industriales SA (Ocisa)	Madrid	57	none
Construcciónes y Contratas SA (CONYCON)	Madrid	73	none

Source: Building

The complexity of requirements for the operations of construction firms and companies has led to the development of firms of *gestores administradores* which specialize in obtaining and filling in forms and obtaining permits.

Clients and finance

Public sector projects, including virtually all infrastructure, are financed mainly from public funds. The Ministry of Public Building and Works - *Ministerio de Obras Públicas* y del Urbanismo (MOPU) - is the main financing agency with over 30% of the total. The situation for 1987 is shown in the table below.

Public expenditure on infrastructure, 1987

Authority	*Value of contracts (pta billion)*	*Percentage of total (%)*	*Main types of work*
MOPU	229	32	roads (192), water, ports
Ministry of Transport	121	17	railways, air transport
Other Ministries	55	8	education, health
Autonomous Communities	220	30	local infrastructure
Local Authorities	99	14	housing
Total	724	100	

In March 1989 the Government issued Spain's first Regional Development Plan for 1989 to 1993 which consists of a public investment programme of some Pta 8020 billion. Just under 22% (Pta 1759 billion) is expected to be funded from European Community structural funds, mainly from the European Regional Development Fund. The money will be split roughly equally between central and regional governments. Nearly half the central government total will be spent on roads.

Housing is largely financed by banks and savings banks although there is some small, direct, local authority involvement. Most housing is constructed for sale under government-subsidized loan schemes for those earning less than 2½ times the minimum wage. It is estimated that subsidies on average represent 10% of the cost of this housing.

Nearly all commercial property is built for sale or for end-user clients. The normal pattern of speculative development is for developer companies to buy a site and sell property from plans, commencing construction only when the cash from sales is sufficient to cover financing costs. Construction is financed from purchasers' deposits and bank loans. Recently there has been a shift towards building commercial property for rent, in which case insurance companies and banks are usually the owners.

Selection of design consultants

The first professional to be employed by the client in the private sector is the architect. He appoints the other professionals, except that the client is required by law also to appoint an independent technical architect to carry out the site supervision of quality and measurement. However, this is actually done in only about 20% of projects.

Selection of the architect is almost always determined by past working relations or personal recommendation. Since there is little private owner-occupier development, clients are usually professionally competent private developers or in the public sector. Because of the regulatory powers of the COA, once an architect has been appointed it is very difficult to change the appointment. Any extensions or rehabilitation work have to be offered to the architect for the original building.

On public civil engineering projects, works are designed within the client body or by a consultant firm acting as a design bureau to whom the design package is let, usually by competitive tender. The contracting authority's own representative engineer is responsible for supervising the works and appoints a quality control company to carry out tests on materials and to monitor sites.

Contractual arrangements

Contracts are awarded to a general contractor who then sub-contracts as necessary. Public contracts are regulated by:

- the Law of State Contracts (LCE) set out in decree 923/1965, law 51/1973, law 50/1984 and decree 931/1986
- the General Regulations for State Contracts (RCE) set out in decree 2410/1975
- the General Administrative Clauses (PCAG) set out in decree 3854/1970.

These provisions are contained in a book published by *Boletin Oficial del Estado* (BOE). The three methods of tendering for public works are outlined below.

1. *Subasta* (auction): Tenderers are provided with documents which include the official estimate containing quantities, rates, extensions and totals. Each tenderer undertakes his own measurement and provides his own rates and prices in order to arrive at his tender figure which is quoted as a discount on the official estimate. Normally the contract will be awarded to the lowest tenderer though if a tender is more than 10% below the average discount offered by the other tenderers, it is judged to be "imprudently low" and automatically disqualified.

2. *Concurso* (competition): In this case the contract is awarded to the most suitable contractor based on an assessment of his record, experience, personnel, plant, equipment, etc., and his tender figure. This method can be used where the contractor's work is to include design.

3. *Contratación directa* (negotiation): In this method the government body concerned will calculate the contract price and will negotiate with a suitable contractor to carry out the work.

Both *subasta* and *concurso* may be "open" in which any qualified contractor may bid, or "restricted" in which it is confined to those already successful in a prequalification stage. In 1987 *subasta* accounted for 44% of value of contracts, of which about two thirds were restricted. Open *subasta* is mainly used by local authorities for rather small contracts. *Concurso* accounts for over half of the value of all contracts but less than 20% of the number of contracts. *Contratación directa* accounts for only 4% of value but nearly a quarter of the number of contracts. There has been a large increase in the 1980s in competitions at the expense of *subasta.*

The documents which must be included in a public project tender are specified in Decree 2.410 dated 25 November 1975. These are:

- conditions of contract
- specifications
- general description of the works
- priced bill of quantities
- contract programme
- drawings
- contractor's qualification details.

Also covered by regulations within the same decree are procedures to be followed in the awarding of contracts, setting out and approvals, penalties for delays in completion, credits to the contractor, variations, determination, handing over, final account, sub-contracts, dayworks and other conditions of contract. Interim payments are made using the official estimate as a basis, rates being adjusted by the appropriate percentage to take into account the difference between the tender figure and the official estimate total.

For private sector work the selection of a contractor is often influenced by the cross holdings of shares in companies or by other obligations. The architect is also very influential.

Private contracts are not subject to the same degree of regulation as contracts for public works, most contract conditions being to some extent negotiable. Depending on the nature of the project, tenders are usually required to be of one of the following types:

- a simple financial bid indicating the length of the contract period and accepting the project documents and conditions
- a tender which also includes the contractor's priced bill of quantities
- a tender which includes both the above features together with technical proposals for financial savings.

Tenders may be "open" in which case quantities are remeasured and valued on completion of the contract or "closed" where the bill of quantities is taken as correct, the contractor's lump sum only being affected by the value of variations. A contractor will only submit his tender after discussions with the client and the technical personnel involved. For private contracts the contract documents would typically consist of:

- contract conditions
- specifications
- general description of the works
- method of measurement
- bills of quantities (possibly including clients' prices)
- drawings.

Contract disputes which cannot be settled by discussion between the client's representative and the contractor may be referred to an independent arbitrator. More serious disputes are dealt with in the courts of justice.

Liability, insurance and inspection

The Spanish Civil Code determines liability but it applies only to the main contractor and the architect and technical architect who share responsibility between them. There is a ten year liability for serious defects - these include those which pose a threat of serious damage to any part of the building or render it unsuitable for its intended use. Liability does not require any proof of negligence. It is said that, in practice, tribunals usually allocate liability to the professionals unless the contractor has deliberately disobeyed an architect's instruction. Architects believe they carry an unfair share of the risk because they are insured and contractors are not.

Architects take responsibility for all design matters, and the technical architect bears a secondary responsibility for all quality control aspects of the contractor's work and materials on site. The architect is liable for the whole design even if part was carried out by a structural engineer. Recently a developer has also been made responsible with the contractor and has been found liable for defects to the owner when the builder went out of business.

In the case of civil engineering works, the civil engineer normally carries the responsibility assigned to the architect in the Civil Code, but this is a matter of contractual liability and not a Civil Code obligation.

The time limitation for initiating claims is 15 years ie 5 years beyond the liability period. This also applies, unlike the UK and other countries, when there has been fraud or concealment. When a defect has been remedied, the 10-year decennial period begins to run again.

In *viviendes de protección oficial* (VPO), or subsidized housing built by private developers (for sale to low-income households with government-subsidized loans and grants), the developer has special responsibilities and must provide a 5-year guarantee, during which period any repairs resulting from construction defects must be carried out. A claim may then be made against the architect or contractor if appropriate.

Both architects and technical architects have public liability insurance arranged through their COAs. This covers the decennial risk on an annual premium basis but, if the individual stops practising and does not pay premiums, no cover is available and plaintiffs have no way of obtaining redress. All architects pay a standard premium, related to the total limit of cover, that is independent of their claims record or experience and, also independent of their level of fee income. Insurance for architects and technical architects is underwritten by mutual insurance companies and by private insurance companies.

There is no obligation to carry insurance cover, and some professionals who are not in independent practice or have low workloads do not bother to insure.

Contractors do not carry any decennial insurance cover, but generally hold an all-risks policy which covers their public liability on site, fire and accident, and their risks during the guarantee period of 12 or 6 months.

Some insurance companies have also begun offering single-premium project decennial insurance policies along French lines, and large projects by international developers now generally appear to be covered by these. This is complicating the professional roles, since now the insurance companies' risk control bureau takes over part of the quality control role of the client's independent technical architect. The liability is then unclear and some technical architects say that, contrary to the intention, this may lead to a reduction in the quality of building.

Development control and standards

The seventeen autonomous communities have ultimate responsibility for planning and control of construction. Powers are exercised by local authorities of which there are some 8000 grouped into 52 provinces.

Building permits are required for all works including even quite small extensions, and an architect needs to be involved. The operation of the system is not very clear and where there is no plan, the criteria for granting a building permit are not known in advance. The local authority is not concerned with technical regulations or standards - these are the sole responsibility of the architect.

Building regulations in Spain are passed principally by the national government but, since 1980, the autonomous communities also have the power to do so. The legislation relating to building covers technical requirements plus both the functions of professionals and of institutions such as the technical approvals body. In practice, there has been very little regional legislation, and it can be taken that, in general, national legislation applies.

Since 1977 the legislation has been divided explicitly into two classes:

- the basic norms (*normas basicas de la edificación* - NBEs) which are the only obligatory standards. NBEs can only be passed by royal decree and cover such things as structural steelwork, fire protection and seismic resistance

- the technical norms (*normas tecnologicas de la edificación* - NTEs) which are developed by MOPU or the autonomous communities, and are advisory but not obligatory.

An index to the legislation on building, the *Indice de Disposiciónes Relaciónadas con la Edificación* (latest edition 1987), is published by MOPU. There is also a compendium set of volumes of NTEs, but not of other regulations.

The Directorate-General of Architecture and Building Technology of MOPU produces or approves codes of practice (*Soluciónes Homologadas de Edificación* - SHEs) whose use guarantees meeting the minimum requirements of the NBEs and the Catalán Institute of Building Technology (*Instituto de Tecnologica de la Construccio de Cataluña* - ITEC) in Barcelona plays the main part in their development.

Product standards in Spain, for all industries, are set by the Spanish standards institution (*Asociación Español de Normalisación y Certificación* - AENOR), the equivalent of the British Standards Institute (BSI). A product approval mark - the "N" mark - has been established by AENOR. As yet, very few construction products have passed through the certification and approval process for "N" marks. There are, however, two other schemes for quality marks - one for reinforcement bars and one for prefabricated prestressed concrete beams. Both these are administered by the research and *agrément* body, the *Instituto Eduardo Torroja de la Construcción y del Cemento* (IETCC), and not by AENOR.

There is, in general, no legal requirement to use approved products, and no import restriction on products which do not meet or are not approved to the Spanish standards *Una Normas Española* (UNE). Neither do the existing insurance arrangements require the use of approved products. Since product suppliers do not carry any legal liability for building defects, the use or otherwise of approved products has no liability implications, except for the architect and technical architect. Many of the UNE standards for steel and concrete plus those for design and testing procedures are, however, embodied in the obligatory NBE regulations.

Research and development

Research on materials (cement, wood etc) is undertaken in specific institutes under the aegis of a National Body for Scientific Research (*Patronato de Investigaciones*

Cientificas) with state funding. Building standards are prepared by technical committees of experts in each field drawn from government research associations and private industries.

Further reading

S D Reynolds and S Sheppard, The Iberian Construction Industry : A Guide for UK Professionals, London: CIRIA, 1989.

RIBA, Architectural Practice in Europe : Spain, London : RIBA, 1991.

Association of Consulting Engineers et al, Engineering Consultancy in the European Community : Spain, London : Telford, 1990.

CONSTRUCTION COST DATA

Cost of labour

The figures below are typical of labour costs in Barcelona as at January 1991. The "wage rate" is the basis of an employee's income, while the "cost of labour" indicates the cost to a contractor of employing that employee. The difference between the two covers a variety of mandatory and voluntary contributions - a list of items which could be included is given in section 4.

	Wage rate (per hour) Pta	*Cost of labour (per hour) Pta*	*Number of hours worked per year*
Site operatives			
Mason/bricklayer	800	1 671	1850
Carpenter	850	1 766	1750
Plumber	920	1 913	1850
Electrician	920	1 913	1850
Structural steel erector	830	1 750	1850
HVAC installer	920	1 913	1850
Semi-skilled worker	780	1 591	1850
Unskilled labourer	700	1 428	1850
Equipment operator	830	1 750	1850
Watchman/security	780	1 591	1850
	(per year)	(per year)	
Site supervision			
General foreman	3 500 000	5 250 000	1950
Trades foreman	3 000 000	4 500 000	1950
Clerk of works	3 000 000	4 500 000	1950

	Wage rate (per year) Pta	*Cost of labour (per year) Pta*	*Number of hours worked per year*
Contractors' personnel			
Site manager	4 500 000	6 750 000	1950
Resident engineer	4 500 000	6 750 000	1950
Resident surveyor	4 000 000	6 000 000	1950
Junior engineer	3 000 000	4 500 000	1950
Junior surveyor	2 400 000	3 600 000	1950
Planner	3 000 000	4 500 000	1950
Consultants' personnel			
Senior architect	6 000 000	9 000 000	1850
Senior engineer	6 000 000	9 000 000	1850
Senior surveyor	4 500 000	6 750 000	1850
Qualified architect	4 000 000	6 000 000	1850
Qualified engineer	4 000 000	6 000 000	1850
Qualified surveyor	3 000 000	4 500 000	1850

Cost of materials

The figures that follow are the costs of main construction materials, delivered to site in the Barcelona area, as incurred by contractors in January 1991. These assume that the materials would be in quantities as required for a medium sized construction project and that the location of the works would be neither constrained nor remote.

All costs exclude value added tax (VAT - see below).

	Units	*Cost Pta*
Cement and aggregate		
Ordinary portland cement in 50kg bags	tonne	13 000
Coarse aggregates for concrete	m^3	2400
Fine aggregates for concrete	m^3	2200
Ready mixed concrete	m^3	8150
Steel		
Mild steel reinforcement	tonne	82 000
Structural steel sections	tonne	110 000
Bricks and blocks		
Common bricks (100 x 130 x 270mm)	1000	24 000
Good quality facing bricks (50 x 130 x 270mm)	1000	50 000
Hollow concrete blocks (200 x 400 x 200mm)	1000	110 000
Solid concrete blocks (210 x 600 x 200mm)	1000	160 000
Precast concrete cladding units with exposed aggregate finish	m^2	2750

	Units	*Cost Pta*
Timber and insulation		
Softwood sections for carpentry	m^3	72 000
Softwood for joinery	m^3	110 000
Hardwood for joinery	m^3	150 000
Exterior quality plywood (10mm)	m^2	500
Plywood for interior joinery (10mm)	m^2	280
Softwood strip flooring (18mm)	m^2	5500
Chipboard sheet flooring (22mm)	m^2	1250
100 mm thick quilt insulation	m^2	440
100 mm thick rigid slab insulation	m^2	2980
Softwood internal door complete with frames and ironmongery	nr	16 300
Glass and ceramics		
Float glass (6mm)	m^2	1950
Sealed double glazing units	m^2	7930
Good quality ceramic wall tiles (200 x 300mm)	m^2	2400
Plaster and paint		
Plaster in 50 kg bags	tonne	6500
Plasterboard (15mm thick)	m^2	590
Emulsion paint in 5 litre tins	litre	325
Gloss oil paint in 5 litre tins	litre	575
Tiles and paviors		
Clay floor tiles (300 x 300 x 10mm)	m^2	1850
Vinyl floor tiles (300 x 300 x 2mm)	m^2	2370
Precast concrete paving slabs (600 x 600 x 80mm)	m^2	2430
Clay roof tiles	1000	36 400
Precast concrete roof tiles	1000	53 000
Drainage		
WC suite complete	nr	20 000
Lavatory basin complete	nr	18 400
100mm diameter clay drain pipes	m	960
150mm diameter cast iron drain pipes	m	4420

Unit rates

The descriptions below are generally shortened versions of standard descriptions listed in five languages (English, French, Italian, German and Spanish) in section 6. Where an item has a two digit reference number (e.g. 05 or 33), this relates to the full description against that number in section 6. Where an item has an alphabetic suffix (e.g. 12A or 34B) this indicates that the standard description has been modified. Where a modification is major the complete modified description is included here and the standard description should be ignored; where a modification is minor (e.g. the insertion of a named hardwood) the shortened description has been modified here but, in general, the full description in section 6 prevails.

The unit rates below are for main work items on a typical construction project in the Barcelona area in January 1991. The rates include all necessary labour, materials, equipment and, where appropriate, allowances for contractors overheads and profit, preliminary and general items and contractors profit and attendance on specialist trades. All the rates in this section exclude value added tax (VAT - see below).

		Unit	*Rate Pta*
Excavation			
01	Mechanical excavation of foundation trenches	m^3	1450
02	Hardcore filling making up levels	m^2	360
03	Earthwork support	m^2	1550
Concrete work			
04	Plain insitu concrete in strip foundations in trenches	m^3	11 340
05	Reinforced insitu concrete in beds	m^3	12 420
06	Reinforced insitu concrete in walls	m^3	11 750
07	Reinforced insitu concrete in suspended floor or roof slabs	m^3	12 610
08	Reinforced insitu concrete in columns	m^3	11 230
09	Reinforced insitu concrete in isolated beams	m^3	11 630
10	Precast concrete slab	m^2	3630
Formwork			
11	Softwood formwork to concrete walls	m^2	2750
12	Softwood or metal formwork to concrete columns	m^2	1880
13	Softwood or metal formwork to horizontal soffits of slabs	m^2	2430
Reinforcement			
14	Reinforcement in concrete walls	tonne	130 000
15	Reinforcement in suspended concrete slabs	tonne	126 000
16	Fabric reinforcement in concrete beds	m^2	440
Steelwork			
17	Fabricate, supply and erect steel framed structure	tonne	178 000
Brickwork and blockwork			
18	Precast lightweight aggregate hollow concrete block walls	m^2	3250
19	Solid (perforated) concrete blocks	m^2	2650
20	Solid (perforated) sand lime bricks	m^2	3850
21	Facing bricks	m^2	3810

		Unit	Rate Pta
Roofing			
22	Concrete interlocking roof tiles 430 x 380mm	m^2	6950
23	Plain clay roof tiles 260 x 160mm	m^2	7850
24	Fibre cement roof slates 600 x 300mm	m^2	2950
25	Sawn softwood roof boarding	m^2	3750
26	Particle board roof coverings	m^2	3140
27	3 layers glass-fibre based bitumen felt roof covering	m^2	3430
28	Bitumen based mastic asphalt roof covering	m^2	2420
29	Glass-fibre mat roof insulation 160mm thick	m^2	1250
30	Rigid sheet loadbearing roof insulation 75mm thick	m^2	1850
31	Troughed galvanized steel roof cladding	m^2	4450
Woodwork and metalwork			
32	Preservative treated sawn softwood 50mm x 100mm	m	1440
33	Preservative treated sawn softwood 50 x 150mm	m	1420
34A	Single glazed casement window in Tea Pine hardwood, size 650 x 900mm	each	15 300
35A	Two panel glazed door in Tea Pine hardwood, size 850 x 2000mm	each	35 450
36A	Solid core half hour fire resisting hardwood internal flush doors, size 800 x 2000mm	each	34 200
37A	Aluminium double glazed window, size 1200 x 1200mm	each	26 400
38A	Aluminium double glazed door, size 850 x 2100mm	each	31 300
39A	Hardwood skirtings	m	1360
40	Framed structural steelwork in universal joist sections	tonne	190 000
41	Structural steelwork lattice roof trusses	tonne	185 000
Plumbing			
42	UPVC half round eaves gutter	m	2640
43	UPVC rainwater pipes	m	2430
44	Light gauge copper cold water tubing	m	964
45	High pressure plastic for cold water supply	m	430
46	Low pressure plastic pipes for cold water distribution	m	290
47	UPVC soil and vent pipes	m	2480
48	White vitreous china WC suite	each	29 400
49	White vitreous china lavatory basin	each	24 600
50	Glazed fireclay shower tray	each	21 700
51	Stainless steel single bowl sink and double drainer	each	30 600
Electrical work			
52	PVC insulated and copper sheathed cable	m	2210
53	13 amp unswitched socket outlet	each	5625
54	Flush mounted 20 amp, 1 way light switch	each	6430

	Unit	Rate Pta
Finishings		
55 2 coats gypsum based plaster on brick walls	m^2	950
56 White glazed tiles on plaster walls	m^2	2450
57 Red clay quarry tiles on concrete floor	m^2	4150
58 Cement and sand screed to concrete floors	m^2	1950
59 Thermoplastic floor tiles on screed	m^2	3450
60A Mineral fibre tiles on concealed suspension system	m^2	3760
Glazing		
61 Glazing to wood	m^2	2950
Painting		
62 Emulsion on plaster walls	m^2	510
63 Oil paint on timber	m^2	1750

Approximate estimating

The building costs per unit area given below are averages incurred by building clients for typical buildings in the Barcelona area as at January 1991. They are based upon the total floor area of all storeys, measured between external walls and without deduction for internal walls.

Approximate estimating costs generally include for mechanical and electrical installations but exclude furniture, loose or special equipment, and external works; they also exclude fees for professional services. The costs shown are for specifications and standards appropriate to Spain and this should be borne in mind when attempting comparisons with similarly described building types in other countries. A discussion of this issue is included in section 4. Comparative data for countries covered in this publication, including construction cost data are presented in Part Three.

Approximate estimating costs must be treated with reserve; they cannot provide more than a rough guide to the probable cost of building.

	Cost m^2 Pta	Cost ft^2 Pta
Industrial buildings		
Factories for letting	29 000	2700
Factories for owner occupation (light industrial use)	30 000	2800
Factories for owner occupation (heavy industrial use)	33 000	3100
Factory/office (high-tech) for letting (shell and core only)	42 000	3900
Factory/office (high-tech) for letting (ground floor shell, first floor offices)	45 000	2700

	Cost m^2 Pta	Cost ft^2 Pta
Factory/office (high tech) for owner occupation (controlled environment, fully finished)	47 000	4400
High tech laboratory (air conditioned)	85 000	7900
Warehouses, low bay (6 to 8m high) for letting (no heating)	29 000	2700
Warehouses, low bay for owner occupation (including heating)	35 000	3300
Warehouses, high bay for owner occupation (including heating)	40 000	3700
Cold stores/refrigerated stores	50 000	4600
Administrative and commercial buildings		
Civic offices, non air conditioned	70 000	6500
Civic offices, fully air conditioned	80 000	7400
Offices for letting, 5 to 10 storeys, non air conditioned	55 000	5100
Offices for letting, 5 to 10 storeys, air conditioned	65 000	6000
Offices for letting, high rise, air conditioned	70 000	6500
Offices for owner occupation 5 to 10 storeys, non air conditioned	62 000	5800
Offices for owner occupation 5 to 10 storeys, air conditioned	78 000	7200
Offices for owner occupation, high rise, air conditioned	85 000	7900
Prestige/headquarters office, 5 to 10 storeys, air conditioned	100 000	9300
Prestige/headquarters office, high rise, air conditioned	110 000	10 000
Health and education buildings		
General hospitals (200 beds)	100 000	9300
Teaching hospitals (200 beds)	100 000	9300
Private hospitals (200 beds)	130 000	12 100
Health centres	90 000	8400
Nursery schools	85 000	7900
Primary/junior schools	75 000	7000
Secondary/middle schools	78 000	7200
University (arts) buildings	85 900	7900
University (science) buildings	85 000	7900
Management training centres	100 000	9300
Recreation and arts buildings		
Theatres (over 500 seats) including seating and stage equipment	85 000	7900
Theatres (less than 500 seats) including seating and stage equipment	90 000	8400
Concert halls including seating and stage equipment	110 000	10 200
Sports halls including changing and social facilities	85 000	7900
Swimming pools (international standard) including changing and social facilities	97 000	9000
Swimming pools (schools standard) including changing facilities	75 000	7000
National museums including full air conditioning and standby generator	180 000	12 100

	Cost m^2 Pta	Cost ft^2 Pta
Local museums including air conditioning	95 000	8800
City centre/central libraries	80 000	7400
Branch/local libraries	75 000	7000
Residential buildings		
Social/economic single family housing (multiple units)	48 000	4500
Private/mass market single family housing 2 storey detached/semidetached (multiple units)	55 000	5100
Purpose designed single family housing 2 storey detached (single unit)	75 000	7000
Social/economic apartment housing, low rise (no lifts)	40 000	3700
Social/economic apartment housing, high rise (with lifts)	45 000	4200
Private sector apartment building (standard specification)	65 000	6000
Private sector apartment buildings (luxury)	80 000	7400
Student/nurses halls of residence	70 000	6500
Homes for the elderly (shared accommodation)	65 000	6000
Homes for the elderly (self contained with shared communal facilities)	75 000	7000
Hotel, 5 star, city centre	125 000	11 600
Hotel, 3 star,city/provincial	80 000	7400
Motel	85 000	7900

Regional variations

The approximate estimating costs are based on projects in Barcelona. Adjust these costs by the following factors for regional variations:

Barcelona, Córdoba.	: 0%
Asturias, Salamanca, Murcia, Granada, Jaén, Badajos, Huelva, Cádiz, Almeria.	: -2%
Coruña, Lugo, Pontevedra, Orense, León, Palencia, La Rioja, Alava, Navarra, Huesca, Lérida, Gerona, Tarragona, Zaragoza, Teruel, Guadalajara, Guipuzcoa, Madrid, Avila, Cáceres, Toledo, Valencia, Cuidad Real, Albacete, Alicante, Sevilla, Málaga, Baleares.	: -4%
Zamora, Valladolid, Sora, Segovia, Burgos, Cantabría, Cuenca, Santa Cruz de Tenerife, Castellón.	: -8%

Value added tax (VAT)

The standard rate of value added tax (VAT) is currently 12%. VAT is chargeable at the standard rate on new construction work and consultants fees. A reduced rate of 6% is chargeable on the sale of private housing.

EXCHANGE RATES AND INFLATION

The combined effect of exchange rates and inflation on prices within a country and price comparisons between countries is discussed in section 4.

Exchange rates

The graph below plots the movement of the Spanish peseta against sterling and the US dollar since 1980. The figures used for the graph are quarterly and the method of calculating these, and other related issues are discussed in section 4. The exchange rate at January 1991 was Pta 184 to the pound sterling and Pta 95 to the US dollar.

The Spanish Peseta against Sterling and the US Dollar

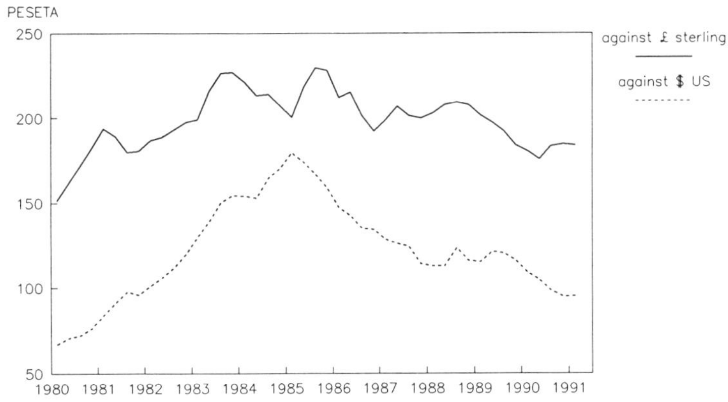

Inflation

The graph and the table below presents general inflation in Spain since 1980. The index has been rebased to 1980=100.

Consumer price inflation

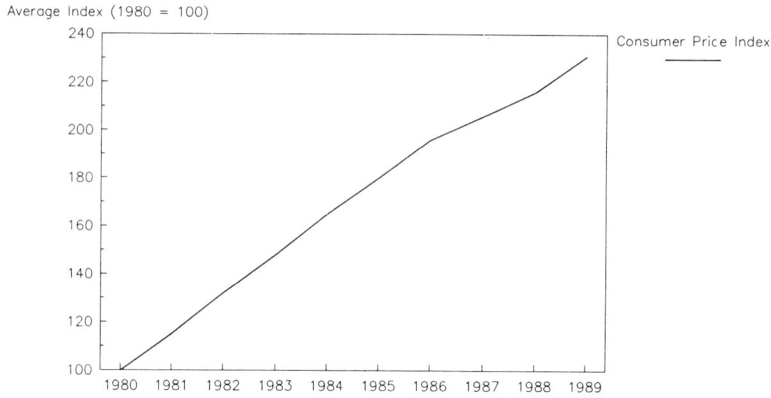

Consumer price inflation

The basis of these figures is the official consumer price index coverted to quarterly averages. The annual change is the percentage change between the last quarters of consecutive years. Inflation peaked in 1981; progress since then in reducing inflation has lagged behind the OECD area as a whole.

Consumer price inflation 1980 = 100

	1981	1982	1983	1984	1985	1986	1987	1988	1989	1990
average index	115	132	148	165	180	196	206	216	231	n.a.
annual change (%)	14.6	14.4	12.2	11.3	8.8	8.8	5.2	4.8	6.8	n.a.

USEFUL ADDRESSES

Ministerio de Obras Publicas y del Urbanismo, Nuevo Ministerios
Pasco de la Castellana 67
Madrid
Tel: 010 34 253 1600

Ministry of Housing,
Plaza de San Juan de la Cruz
Madrid 3

Instituto Eduardo Torroja (IETCC)
Costiliares - Chemartin
Madrid 33

Consejo Superior de Los Colegios de Arquitectos
Paseo de la Castellana 12
Madrid
Tel: 010 34 1 435 2200

Colegio General de Colegios Oficiales de Ingenieros Industriales
General Arrando 38
Madrid
Tel: 010 34 1 419 7428

Consejo General de Colegios Oficiales de Aparejadores y
Arquitectos Tecnicos
Paseo de la Castellana 155
Madrid.
Tel: 010 34 1 222 9060

Confederación Nacional de la Construcción (CNC)
Diego de Leon 50
28006 Madrid
Tel: 010 34 1 261 9715

Sindicato de Empresas de Obra Publicas de Ambito Nacional (SEOPAN)
Serrano 174
28002 Madrid
Tel: 010 34 1 262 9204

Asociacion Española de Consultores en Ingenieria
Paseo de la Castellana 157
28046 Madrid

Asociacion Española de Consultores en Ingenieria
Paseo de la Castellana 157
28046 Madrid.

Sweden

KEY DATA

Population

Population (1989)	8.44m
Urban population	84%
Population under 15	16.5%
Population over 65	18.3%
Average annual growth rate (1983 to 1988)	0.3%

Geography

Land area	450 000 km^2
Agricultural area	35 000 km^2
Capital city	Stockholm (population 1 435 474)

Economy

Monetary unit	Swedish Krona (SKr)
Exchange rate (January 1991) to	
the ECU	SKr 7.64
the pound sterling	SKr 10.9
the US dollar	SKr 5.65
Average annual inflation (1984 to 1989)	5.6%
Gross Domestic Product (GDP)	SKr 1097bn
GDP per capita (1989)	SKr 129 976
GDP growth (real) (1980 to 1988)	1.8%
Private consumption (% of total)	51.0%
Public consumption (% of total)	28.6%
Investment (% of total)	20.2%
Central government expenditure as percentage of GDP	57.8%

Construction

Total value of construction output (1989)	SKr 187bn
Construction output per capita	SKr 22 150
Construction output as a percentage of GDP	17%

THE CONSTRUCTION INDUSTRY

Construction output

In 1989 the value of gross output of the Swedish construction industry was about SKr 187 billion or 26.5 billion ECUs. This represents about 17% of GDP. The breakdown by type of work is shown below.

Construction output 1989

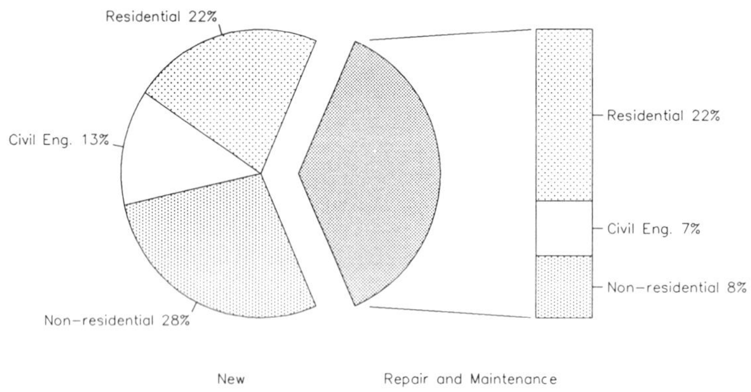

A period of growth in construction output is expected to have come to an end in 1991. Expenditure on residential repair and maintenance is equal to that on new residential building.

In 1987 Swedish construction companies had a turnover abroad of SKr 6 billion, notably in the Middle East and Africa, but also in other parts of the world including Europe. Output abroad, however, is only 3% of total work done and Sweden is small in the world market.

Characteristics and structure of the industry

The National Association of Swedish Architects (SAR) has some 4000 members. The Swedish association of Building Engineers has some 2300 members. In addition there is the Swedish Association of Consulting Architects with 271 member firms.

On large schemes the *byggnadskontrollant* (building controller) has replaced the architect as project co-ordinator and is responsible for financial control, site co-ordination, progress payments and variations. Often, he is a building engineer with special training in cost accounting as well as in structures and architecture. The building controller has evolved because the Swedish practice of separate trade contracting left all site co-ordination responsibility in the hands of the architect whose job became more complicated and far reaching because of the increase in the number of consultants and the greater degree of planning required by more sophisticated buildings. The fact that a considerable proportion of building work is carried out by labour-only gangs further aggravates the problem.

At one time most contracts in Sweden were let on the separate trades system, but general contracting is now common with the general contractor being given responsibility for site management and liability for all sub-contractors' work. However, because of the specialization of contractors in the past, it is also common for the main contractor to be responsible for completing only the structure, sub-letting all other tasks.

There are some 10 000 building and civil engineering contracting firms employing at least two persons. In 1987 there were 28 firms employing over 500 persons and accounting for 44% of the work. This illustrates that the Swedish industry, like the Finnish, is dominated by large firms - to a greater extent than many other countries in Europe, with the exception of the Eastern European former planned economies. Fifteen Swedish companies are in Building magazine's "Top 300 European Contractors" and nine are in the top 100. Only NCC is listed in Engineering News Record's "Top 100 International Contractors", though it would seem that Skanska should also be included there.

The Swedish construction industry has invested heavily in construction technology. Prefabricated components are extensively used - including timber and timber framed housing - and productivity is high. There have been substantial investments in special facilities to enable construction to continue during the winter months. Steam is used to rid moulds and formwork of snow; concrete is delivered warm to site with additives which accelerate hardening; window openings are covered; newly cast concrete is insulated and driers are used to attain the required hardening heat.

Major Swedish contractors

Major contractors	*Place in Building's "Top 300 European Contractors"*	*Exports (% of turnover)*
Skanska	6	15-25
NCC	17	15-25
BPA	27	No
Siab	46	less than 5%
DIOS	53	n.a.
JM Byggnads Och Fastighets	65	less than 5%
Andersons	89	none
Conata	93	less than 5%
PEAB	98	none

Source: Building

Clients and finance

Some 25% to 30% of new dwellings are constructed by the public sector, mostly for rent. In the private sector, over half are built by private persons but co-operatives are also very important. However, the public sector is involved in all housing. Most people (up to 99%) buying a new dwelling can benefit from a loan direct from the National Council of Housing which can amount to 30% of the cost. The remainder of the finance is private but usually at preferential rates. Private owners can also obtain tax relief on improvement expenditures.

Public non-residential construction accounts for less than a third of total non-residential building.

Contractual arrangements

The majority of Government contracts are placed as a result of public advertisment, but for Local Authority and private work, selective tendering is more common.

The Statute of Tendering Procedure lays down official procedures to be followed in obtaining and accepting tenders. It is the basis for all Government construction procurement. Whilst it is not compulsory for work carried out by Local Authorities or private developers, it is commonly used by them as a guide. There is no requirement that the lowest tender should be accepted even for publicly advertised government tenders.

The principal tender documents are:

- general conditions of contract supplemented by specific conditions
- instructions relating to the particular project
- a comprehensive specification
- drawings
- bills of approximate "unit" quantities.

A handbook of typical standard specification clauses exists (AMA 72) as well as a national building products catalogue. There are also recommended rules for contract drawings and project specifications. A standard form of building contract (issued by the Swedish Association of Engineers and Architects) is used for about 80% of contracts. The tender documents, together with this contract, when signed, become the contract documents, with copies being held by both the client and the contractor.

If quantities form part of the contract, they are usually measured by a *massberaknar*' (quantity reckoner). It is accepted in Sweden that, within a stipulated period, the successful contractor can check these quantities; thereafter only authorized variations will be accepted.

Interim payments made to the contractor during the contract period are calculated either on the basis of a valuation of the completed work or by agreed stage payments.

Adjustments are made to the contract sum for fluctuations in wage rates and material prices. This is normally by means of an agreed index, provided the appropriate clauses are included in the contract documents.

Variations are usually permitted and these are either valued on the basis of the rates submitted with the tender or for an agreed lump sum. On completion of the contract and until final settlement, 10% of the contract sum is retained and only released at final settlement or when any defects have been made good. It is the duty of the contractor to prepare the final account.

Development control and standards

In 1987, the older legislation covering physical planning was replaced by two acts : the Management of Natural Resources Act and the Planning and Building Act. The first deals with resolution of conflicting interests in land and water use. The second concerns the planning system. Every municipality is required to have an up-to-date Structure Plan. This is not binding because it is basically a statement of broad intentions. To complement the structure plan, the Municipality is, within certain limits, able to use Special Area Regulations to determine land use including the siting and design of buildings. In addition, Detailed Plans may be prepared to cover:

- new areas of building development
- new individual buildings with considerable impact on their surroundings or where there is great demand for land for building development
- building development which is to be altered or conserved and needs to be regulated within a broader context.

The implementation period must be specified and must be between five and fifteen years with the possibility of a five year extension. The plan must not be changed in this period, except in exceptional circumstances, in which case land owners may be compensated. It can be changed after the implementation period with no compensation. Landowners have the right to build according to the plan but cannot force through changes in it.

Overall building control, supervision, and enforcement of building legislation in Sweden rests with *Statens Planverk* (The National Board of Physical Planning and Buildings). Regulations are laid down in the Swedish Building Code, *Svensk Byggnorm* (SBN), and apply to the whole country for various building types - hotels, places of assembly, and certain work premises. At a regional level, the County Administration exercises control and supervision. Every municipality has a local building committee which examines drawings and calculations to check that a proposed building complies with the requirements of the regulations. The committee grants a building permit and supervises the construction. For every building project there must be a site supervisor who is responsible for obtaining the necessary tests and inspections.

Swedish standards for building, published by the Swedish Standards Institute, are not mandatory. References are made to standards in SBN. Solutions are deemed to satisfy the requirements in the regulations.

The Swedish Institute of Building Documentation publishes an annual Building Codes and Standards Index.

Research and development

The Swedish Council for Building Research is the central agency for co-ordination of research in housing, building and planning. Its research funds are mainly distributed in the form of project grants to a large number of research organisations including the National Institute for Building Research, universities, institutes of technology and specialized materials institutes. The Council stresses the need for research in order to remain internationally competitive.

CONSTRUCTION COST DATA

Cost of labour

The figures below are typical of labour costs in Stockholm as at January 1991. The "wage rate" is the basis of an employee's income, while the "cost of labour" indicates the cost to a contractor of employing that employee. The difference between the two covers a variety of mandatory and voluntary contributions - a list of items which could be included is given in section 4.

	Wage rate (per hour) SKr	*Cost of labour (per hour) SKr*
Site operatives		
Skilled labour	92.00	190.00
Unskilled labour	89.00	185.00
Semi-skilled labour	82.00	180.00

Cost of materials

The figures that follow are the costs of main construction materials, delivered to site in the Stockholm area, as incurred by contractors in January 1991. These assume that the materials would be in quantities as required for a medium sized construction project and that the location of the works would be neither constrained nor remote.

All the costs in this section exclude value added tax (VAT - see below).

	Units	*Cost SKr*
Cement and aggregate		
Ready mixed concrete (K25)	m^3	680.40
Ready mixed concrete (K30)	m^3	705.60
Steel		
Mild steel reinforcement	tonne	5985.00
High tensile steel reinforcement	tonne	6750.00
Structural steel sections	tonne	12 075.00
Bricks and blocks		
Precast concrete cladding units with exposed aggregate finish	m^2	129.73
Timber and insulation		
Exterior quality plywood (12.5mm)	m^2	71.93
Plywood for interior joinery (7.5mm)	m^2	56.44
Softwood strip flooring (24mm)	m^2	65.45
Chipboard sheet flooring (3.2mm)	m^2	29.50
100mm thick quilt insulation	m^2	34.02
100mm thick rigid slab insulation	m^2	60.90
Softwood internal door complete with frames and ironmongery	nr	810.00
Glass and ceramics		
Float glass	m^2	243.23
Sealed double glazing units	m^2	593.25
Plaster and paint		
Good quality ceramic wall tiles (100 x 100mm)	m^2	355.95
Plasterboard (13mm thick)	m^2	38.27
Tiles and paviors		
Clay floor tiles (150 x 150mm)	m^2	301.33
Vinyl floor tiles (200 x 200 x 2.5mm)	m^2	128.82
Precast concrete paving slabs (600 x 600 x 50mm)	m^2	227.85
Precast concrete roof tiles	m^2	56.89
Drainage		
WC suite complete	nr	5000.00
Lavatory basin complete	nr	4900.00

Unit rates

The descriptions below are generally shortened versions of standard descriptions listed in five languages (English, French, Italian, German and Spanish) in section 6. Where an item has a two digit reference number (e.g. 05 or 33), this relates to the full description against that number in section 6. Where an item has an alphabetic suffix (e.g. 12A or 34B) this indicates that the standard description has been modified. Where a modification is major the complete modified description is included here and the standard description should be ignored; where a modification is minor (e.g. the insertion of a named hardwood) the shortened description has been modified here but, in general, the full description in section 6 prevails.

The unit rates below are for main work items on a typical construction project in the Stockholm area in January 1991. The rates include all necessary labour, materials, equipment and, where appropriate, allowances for contractors overheads and profit, preliminary and general items and contractors profit and attendance on specialist trades. All the rates in this section exclude value added tax (VAT - see below).

		Unit	*Rate SKr*
Excavation			
01	Mechanical excavation of foundation trenches	m^3	89.83
02	Hardcore filling making up levels	m^2	29.71
Concrete work			
04	Plain insitu concrete in strip foundations in trenches	m^3	680.40
05	Reinforced insitu concrete in beds	m^3	824.25
06	Reinforced insitu concrete in walls	m^3	783.72
07	Reinforced insitu concrete in suspended floor or roof slabs	m^3	912.70
08	Reinforced insitu concrete in columns	m^3	1150.47
09	Reinforced insitu concrete in isolated beams	m^3	1251.77
Formwork			
11	Softwood formwork to concrete walls	m^2	199.50
12	Softwood or metal formwork to concrete columns	m^2	264.60
13	Softwood or metal formwork to horizontal soffits of slabs	m^2	230.96
Reinforcement			
14	Reinforcement in concrete walls	tonne	7035.00
15	Reinforcement in suspended concrete slabs	tonne	8000.00
16	Fabric reinforcement in concrete beds	m^2	18.48

		Unit	Rate SKr
Brickwork and blockwork			
18	Precast lightweight aggregate hollow concrete block walls	m^2	194.46
19	Solid (perforated) concrete blocks	m^2	305.76
21	Facing bricks	m^2	318.68
Roofing			
22	Concrete interlocking roof tiles 430 x 380mm	m^2	126.00
23	Plain clay roof tiles 260 x 160mm	m^2	131.40
25	Sawn softwood roof boarding	m^2	103.20
29	Glass-fibre mat roof insulation 160mm thick	m^2	51.24
30	Rigid sheet loadbearing roof insulation 75mm thick	m^2	22.25
31	Troughed galvanized steel roof cladding	m^2	157.34
Woodwork and metalwork			
32	Preservative treated sawn softwood 50mm x 100mm	m	33.78
33	Preservative treated sawn softwood 50 x 150mm	m	25.50
34	Single glazed casement window in hardwood, size 650 x 900mm	each	1255.30
35	Two panel glazed door in hardwood, size 850 x 2000mm	each	3136.23
36	Solid core half hour fire resisting hardwood internal flush doors, size 800 x 2000mm	each	1367.31
37A	Aluminium triple glazed window, size 1200 x 1200mm	each	2552.13
38	Aluminium double glazed door, size 850 x 2100mm	each	8384.36
39A	Hardwood skirtings 12 x 44mm	m	1239.00
Plumbing			
48	White vitreous china WC suite	each	6750.00
49	White vitreous china lavatory basin	each	6500.00
50	Glazed fireclay shower tray	each	3000.00
Finishings			
55	2 coats gypsum based plaster on brick walls	m^2	37.90
56	White glazed tiles on plaster walls	m^2	226.00
57	Red clay quarry tiles on concrete floor	m^2	559.35
58	Cement and sand screed to concrete floors	m^2	67.10
59	Thermoplastic floor tiles on screed	m^2	204.53
60	Mineral fibre tiles on concealed suspension system	m^2	185.60
Glazing			
61	Glazing to wood	m^2	231.65
Painting			
62	Emulsion on plaster walls	m^2	56.50
63	Oil paint on timber	m^2	93.79

Approximate estimating

The building costs per unit area given below are averages incurred by building clients for typical buildings in the Stockholm area as at January 1991. They are based upon the total floor area of all storeys, measured between external walls and without deduction for internal walls.

Approximate estimating costs generally include for mechanical and electrical installations but exclude furniture, loose or special equipment, and external works; they also exclude fees for professional services. The costs shown are for specifications and standards appropriate to Sweden and this should be borne in mind when attempting comparisons with similarly described building types in other countries. A discussion of this issue is included in section 4. Comparative data for countries covered in this publication, including construction cost data are presented in Part Three.

Approximate estimating costs must be treated with caution; they cannot provide more than a rough guide to the probable cost of building.

All the rates in this section exclude value added tax (VAT - see below).

	Cost m^2 *SKr*	*Cost* ft^2 *SKr*
Industrial buildings		
Factories for letting	3500	330
Factories for owner occupation (light industrial use)	5500	510
Factory/office (high-tech) for letting (shell and core only)	4250	395
Factory/office (high tech) for owner occupation (controlled environment, fully finished)	6750	630
Warehouses, low bay (6 to 8m high) for letting (no heating)	3130	290
Warehouses, low bay for owner occupation (including heating)	4910	460
Administrative and commercial buildings		
Offices for letting, non air conditioned	8350	780
Prestige/headquarters office, air conditioned	11 110	1035
Health and education buildings		
General hospitals	12 260	1140
Long term hospitals	9480	880
Nursery schools	9280	860
Primary/secondary schools	8160	760
Recreation and arts buildings		
Sports halls including changing and social facilities	6520	610
Swimming pools (international standard) including changing and social facilities	10 500	980

	Cost m^2 SKr	Cost ft^2 SKr
Residential buildings		
Private/mass market single family housing 2 storey detached/semidetached (multiple units)	8450	780
Private sector apartment building (standard specification)	6600	615
Multi storey car park	4460	415
Underground car park	6760	630

Value added tax (VAT)

The standard rate of value added tax (VAT) is currently 25%, chargeable on general building work.

EXCHANGE RATES AND INFLATION

The combined effect of exchange rates and inflation on prices within a country and price comparisons between countries is discussed in section 4.

Exchange rates

The graph below plots the movement of the Swedish Krona against the American dollar and sterling since 1980. The figures used for the graph are quarterly and the method of calculating these, and other related issues are discussed in section 4. The exchange rate at January 1991 was SKr5.65 to the US$ and SKr10.9 to the pound sterling.

The Swedish Krona against the Sterling and the US Dollar

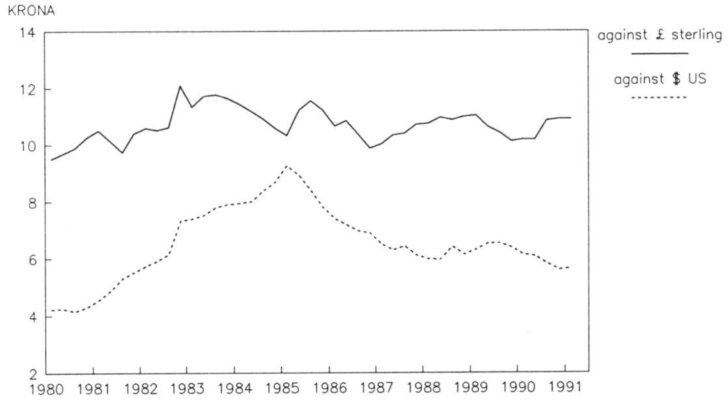

Inflation

The graph and the table below present the consumer price index showing the trend in inflation since 1980. The index has been rebased to 1980=100. Notes on inflation and the distinction between costs and prices are included in section 4.

Consumer price inflation

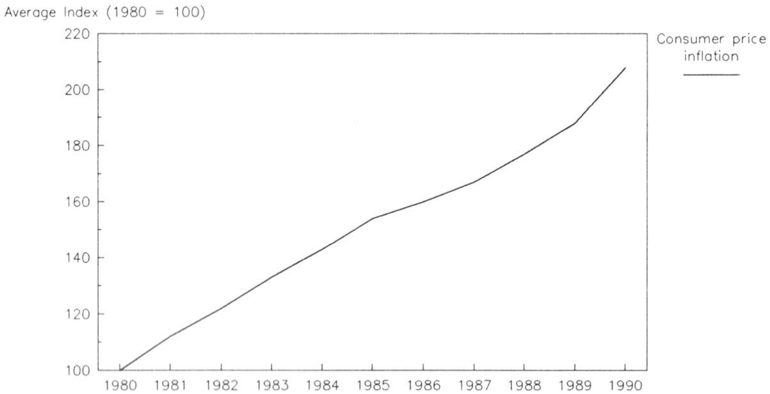

The basis of the figures is the official consumer price index converted to quarterly averages. Inflation peaked in 1981 but following a fall in the mid 1980s (to a low of 4.2%) it has began to increase again and by 1990 it was just over 10%.

Consumer price inflation 1980 = 100

	1981	1982	1983	1984	1985	1986	1987	1988	1989	1990
average index	112	122	133	143	154	160	167	177	188	208
annual change (%)	12.1	8.6	9.0	8.0	7.4	4.2	4.2	5.8	6.5	10.4

USEFUL ADDRESSES

Ministry of Transport and Communications
S-103 33 Stockholm
Tel: 08-763 1000

Swedish National Road Administration
S-781 87 Borlange
Tel: 0243-750 00

The Swedish Civil Aviation Administration
S-601 79 Norrköping
Tel: 011-19 20 00

Swedish National Rail Administration
S-105 50 Stockholm
Tel: 08-762 2000

SIS Standardiseningskommission i Sveriga
(National Standards Institution)
Tegnérgatan 11, Box 3295
S-103 66 Stockholm
Tel: 08-613 52 00

Byggforskningsradet
(Building Research Body)
S:t Göransgatan 66
S-112 33 Stockholm
Tel: 08-617 7300

Byggentreprenörerna
(Building Contractors Federation)
Narvavagen 12, Box 27308
S-102 54 Stockholm
Tel: 08-665 3500

Svenska Arkitekters Riksforund
(Professional Architectural Assoc.)
Norrlandsgatan 18
Stockholm
Tel: 08-679 7230

Svenska Vag och Vattenbyggares Riksforbund
(Professional Engineering Assoc.)
Drottninggatan 88c, Box 1334
S-111 83 Stockholm
Tel: 08-24 5450

Statistical Office
Statistics Sweden
S-115 81 Stockholm
Tel: 08-783 4000

Switzerland

KEY DATA

Population

Population (1989)	6.62m
Urban population	61%
Population under 15	16.4%
Population over 65	15.3%
Average annual growth rate (1983 to 1988)	0.4%

Geography

Land area	41 000 km^2
Agricultural area	20 000 km^2
Capital city	Berne (population 135 000)

Economy

Monetary unit	Swiss Franc (Swfr)
Exchange rate (January 1991) to	
the ECU	Swfr 1.72
the pound sterling	Swfr 2.46
the US dollar	Swfr 1.27
Average annual inflation (1984 to 1989)	2.1%
Gross Domestic Product (GDP)	Swfr 270bn
GDP per capita (1989)	Swfr 40 785
GDP growth (real) (1980 to 1988)	1.9%
Private consumption (% of total)	57.9%
Public consumption (% of total)	12.9%
Investment (% of total)	29.1%
Central government expenditure as percentage of GDP	30.4%

Construction

Total value of construction output (1989)	Swfr 50.8bn
Construction output per capita	Swfr 7670
Construction output as a percentage of GDP	18.8%

THE CONSTRUCTION INDUSTRY

Construction output

The value of the gross output of the construction industry in 1989 in Switzerland was Swfr 50.8 billion or 28.2 billion ECUs. This represents 18.8% of GDP. It is expected that the rate of growth in output, which has been buoyant in the last three years, will slow down and even fall in 1991. A breakdown of output by type of work is shown below.

Construction output 1989

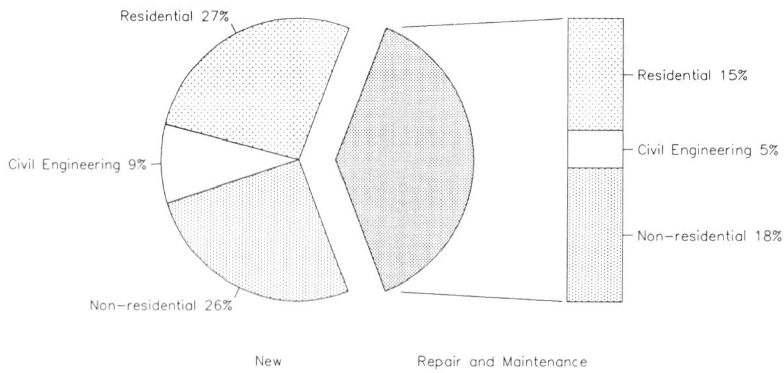

Although it might be expected that Switzerland - a country with a well developed infrastructure - would require little additional construction, the level of activity is, in fact, relatively high both in terms of percentage of GDP and in terms of construction output per capita. However, because infrastructure is well developed, most expenditure is on buildings rather than civil works. About 65% of non-residential building is industrial, office or commercial buildings including hotels.

Characteristics and structure of the industry

Swiss cantons and municipalities enjoy a high degree of autonomy and this has implications for the construction industry. The cantons and municipalities impose their own taxes and arrive at their own spending policy and budgets, though moves are afoot to co-ordinate policies and harmonize taxation.

The construction industry in Switzerland comprises a large number of small firms, many of whom are specialist single trade contractors who operate only on a local basis, and large general contractors who operate throughout Switzerland and undertake a large proportion of the work. Many of the larger companies have contributed significantly to the country's external trade position, though since 1983 the importance of construction exports has decreased. There are 19 Swiss contractors in Building magazine's "Top

300 European Contractors" and three are in the top 100. There are three in Engineering News Record's "Top 250 International Contractors".

Major Swiss contractors

Major contractors	*Head office*	*Place in Building's "Top 300 European Contractors"*	*Place in ENR's "Top 250 International Contractors"*	*Exports (% of turnover)*
Marti Holding AG		62		n.a.
Zschoke Holding AG	Geneva	75	188	5%
Losinger AG	Bern	82	79	40%
Preiswerk Holding AG	Basel	119		less than 10%
Frutinger Unternehmungen AG	Thun	134		less than 10%
APROFIM TP	Geneva		62	75-80%

Source: Building; Engineering News Record

A large number of small specialist contractors have foreign origins. For example, many specialists in high quality finishings such as plastering, marble work, and terrazzo have entered Switzerland from Italy. Construction philosophy and techniques tend to be allied to those of West Germany. German trade press and publications are widely read and many German materials and components are available.

At the end of September 1989 about 171 000 workers were employed in the construction industry, but of these about 112 000 were foreigners and about 59 000 Swiss. Of the foreigners, about 53 000 were migrant seasonal workers and about 12 000 commuters from Italy, France, Germany and Austria. Most of the migrants are Italian but the proportion of Spaniards, Yugoslavs and Turks has grown over the last decade.

Clients and finance

The public sector as a client of the construction industry is not very important in Switzerland. Only a small proportion (about 3% in 1988) of housing is constructed by the public sector. Of private sector housing in 1988, 51% was constructed by individuals, 8% by co-operatives and 41% by other private bodies.

Credit in Switzerland is relatively cheap, especially for long-term loans and mortgages. Loan conditions are generally more favourable than in many other European countries. Normally 50% of the total investment value (ie. market value or construction costs including machinery and equipment) can be favourably financed.

Some public sector support for investment is available from cantons which may procure credits, grant guarantees and direct loans, and arrange for contributions towards interest payments and other services. The Federal Government may also support investment plans for specific assisted regions.

The main tax affecting construction in Switzerland is a "Trade Turnover Tax"; there is no value added tax (VAT) and customs duties on imported materials are minimal.

Tendering and contract procedures

In Switzerland the architect has the responsibility of designer and project manager, planning and co-ordinating the activities of a number of (most commonly) single trade contractors. Larger contracts are often carried out by general contractors on the basis of bills of approximate quantities, with the onus of management transferred to a great extent to the contractor. A similar transfer of responsibility takes place in the case of design and build contracts.

Competitive tendering based on drawings, specifications and bills of quantities is the norm. Bills of quantities are approximate and subject to remeasurement on completion of the work. Prequalification is not necessary though performance bonds are required from general contractors.

Development controls, standards and liability

Planning and development control are based on the principle that a right to build exists under the constitution, provided that no valid legal objection is raised. However, cantonal legislation covers the legal and procedural rules with respect to zoning, designation of development land and ownership. Such legislation may be supplemented at municipal level by further "building ordinances" (*Bauordnungen*) mainly concerned with planning, but also governing building heights, spacings etc. and sometimes specific constructional requirements in respect of safety and health.

A construction permit procedure must be completed before beginning a construction project. Applications must be filed with the political community. The decisions on planning are very much a community matter in rural areas. It is not unusual for a framed structure to be erected on the site so that the public can see the effect of the dimensions of the proposed building.

Industrial buildings also require operations permits in the interest of work safety. These applications are filed with cantonal authorities.

Structural design and properties of building materials are covered by the standards of the Association of Engineers and Architects (SIA) which virtually serve as national regulations.

The building regulation system tends to be complex due to involvement, in varying degrees, of federal, cantonal and municipal government. The power of federal government in this context is restricted by constitution to a limited number of areas of national importance, for example in civil defence where there is a requirement for shelters to be provided in all buildings. Other federal government regulations cover the safety of electrical systems and anti pollution measures.

Although smaller authorities exercise minimal supervision and inspection of building work, in larger cities, including Zurich and Basle, organized building inspectorates exist to check applications for permits and carry out on-site inspections. Resulting technical standards are generally high.

Much of the responsibility for technical standards in construction lies with the building engineer whose personal liability in civil and criminal law is regarded as an important safety factor, ensuring continuing high standards.

CONSTRUCTION COST DATA

Cost of labour

The figures below are typical of labour costs in Berne as at January 1991. The "wage rate" is the basis of an employee's income, while the "cost of labour" is the cost to a contractor of employing that employee. The difference between the two covers a variety of mandatory and voluntary contributions - a list of items which could be included is given in section 4.

	Wage rate (per day) Swfr
Site operatives	
Mason/bricklayer	121.20
Carpenter	136.80
Plumber	155.60
Electrician	161.60
Semi-skilled worker	97.80
Site supervision	
General foreman	197.20
Trades foreman	141.40

Cost of materials

The figures that follow are the costs of main construction materials, delivered to site in the Berne area, as incurred by contractors in January 1991. These assume that the materials would be in quantities as required for a medium sized construction project and that the location of the works would be neither constrained nor remote.

	Units	*Cost Swfr*
Cement and aggregate		
Ordinary portland cement in 50kg bags	tonne	275
Coarse aggregates for concrete	m^3	50
Ready mixed concrete (20N/mm^2)	m^3	165
Ready mixed concrete (30N/mm^2)	m^3	185
Steel		
Mild steel reinforcement	tonne	1500
Bricks and blocks		
Good quality facing bricks (225 x 115 x 75mm)	1000	1200
Timber and insulation		
Softwood sections for carpentry	m^3	675
Softwood for joinery	m^3	1210
Hardwood for joinery	m^3	2750
Softwood internal door complete with frames and ironmongery	nr	375
Glass and ceramics		
Float glass (6mm)	m^2	30
Sealed double glazing units	m^2	300

Unit rates

The descriptions below are generally shortened versions of standard descriptions listed in five languages (English, French, Italian, German and Spanish) in section 6. Where an item has a two digit reference number (e.g. 05 or 33), this relates to the full description against that number in section 6. Where an item has an alphabetic suffix (e.g. 12A or 34B) this indicates that the standard description has been modified. Where a modification is major the complete modified description is included here and the standard description should be ignored; where a modification is minor (e.g. the insertion of a named hardwood) the shortened description has been modified here but, in general, the full description in section 6 prevails.

The unit rates below are for main work items on a typical construction project in the Berne area in January 1991. The rates include all necessary labour, materials, equipment and, where appropriate, allowances for contractors overheads and profit, preliminary and general items and contractors profit and attendance on specialist trades.

		Unit	*Rate Swfr*
Excavation			
01	Mechanical excavation of foundation trenches	m^3	19.00
02	Hardcore filling making up levels	m^3	10.50
03	Earthwork support	m^2	50.00
Concrete work			
04	Plain insitu concrete in strip foundations in trenches	m^3	235.00
05	Reinforced insitu concrete in beds	m^3	259.00
06	Reinforced insitu concrete in walls	m^3	304.00
07	Reinforced insitu concrete in suspended floor or roof slabs	m^3	244.00
08	Reinforced insitu concrete in columns	m^3	356.00
09	Reinforced insitu concrete in isolated beams	m^3	305.00
10	Precast concrete slab	m^2	223.21
Formwork			
11	Softwood formwork to concrete walls	m^2	44.00
12	Softwood or metal formwork to concrete columns	m^2	76.00
13	Softwood or metal formwork to horizontal soffits of slabs	m^2	42.00
Reinforcement			
14	Reinforcement in concrete walls	tonne	2200.00
15	Reinforcement in suspended concrete slabs	tonne	1950.00
16	Fabric reinforcement in concrete beds	m^2	2.60
Brickwork and blockwork			
18	Precast lightweight aggregate hollow concrete block walls	m^2	78.00
19	Solid (perforated) concrete blocks	m^2	75.19
20	Solid (perforated) sand lime bricks	m^2	74.01
21	Facing bricks	m^2	150.00
Roofing			
22	Concrete interlocking roof tiles 430 x 380mm	m^2	38.00
23	Plain clay roof tiles 260 x 160mm	m^2	39.00
24	Fibre cement roof slates 600 x 300mm	m^2	63.00
26	Particle board roof coverings	m^2	52.00
27	3 layers glass-fibre based bitumen felt roof covering	m^2	40.00
28	Bitumen based mastic asphalt roof covering	m^2	50.52
29	Glass-fibre mat roof insulation 160mm thick	m^2	47.00
31	Troughed galvanized steel roof cladding	m^2	32.00

		Unit	Rate Swfr
Woodwork and metalwork			
32	Preservative treated sawn softwood 50 x 100mm	m	199.72
33	Preservative treated sawn softwood 50 x 150mm	m	24.67
34A	Double glazed casement window in hardwood, size 1000 x 1000mm	each	509.00
35	Two panel glazed door in hardwood, size 850 x 2000mm	each	3000.00
36	Solid core half hour fire resisting hardwood internal flush doors, size 800 x 2000mm	each	1900.00
37A	Aluminium double glazed window, size 1000 x 1600mm	each	1066.00
38	Aluminium double glazed door, size 850 x 2100mm	each	1591.00
39	Hardwood skirtings	m	14.00
40	Framed structural steelwork in universal joist sections	tonne	2643.30
Plumbing			
42A	Copper half round eaves gutter	m	72.00
43	UPVC rainwater pipes	m	45.00
44	Light gauge copper cold water tubing	m	27.02
46	Low pressure plastic pipes for cold water distribution	m	20.56
47	UPVC soil and vent pipes	m	45.82
48	White vitreous china WC suite	each	950.00
49	White vitreous china lavatory basin	each	1000.00
50	Glazed fireclay shower tray	each	751.87
51	Stainless steel single bowl sink and double drainer	each	646.14
Electrical work			
52	PVC insulated and copper sheathed cable	m	19.97
53	13 amp unswitched socket outlet	each	43.47
54	Flush mounted 20 amp, 1 way light switch	each	70.49
Finishings			
55	2 coats gypsum based plaster on brick walls	m^2	26.00
56	White glazed tiles on plaster walls	m^2	99.00
57	Red clay quarry tiles on concrete floor	m^2	111.00
58	Cement and sand screed to concrete floors	m^2	25.00
59	Thermoplastic floor tiles on screed	m^2	48.00
60	Mineral fibre tiles on concealed suspension system	m^2	50.00
Glazing			
61	Glazing to wood	m^2	41.12
Painting			
62	Emulsion on plaster walls	m^2	17.00
63	Oil paint on timber	m^2	32.00

Approximate estimating

The building costs per unit area given below are averages incurred by building clients for typical buildings in the Berne area as at January 1991. They are based upon the total floor area of all storeys, measured between external walls and without deduction for internal walls.

Approximate estimating costs generally include for mechanical and electrical installations but exclude furniture, loose or special equipment, and external works; they also exclude fees for professional services. The costs shown are for specifications and standards appropriate to Switzerland and this should be borne in mind when attempting comparisons with similarly described building types in other countries. A discussion of this issue is included in section 4. Comparative data for countries covered in this publication, including construction cost data are presented in Part Three.

Approximate estimating costs must be treated with reserve; they cannot provide more than a rough guide to the probable cost of building.

	Cost m^2 *Swfr*	*Cost* ft^2 *Swfr*
Industrial buildings		
Factories for letting	1200	110
Factory/office (high tech) for owner occupation (controlled environment, fully finished)	2000	185
Warehouses, low bay (6 to 8m high) for letting (no heating)	1150	110
Administrative and commercial buildings		
Civic offices, air conditioned	2000	185
Offices for letting, 5 to 10 storeys, non air conditioned	1800	170
Offices for letting, 5 to 10 storeys, air conditioned	2000	185
Offices for owner occupation 5 to 10 storeys, non air conditioned	1900	180
Offices for owner occupation 5 to 10 storeys, air conditioned	2150	200
Health and education buildings		
General hospitals (100 beds)	2600	240
Nursery schools	1600	150
Primary/secondary schools	2050	190
University (arts) buildings	2200	200
Recreation and arts buildings		
Theatres (over 500 seats) including seating and storage equipment	1900	180
Sports halls including changing and social facilities	1800	170
Local museums including air conditioning	2500	230

	Cost m^2 Swfr	Cost ft^2 Swfr
Residential buildings		
Private/mass market single family housing 2 storey detached/semidetached (multiple units)	1375	130
Private sector apartment building (standard specification)	1800	170
Private sector apartment building (luxury)	2100	195
Homes for elderly (self contained with shared communal facilities)	2600	240
Hotel, 5 star, city centre	2600	240
Hotel, 3 star, city/provincial	2000	185

EXCHANGE RATES AND INFLATION

The combined effect of exchange rates and inflation on prices within a country and price comparisons between countries is discussed in section 4.

Exchange rates

The graph below plots the movement of the Swiss Franc against the American dollar and sterling since 1980. The figures used for the graph are quarterly and the method of calculating these, and other related issues are discussed in section 4. The exchange rate at January 1991 was Swfr1.27 to the US$ and Swfr2.46 to the pound sterling.

The Swiss Franc against the Sterling and the US Dollar

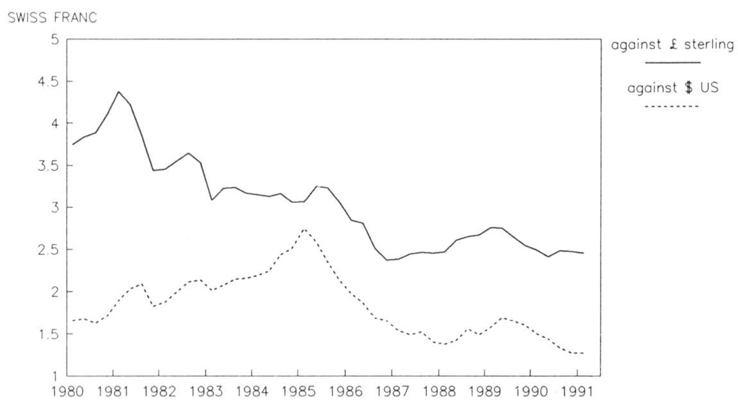

Inflation

The graph and the table below present the consumer price index showing the trend in inflation since 1983. The index has been rebased to 1982=100. Notes on inflation and the distinction between costs and prices are included in section 4.

Consumer price inflation

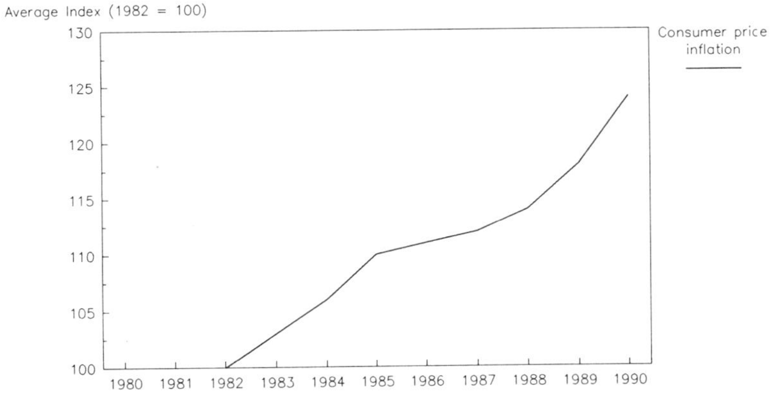

Consumer price inflation

The basis of these figures is the official consumer price index converted to quarterly averages. Prior to 1990 inflation had been below the average for the OECD area as a whole.

Consumer price inflation 1982 = 100

	1983	1984	1985	1986	1987	1988	1989	1990
average index	103	106	110	111	112	113	118	124
annual change (%)	3.0	3.0	3.4	0.8	1.4	1.9	3.2	5.4

USEFUL ADDRESSES

Schweizerischer Baumeisterverband
(Swiss Contractors' Association)
Weinbergstrasse 49
CH-8035 Zürich
Tel: 01 258 8111
Fax: 01 204 9426

Schweizerischer Ingenieur und Architekten Verein
(Specialized Group of Industrialized Construction:
Swiss Society of Engineers and Architects)
Selnaustrasse 16
8039 Zürich
Tel: 01 201 1570
Fax: 01 201 6335

Bund Schweizer Architekten
Fröbelstrasse 33
8032 Zurich

Verband Schweizerischer Generalunternehmer, (VSGU)
(Swiss General Contractors Association)
Weisenstrasse 7
CH-8023 Zürich
Tel: 01 383 8614
Fax: 01 383 2593

Vereinigung Schweizerischer Beratender Ingenieur - und
Architekturgesellschaften (USSI)
(Association of Swiss Consulting Engineering and Architecture Firms)
Seefeldstrasse 9
CH-8008 Zürich
Tel: 01 252 9880
Fax: 01 252 9896

Schweizerischer Technischer Verband (STA)
(Swiss Technical Association)
Weinbergstrasse 41
Postfach 6383
CH-8023 Zürich
Tel: 01 261 3794
Fax: 01 251 4802

Turkey

KEY DATA

Population

Population (1989)	53.8m
Urban population	47%
Population under 15	35.4%
Population over 65	4.3%
Average annual growth rate (1983 to 1988)	2.3%

Geography

Land area	779 000 km²
Agricultural area	366 000 km²
Capital city	Ankara (population 3 196 460)

Economy

Monetary unit	Turkish Lira (TL)
Exchange rate (January 1991) to	
the ECU	TL 4219
the pound sterling	TL 6017
the US dollar	TL 3123
Average annual inflation (1984 to 1989)	
Gross Domestic Product (GDP)	TL 166 700bn
GDP per capita (1989)	TL 3.10m
GDP growth (real) (1980 to 1988)	5.3%
Private consumption (% of total)	65.0%
Public consumption (% of total)	9%
Investment (% of total)	24%
Central government expenditure as percentage of GDP	24%

Construction

Total value of construction output (1989)	TL 20 000bn
Construction output per capita	TL 371 700
Construction output as a percentage of GDP	12%

THE CONSTRUCTION INDUSTRY

Construction output

The gross value of output of new construction in Turkey in 1989 is estimated at about TL 20 000 billion, equivalent to 8.8 billion ECUs (at mid-1989 rates). However, because of high inflation rates in Turkey and changing exchange rates, any figure is subject to considerable uncertainty. Construction output was probably about 12% of GDP in 1989. Of the total, over 50% is in housing construction of which about 80% is privately constructed. The second most important component is civil engineering construction which may account for some 30% or more of the total, leaving less than 20% for other new work. Expenditure on repair and maintenance is not known but is almost certainly a low proportion of the total.

Turkey needs massive infrastructure investment, but the high rate of inflation in the economy is one of the reasons why the 1990 investment programme, drawn up by the State Planning Organization, is expected to show a substantial shortfall.

A boom in housing construction started in 1985 and was a useful cushion for the fall in external work, particularly in the Middle East, where Turkish contractors were active. One of the notable features of the Turkish construction industry is the substantial output abroad. This rose from US$1.6 billion in 1978 to US$17 billion in 1987. Nearly 60% of this work was in Libya and 31% was in Saudi Arabia. Since 1987 work in other Middle Eastern and North African countries has increased in importance, but Libya and Saudi Arabia are still dominant. Workers remittances and company profits constitute important components in the Turkish balance of payments.

Characteristics and structure of the industry

Turkish contractors, in general, are not large by international standards although in 1989 there were six Turkish companies in Engineering News Record's "Top 250 International Contractors". Turkish firms tend to be family-run in a centralized, informal way by the founder. This gives flexibility in overseas activities and over 300 firms were operating abroad in 1987. Firms tend to specialize by type of work using directly employed labour and traditional trade subcontractors. The strongest selling point of Turkish firms abroad has been their large, flexible labour force.

There are about 10 000 registered contractors in the domestic market. Most of the larger contractors are conglomerates with activities in other countries. The Ministry of Public Works and Resettlement groups contractors into three categories:

- Group A(capable of undertaking large and technical projects)
- Group B (capable of undertaking medium and small projects)
- Group C (capable of undertaking electrical and mechanical).

In addition there are some 20 000 subcontractors.

Major Turkish contractors

Major contractors	*Head office*	*Place in ENR's "Top 250 International Contractors"*	*Exports (% of turnover)*	*Main work types*
Dogus Insaat	Istanbul	75	40	Dams, Water supply
Mesa Housing Industries Inc.	Ankara	76	55	Building
Enka Insaat ve Sanayi AS	Istanbul	80	65	Building, Power Water, Sewerage
Tekfen Construction and installation Co. Inc.	Istanbul	149	10	Pipelines
SOYAK Construction Company	Istanbul	159	70	Building
Nurol Construction trading Co Inc	Ankara	225	2	Building and general civil engineering

Source: Engineering News Record

There are about 125 000 graduate engineers and about 85 000 architects. Architects and engineers produce detailed designs, but working drawings are the responsibility of the contractor.

Clients and finance

The public sector is an important client of the construction industry for work other than housing and especially for civil engineering work. It also undertakes significant investment in manufacturing industry, agriculture, mining, education, health and tourism. The major public sector client is the Ministry of Public Works and Resettlement; others include the Ministry of Energy and Natural Resources, the Ministry of Communications and Transportation, the Ministry of National Defence and the Ministry of Tourism.

In 1984 the Government set up a Mass Housing Fund to build new residential areas in most large cities. It is financed by a surcharge on luxury imports. It provided finance for 120 000 of the 255 000 housing units built in 1988.

Design and build contracts are being used increasingly. A variant is the build-operate-transfer project in which the contractor operates the building or works for a specified period of time and retains the profits. At the end of the period the building or other work is transferred to the Government which therefore receives it free in exchange for the original concession to build and operate. There is also a mutual construction contract whereby the landowner enters into partnership with a contractor who will develop the land. This is used mainly in private residential and commercial developments.

Contractor selection

Apart from the arrangements mentioned under "Clients and finance", competitive bids are normal in the public sector while negotiation is more usual in the private sector. For public sector projects, bidding takes place on specification, drawings, schedules of quantities (on major projects) and a priced schedule of rates (for smaller works). The contractor quotes on the basis of those rates.

Before 1983, acceptance of the lowest price led to unprofitable contracts, high levels of bankruptcy and insolvencies. The public sector now selects a bid near to the average on the basis of financial strength, reputation, experience and reliability.

The normal contract is a lump sum but for larger projects there is cost escalation reimbursement based on cost indices. Bid bonds and performance bonds are normal.

Development control

The Town Development Act of 1985 gave most of the planning and development control powers to the municipalities over 10 000 population. Municipalities with less than 10 000 population may make plans but are not obliged to do so. Countryside planning, where it exists, is under the provincial local administration. Regional plans are optional.

Municipal councils act as approval authorities for all levels of plans. In major cities the mayor has the final say in the approval and amendment of plans. The development of the system is still underway and the full effects of the 1985 act are not yet evident.

Further reading

Economist Intelligence Unit Country Profiles and Country Reports Turkey, various issues EIU.

Amir Tavokoli and Sevket Con Tulumen, "Construction Industry in Turkey", Construction Management and Economics, Vol 8, pp 77-87, 1990.

Heyecan Giritli, Zeynep Sozen, Roger Flanagan and Peter Lansley, International Contracting: a Turkish Perspective, Construction Management and Economics, Vol 8, pp 415-430, 1990.

CONSTRUCTION COST DATA

Cost of labour

The figures below are typical of labour costs in Ankara as at January 1991. The "wage rate" is the basis of an employee's income, while the "cost of labour" indicates the cost to a contractor of employing that employee. The difference between the two covers a variety of mandatory and voluntary contributions - a list of items which could be included is given on in section 4.

	Wage rate (per day) TL	*Cost of labour (per day) TL*	*Number of hours worked per year*
Site operatives			
Mason/bricklayer	4444	9222	2700
Carpenter	4444	9222	2700
Plumber	3889	8111	2700
Electrician	3333	7111	2700
Structural steel erector	5556	11 556	2700
HVAC installer	5556	11 556	2700
Semi-skilled worker	3333	7111	2700
Unskilled labourer	2778	5778	2700
Equipment operator	6250	13 000	2400
Watchman/security	3000	6250	2400
Site supervision			
General foreman	6666	13 333	2700
Trades foreman	5778	12 000	2700
Clerk of works	2778	5778	2700
Contractors' staff			
Site manager	20 000	31 400	2700
Resident engineer	15 000	22 400	2700
Resident surveyor	15 000	22 400	2700
Junior engineer	8000	15 000	2700
Junior surveyor	6000	11 600	2700
Planner	12 000	21 000	2700
Consultants' personnel			
Senior architect	25 000	39 250	2160
Senior engineer	13 750	24 500	2160
Senior surveyor	13 750	24 500	2160
Qualified architect	13 750	24 500	2160
Qualified engineer	13 750	24 500	2160
Qualified surveyor	16 250	28 000	2160

Cost of materials

The figures that follow are the costs of main construction materials, delivered to site in the Ankara area, as incurred by contractors in January 1991. These assume that the materials would be in quantities as required for a medium sized construction project and that the location of the works would be neither constrained nor remote.

	Units	*Cost TL*
Cement and aggregate		
Ordinary portland cement in 50kg bags	tonne	220 000
Coarse aggregates for concrete	m^3	27 000
Fine aggregates for concrete	m^3	27 000
Ready mixed concrete (mix B160)	m^3	150 000
Ready mixed concrete (mix B225)	m^3	165 000
Steel		
Mild steel reinforcement	kg	1250
High tensile steel reinforcement	kg	1350
Structural steel sections	tonne	1 300 000
Bricks and blocks		
Common bricks (19 x 19 x 13.5)	each	250
Good quality facing bricks (19 x 19 x 5)	each	440
Hollow concrete blocks (20 x 19 x 39)	each	500
Timber and insulation		
Softwood for joinery	m^3	1 500 000
Hardwood for joinery	m^3	2 000 000
Plywood for interior joinery (2mm)	m^3	2 100 000
Softwood strip flooring	m^2	65 000
50mm/18kg thick quilt insulation	m^3	10 150
50mm/30kg thick rigid slab insulation	m^3	10 200
Softwood internal door complete with frames and ironmongery	m^2	1 250 000
Glass and ceramics		
Float glass (6mm)	m^2	26 379
Sealed double glazing units (3+3mm)	m^2	75 623
Plaster and paint		
Good quality ceramic wall tiles (15 x 15)	each	750
Emulsion paint in 5 litre tins	kg	6560
Gloss oil paint in 5 litre tins	kg	11 000

	Units	Cost TLira
Tiles and paviors		
Clay floor tiles (20cm x 20cm x 9cm)	m^2	31 000
Vinyl floor tiles (25cm x 25cm x 2cm)	m^2	8000
Clay roof tiles	each	700
Drainage		
WC suite complete	nr	716 000
Lavatory basin complete	nr	423 750
150mm diameter cast iron drain pipes	m	65 420
Steel butt hinges	nr	1500
Espagnolette bolt	nr	9200

Unit rates

The descriptions below are generally shortened versions of standard descriptions listed in five languages (English, French, Italian, German and Spanish) in section 6. Where an item has a two digit reference number (e.g. 05 or 33), this relates to the full description against that number in section 6. Where an item has an alphabetic suffix (e.g. 12A or 34B) this indicates that the standard description has been modified. Where a modification is major the complete modified description is included here and the standard description should be ignored; where a modification is minor (e.g. the insertion of a named hardwood) the shortened description has been modified here but, in general, the full description in section 6 prevails.

The unit rates below are for main work items on a typical construction project in the Ankara area in January 1991. The rates include all necessary labour, materials, equipment and, where appropriate, allowances for contractors overheads and profit, preliminary and general items and contractors profit and attendance on specialist trades.

		Unit	Rate TL
Excavation			
01	Mechanical excavation of foundation trenches	m^3	2398
02	Hardcore filling making up levels	m^3	12 154
03	Earthwork support	m^2	27 426
Concrete work			
04	Plain insitu concrete in strip foundations in trenches	m^3	120 719
05	Reinforced insitu concrete in beds	m^3	191 294
06	Reinforced insitu concrete in walls	m^3	203 794
07	Reinforced insitu concrete in suspended floor or roof slabs	m^3	203 794
08	Reinforced insitu concrete in columns	m^3	203 794
09	Reinforced insitu concrete in isolated beams	m^3	203 794

		Unit	Rate TL
Formwork			
11	Softwood formwork to concrete walls	m^2	19 194
12	Softwood or metal formwork to concrete columns	m^2	19 194
Reinforcement			
14	Reinforcement in concrete walls	tonne	1 607 094
15	Reinforcement in suspended concrete slabs	tonne	1 607 094
16A	Fabric reinforcement in concrete beds	tonne	2 055 325
Brickwork and blockwork			
18	Precast lightweight aggregate hollow concrete block walls	m^2	17 921
19	Solid (perforated) concrete blocks	m^2	25 347
21	Facing bricks	m^2	39 108
Roofing			
22A	Concrete interlocking roof tiles 420 x 335mm	m^2	13 678
23	Plain clay roof tiles 260 x 160mm	m^2	35 728
25	Sawn softwood roof boarding	m^2	39 490
27	3 layers glass-fibre based bitumen felt roof covering	m^2	37 156
28	Bitumen based mastic asphalt roof covering	m^2	19 075
29A	Glass-fibre mat roof insulation 50mm thick	m^2	9848
31	Troughed galvanized steel roof cladding	m^2	71 848
Woodwork and metalwork			
34A	Single glazed casement window in hardwood, size 650 x 900mm	m^2	154 942
35A	Two panel glazed door in hardwood, size 850 x 2000mm	m^2	334 686
36A	Solid core half hour fire resisting hardwood internal flush doors, size 800 x 2000mm	m^2	251 046
37	Aluminium double glazed window, size 1200 x 1200mm	m^2	398 288
38A	Aluminium double glazed door, size 850 x 2100mm	m^2	341 251
39A	Hardwood skirtings	m	8411
40	Framed structural steelwork in universal joist sections	tonne	1 934 103
41	Structural steelwork lattice roof trusses	tonne	2 989 419
Plumbing			
42A	UPVC half round eaves gutter	m	21 165
43A	UPVC rainwater pipes	m	16 570
44A	Light gauge copper cold water tubing	m	12 970
45A	High pressure plastic for cold water supply	m	1760
46	Low pressure plastic pipes for cold water distribution	m	1300
47A	UPVC soil and vent pipes	m	16 570
48A	White vitreous china WC suite	each	753 000
49A	White vitreous china lavatory basin	each	423 750

		Unit	Rate TL
50A	Glazed fireclay shower tray	each	430 000
51A	Stainless steel single bowl sink and double drainer	each	412 000
Electrical work			
52	PVC insulated and copper sheathed cable	m	7650
53	13 amp unswitched socket outlet	each	84 000
54	Flush mounted 20 amp, 1 way light switch	each	79 900
Finishings			
55	2 coats gypsum based plaster on brick walls	m^2	13 997
56A	White glazed tiles on plaster walls	m^2	82 679
58A	Cement and sand screed to concrete floors	m^2	14 108
59A	Thermoplastic floor tiles on screed	m^2	20 564
Glazing			
61	Glazing to wood	m^2	31 087
Painting			
62	Emulsion on plaster walls	m^2	6466
63	Oil paint on timber	m^2	14 927

Approximate estimating

The building costs per unit area given below are averages incurred by building clients for typical buildings in the Ankara area as at January 1991. They are based upon the total floor area of all storeys, measured between external walls and without deduction for internal walls.

Approximate estimating costs generally include for mechanical and electrical installations but exclude furniture, loose or special equipment, and external works; they also exclude fees for professional services. The costs shown are for specifications and standards appropriate to Turkey and this should be borne in mind when attempting comparisons with similarly described building types in other countries. A discussion of this issue is included in section 4. Comparative data for countries covered in this publication, including construction cost data are presented in Part Three.

Approximate estimating costs must be treated with caution; they cannot provide more than a rough guide to the probable cost of building.

	Cost m^2 TL	Cost ft^2TL
Industrial buildings		
Industrial building (Basic)	579 700	53 900
Industrial building (standard)	662 000	61 600
Small industrial centres	662 000	61 600
Organised industrial development	828 200	76 900
Cold stores/refrigerated stores	993 800	92 300
Administrative and commercial buildings		
Offices excluding lifts and heating	579 700	53 900
Offices, standard specification	725 000	67 500
Local government offices	725 000	67 500
City Hall offices	745 400	69 300
Large court house	993 800	92 300
Health and education buildings		
Hospitals	993 800	92 300
Health centres	662 500	61 600
Nursery schools	745 000	69 200
Schools	662 500	61 600
School including gymnasium, conference hall	745 400	69 300
University buildings	828 200	76 900
Recreation and arts buildings		
Theatres	2 038 600	189 400
Concert halls, congress centre	1 490 700	138 500
Sport halls	579 700	53 900
Stadium and sports halls	828 200	76 900
Swimming pools (covered)	993 800	92 300
Museum, exhibition and library complex	1 490 700	138 500
Large library and cultural buildings	993 800	92 300
Residential buildings		
Social/economic single family housing (multiple units)	344 000	32 000
Social/economic apartment housing, excluding heating	579 900	53 900
Social/economic apartment housing, including heating	662 500	61 600
Private sector apartment buildings (luxury)	828 000	76 900
Student halls of residence	993 800	92 300
Homes for the elderly (shared accommodation)	745 000	69 200
Hotel, 5 star, city centre	1 739 200	161 600
Hotel, 3 star, city/provincial	993 800	92 300

EXCHANGE RATES AND INFLATION

The combined effect of exchange rates and inflation on prices within a country and price comparisons between countries is discussed in section 4.

Exchange rates

The graph below plots the movement of the Turkish Lira against sterling and the US dollar since 1979. The figures used for the graph are quarterly and the method of calculating these, and other related issues are discussed in section 4. The exchange rate at January 1991 was Turkish Lira 6017 to the pound sterling and Turkish Lira 3123 to the US dollar.

The Turkish Lira against Sterling and the US Dollar

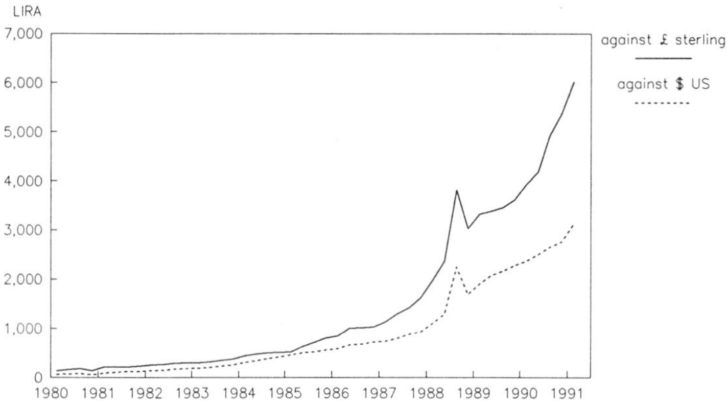

Inflation

The graph below presents a number of indices showing the trends in inflation, prices and costs. All the indices have been rebased to 1980=100. Notes on inflation and the distinction between costs and prices are included in section 4.

Cost and price indices

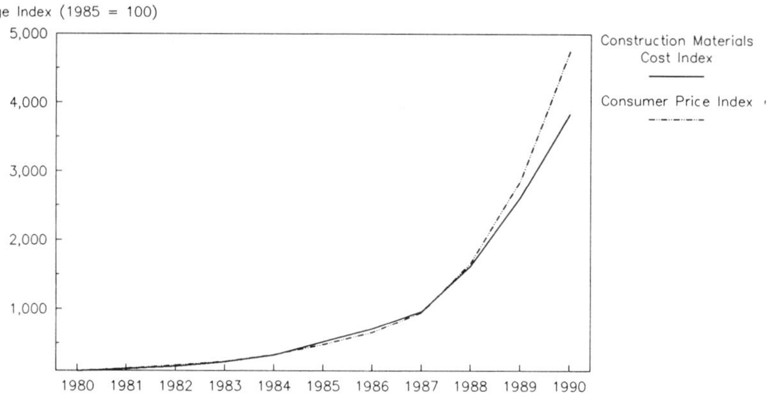

Consumer price inflation

The table below presents general inflation in Turkey since 1980. The annual change is the percentage change between the last quarters of consecutive years.

The basis of the figures is the official consumer price index converted to quarterly averages. Inflation has been high in the 1980s reaching an annual rate of 76% in 1988.

Consumer price inflation 1980 = 100

	1981	1982	1983	1984	1985	1986	1987	1988	1989	1990
average index	135	182	233	335	481	654	943	1664	2862	4748
annual change (%)	35	35	28	44	44	36	44	76	72	66

Building materials cost inflation

The table below presents an average index for the movement of costs of building materials since 1980.

Building materials cost index 1980 = 100

	1981	1982	1983	1984	1985	1986	1987	1988	1989	1990
average index	128	162	228	330	524	708	966	1621	2624	3838
annual change (%)	28	27	41	45	59	35	36	68	62	46

United Kingdom

KEY DATA

Population

Population (1989)	57.1m
Urban population	92%
Population under 15	18.9%
Population over 65	15.5%
Average annual growth rate (1983 to 1988)	0.3%

Geography

Land area	244 000 km^2
Agricultural area	186 000 km^2
Capital city	London (population 6 735 000)

Economy

Monetary unit	Pound Sterling (£)
Exchange rate (January 1991) to	
the ECU	£ 0.700
the US dollar	£ 0.518
Average annual inflation (1984 to 1989)	5.3%
Gross Domestic Product (GDP)	£ 464bn
GDP per capita (1989)	£ 8130
GDP growth (real) (1980 to 1988)	2.8%
Private consumption (% of total)	64.6%
Public consumption (% of total)	19.2%
Investment (% of total)	28.4%
Central government expenditure as percentage of GDP	38.75%

Construction

Total value of construction output (1989)	£ 46.2bn
Construction output per capita	£ 810
Construction output as a percentage of GDP	9.95%

THE CONSTRUCTION INDUSTRY

Construction output

In 1989 the value of the gross output of the UK construction industry was £46.2 billion, equivalent to 67.8 billion ECUs. This represents just under 10% of GDP - low by comparison with other European countries. Output has increased over the last few years but will to fall over the next year or so.

The breakdown of output by type of work is shown below. This is based on estimates, because official statistics do not separate civil engineering from building.

Construction output 1989

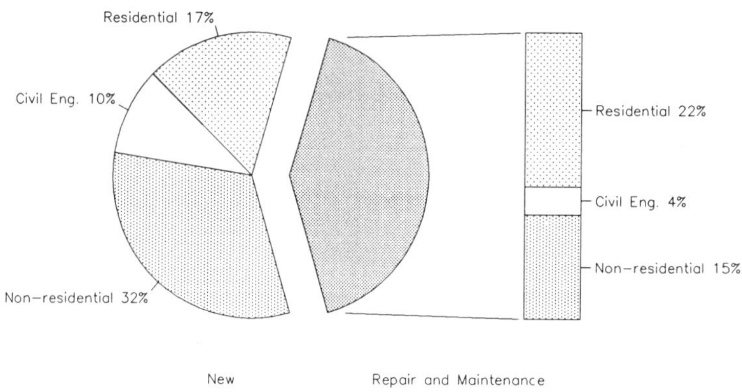

The very high level of output attributed to offices, retail and other commercial building accounts for about 56% of all non-residential building. Industrial building accounted for about 28% of non-residential building in 1989. However, the office and retail boom has now come to an end and construction output is expected to fall dramatically over the next few years. It is noteworthy that residential repair and maintenance is greater than new residential construction. In part this is because new house construction is low compared to previous years. However, total maintenance expenditure is much higher than is indicated by the diagram above. Expenditure on DIY materials (£4.7 billion in 1989) by the residential sector and substantial expenditure on maintenance (undertaken by private direct labour organizations and therefore unrecorded) by the non-residential sector together may increase maintenance expenditure by nearly 40%. In addition there is thought to be uncounted expenditure for work done by those avoiding tax and insurance obligations.

One of the most marked features of construction activity over recent years has been the shift in activity by region. Historically, construction work throughout the country has been closely related to population. The table below lists the standard regions of Great Britain and, against each, percentages showing the 1985 population plus construction new orders for 1979, 1986 and 1989.

Regional population and new orders 1979 to 1989

Region	*Population 1985 (%)*	*New Orders 1979 (%)*	*1986 (%)*	*1989 (%)*
North	5.6	5.5	3.8	4.6
Yorks and Humberside	8.9	9.2	7.2	7.7
East Midlands	7.1	5.8	6.3	6.2
East Anglia	3.6	4.6	4.8	4.8
South East	31.2	32.3	42.5	38.4
South West	8.2	7.7	8.5	9.1
West Midlands	9.4	8.1	6.8	8.5
North West	11.6	10.9	8.4	8.5
Wales	5.1	4.7	4.3	4.0
Scotland	9.3	11.2	7.5	8.4
Total	100	100	100	100

Sources: DoE Housing and Construction Statistics and CSO Regional Trends

In 1979 four regions had construction new orders 10% above or below their share of population. The East and West Midlands each had less than their shares; Scotland and East Anglia had substantially more than their shares. In 1989 no region had new orders within 10% of their share by population. Three regions, all in the South, had more than their share (East Anglia, the South East, and the South West). All other regions had significantly less than their share by population.

East Anglia has more than its share in all sectors; the South West in all except public work (other than housing); and the South East in commercial building, where it had 50% of the total, and in public work (other than housing), where it had 39% of the total.

In 1989 British contractors obtained orders abroad totalling £1426 million as shown below.

UK construction work overseas (£ million - current prices)

Areas of World	*1984/5*	*%*	*1989*	*%*
EEC	45	2	39	3
Rest of Europe	57	2	36	3
ME in Asia	700	28	113	8
ME in Africa	48	2	63	4
Rest of Asia	306	12	140	10
Rest of Africa	482	19	162	11
Americas	611	25	714	50
Oceania	237	10	159	11
All countries	2486	100	1426	100

Source: DoE Housing and Construction Statistics

The total value of work done by British contractors overseas has declined since the mid 1980s. The largest falls have been in work in the Middle East and Africa, the two largest overseas markets for British contractors in 1984/85.

Characteristics and structure of the industry

The bulk of building work is carried out by general contractors who traditionally employed their own labour force but increasingly employ labour-only sub-contractors. Specialist sub-contractors may be nominated by the client's consultant team or employed by the general contractor.

Traditionally, building work in the UK has been administered by professional consultants appointed by the building client. The consultants are responsible for the design and specification of the work, the contractual arrangements and the supervision of the contract. However, over the last 20 years, design-and-build contractors offering a single point of responsibility have also established a sizeable share of the market as have other, less traditional, arrangements (see "Contractual arrangements").

The construction industry in the UK consists of a large number of small firms and several very large firms. In 1989 there were estimated to be 201 000 firms of which 48 employed over 1200 persons and 267 employed over 300. However, these employment figures do not truly indicate the importance of large firms because of the prevalence of labour-only sub-contractors. A better indication is that in Building magazine's "Top 300 European

Contractors" there are 79 UK companies and 30 in the top 100 - more than any other country. The Engineering News Record's "Top International Contractors" lists 13 UK contractors in the top 100. In both the Building and Engineering News Record lists, some contractors are very specialised, for example in engineering or housing development, rather than building and civil engineering. However, the comparison between countries is relevant.

The principal UK contractors with some of their characteristics are shown in the table below.

Major UK contractors

Major contractors	*Place in Building's "Top 300 European Contractors"*	*Place in ENR's "Top 250 International Contractors"*	*Main work types*
BICC	2	118	Largest because conglomerate in cables etc. Balfour Beatty is construction arm
Tarmac	3	187	Materials, property, construction
Trafalgar House	5	-	Shipping, other businesses, construction
Wimpey	12	35	Property, construction
AMEC	13	-	Nearly all construction but includes development. Construction for offshore.
Beazer	14	36	Mainly construction, property and materials
Costain	18	96	Mining, property, construction
John Laing	19	100	Mainly construction, property and plant and mechanical and electrical engineering
Taylor Woodrow	20	53	Mainly construction, property, offshore and mechanical and electrical engineering
Mowlem	21	92	Mainly construction and scaffolding
Bovis	28	5	Construction subsidiary of P&O
Alfred McAlpine	36	-	Mainly construction

Source: Building; Engineering News Record

For consistency, contractors are listed in the order quoted in Building magazine's list which is based on total turnover. This does not, however, really represent their construction size nor their contracting size. Other businesses in which they operate are shown on the table. Thus Wimpey is probably larger than the first three in contracting operations.

One feature of British contractors is the extent of diversification beyond general contracting into property development, speculative housebuilding, material production and other businesses such as mining or airports. This has enabled the larger contractors to weather the fluctuations in construction demand and, in many cases, gives a better return on capital employed.

Construction design work is undertaken mainly by architects and engineers. Architects must register with the Architects Registration Council of the United Kingdom (ARCUK). The designation architect can be used only by persons so registered. There are about 30 500 architects in the UK, about 530 per million inhabitants. About 80% of architects are members of the Royal Institute of British Architects (RIBA). In 1989, 61% of architects were working in private practice - two thirds as principals and one third salaried. Nearly 70% of practices have fewer than 5 architects, but there are 70 to 80 practices employing over 50 architects and some of these are very large. The remaining architects are employed by local authorities or government, in education or in other private employment.

Civil engineers are normally members of the Institution of Civil Engineers with a membership approaching 50 000. Structural engineers are members of the Institution of Structural Engineers, with a membership of about 11 000. Building services engineers are members of the Chartered Institution of Building Service Engineers having a membership of over 7000. But, because engineers may be members of more than one engineering institution, there are in total about 600 engineers per million inhabitants. The title of Chartered Engineer is registered and protected, either by the professional institution or by the Engineering Council. Most construction design work in the UK is undertaken by private practices. The amount of in-house work has shrunk considerably in the last ten years and contractors' design departments are relatively small being mostly concerned with building rather than civil engineering work. Civil and Structural engineers are most important numerically in private practice. In 1989, nine engineering consultancy firms had over 1000 staff and of these, seven had more than 1000 civil and structural staff. There is a long tradition of British consulting engineers working abroad.

The surveying profession is very important in the UK. The Royal Institution of Chartered Surveyors is an umbrella organization for quantity surveyors and building surveyors as well as a number of other surveyors less closely connected to the construction industry. In 1989 there were about 22 000 quantity surveyors and about 5000 building surveyors. The quantity surveyor plays a key role in the UK construction industry and countries influenced by UK practices. Originally his role was to prepare a bill of quantities and measure work on site. The profession has, however, developed cost advisory roles to clients and has a full professional status equivalent to that of designers. Over 54% of quantity surveyors work in private practice, 19% with contractors, 15% in public service and the remainder in commercial organizations and education. Most quantity surveying practices are small but there are a number of very large firms employing several hundred staff.

Clients and finance

Historically, the UK construction industry maintained a fairly even split between orders in the public and private sectors. However, since 1979 there has been a marked decline in public sector investment. In 1979, 41% of new construction output was for public sector clients, by 1985 this had reduced to 32% and by 1989 to 22%. Although this decline has been for all types of public construction it has been most dramatic in the public housing sector. In 1989 public sector housing starts were only 22% of their 1979 levels, whereas private sector housing starts in 1985 were 116% of their 1979 level.

Approximately 80% of investment in housing is in dwellings in the private personal sector, with most of the remainder being by local authorities. A large part of the purchase of housing, both existing and new, is financed by mortgages. In 1989 mortgage finance was provided as follows:

	%
Building Societies	72
Banks	25
Insurance Companies	3
Local Authorities	neg.
	100

Other buildings in the private sector may be financed in a number of ways:

- they may be built and owned by owner occupiers
- they may be built by investors and let.

It is estimated, for example, that owner occupiers account for up to 80% of new construction of industrial buildings. However, the amount of other private building built and owned by the occupiers is much less. Such statistics as are available suggest that the majority of non-industrial building and non-housing building is financed by the banking, finance and insurance sectors or by property developers.

Selection of design consultants

The selection of design consultants is not regulated by law in the UK, even in the public sector. The public sector client, where there is no in-house department, selects the designer - normally an architect for building work and civil engineering consultant for civil work - from a list of designers considered capable of undertaking the work. Interviews may be conducted with a short list of designers who will present their credentials, including relevant experience, and may discuss their ideas for the design. Fees are negotiable and have, in recent years, become an element in the selection process. There are in-house design teams for some civil engineering work but the largest is the Property Services Agency (PSA) which is in charge of government buildings. However the general move towards privatization in the UK means that many in-house teams, including the PSA, are being disbanded.

In the private sector, the procedures are similar to the public sector, but with more variability, for example, the private client may appoint a quantity surveyor, even before an architect, if the project is one where cost is critical.

Contractual arrangements

Probably about two-thirds of contracts are awarded on a lump sum basis as a result of competitive tenders received from a short list of firms that have been selected by client's consultants. The National Joint Consultative Committee, which comprises a group of client and consultant bodies, publishes codes for selective tendering. Generally, prices are based on firm bills of quantities though sometimes approximate bills of quantities or specifications and drawings are used.

The predominant form of contract used in the UK is the Joint Contracts Tribunal (JCT) 1980 Standard Form of Contract with quantities. This contract assumes the use of measured bills of quantities that are normally prepared by quantity surveyors. The JCT also produce the 1984 Intermediate Form of Building Contract for works of simpler content. The tender documentation under both these forms of contract might comprise:

- drawings
- specification
- bills of quantities, schedules of works or schedules of rates.

The JCT building contracts are produced in private and local authority editions. Government buildings and civil engineering contracts are placed using the General Conditions of Government Contracts for building and Civil Engineering Works, otherwise known as Form GC/Works/1.

For civil engineering, local authority and private clients generally use the Institution of Civil Engineers Conditions of Contract, fifth edition, but an amended version of GC/Works/1 is also sometimes used.

Contracts of up to 18 months duration or less are generally let on a fixed price basis. Fluctuations in labour, materials and plant costs on longer term contracts can be adjusted, where the contract permits. Such increased costs on private contracts may be paid on the basis of invoices for materials and plant, and time sheets for labour. Local authority government contracts and some private contracts generally involve the use of an adjustment formula and monthly published indices.

A number of alternative contractual arrangements are available, most notably management contracting, design-and-build contracting and construction management. Under management contracting, the client enters into separate contracts with a designer and a management contractor and the management contractor enters into subcontracts with works contractors. However, construction management, in which the client enters into separate contracts with a designer, a construction manager and works contractors is increasingly being used in preference to management contracting.

There are separate contract forms for the other types of construction arrangements. In addition, there are some contracts notably the Association of Consulting Architects (ACA) British Property Federation Contract, which have been developed as an alternative to the JCT form, but their use is not significant.

Liability and insurance

Since the mid 1970s there has been a marked increase in litigation on professional liability. Liability may arise in contract or in tort. In contract, proof is required of a breach of contract. However, contracts do not normally define clearly the limits of work or duties of the professional and are in any case often informal, thus giving plenty of opportunity for litigation. The claims under tort, though increasing in the 1970s, have more recently declined as a result of various legal judgements.

Largely because of the reliance on case law made by the courts, the great problem of the English system is its uncertainty and complexity. There is no certainty on whether liability exists, on the amount involved, the period of liability or the time lag before any liability is determined.

Not all professional practices carry professional indemnity insurance. However most larger ones are insured. The cost of premiums is now very high and has been estimated at an average of 11% of professional labour costs.

The proposals of the European Commission for a standard liability throughout the Community may substantially alter the UK position and create greater certainty.

Development control and standards

The system for planning and control of development was introduced by the Town and Country Planning Act 1947. Although this has been amended, notably by the Town and

Country Planning Act 1971, the principles remain the same. The main responsibility for planning lies with local authorities: the county councils and metropolitan areas and large towns as well as the district councils. A development plan for each area must be produced and every development (which is very widely defined) must receive permission from the relevant authority.

In general, applications for development must be made to the relevant district council. The total volume of applications is about 400 000 a year, of which about 44% are householder applications for very small alterations to dwellings; 12% are for changes of use not involving building works; 40% are minor developments and only 4% are major developments of more than 10 dwellings or 1000 square metres of floor space. Most planning applications are therefore simple and go through a process which varies from district to district but is straightforward. Judgement of the local planning officer and ultimately of the elected council members has an important role and therefore there is no certainty that an application will be granted. Negotiation on changes to the original application which could be acceptable is often possible. About 6% of applications by householders and nearly 20% of those for change of use are refused. On average 10% of all applications are refused.

Decisions on planning applications have to be given within 8 weeks unless extension of the period is arranged. About 60% to 70% of planning applications are dealt with within this period.

All new construction in the UK has to comply with the 1985 Building Regulations, which are couched in terms of a series of technical requirements. These requirements are backed up by Approved Documents which set out ways in which the requirements can be met. These documents give guidance but are not mandatory. In many areas the Approved Document will refer a designer to an appropriate British Standard on complex issues. At present, approval for construction work is given by local authorities. In the case of private sector housing the National House Builders Council (NHBC) can act as an alternative Approved Inspector. The 1985 Building Act has provision for other persons or bodies to act as Approved Inspectors, but to date this has not happened, primarily because of the problem of liability insurance.

Further reading

Association of Consulting Engineers et al, Engineering Consultancy in the European Community: United Kingdom, London: Telford, 1990.

H W E Davies, "England" in Department of the Environment, Planning Control in Western Europe, London: HMSO, 1989, pp 33-78.

Patricia M Hillebrandt, Analysis of the British Construction Industry, London: Macmillan, 1984.

Department of the Environment, Housing and Construction Statistics, 1979-89, London: HMSO, 1990.

CONSTRUCTION COST DATA

Cost of labour

The figures below are typical of labour costs in London as at January 1991. The "wage rate" is the basis of an employee's income, while the "cost of labour" indicates the cost to a contractor of employing that employee. The difference between the two covers a variety of mandatory and voluntary contributions - a list of items which could be included is given on in section 4.

	Wage rate (per hour) £	*Cost of labour (per hour)* £	*Number of hours worked per year*
Site operatives			
Mason/bricklayer	4.12	5.32	2063
Carpenter	4.12	5.32	2063
Plumber	4.22	6.05	2034
Electrician	4.54	5.76	1755
Structural steel erector	4.75	6.41	2063
HVAC installer	4.68	6.50	1756
Semi-skilled worker	3.61	4.78	2063
Unskilled labourer	3.54	4.63	2063
Equipment operator	3.81	4.94	2063
Watchman/security	3.61	4.78	2063
Site supervision			
General foreman	4.46	5.74	2063
Trades foreman	3.84	4.97	2063
	(per year)	*(per year)*	
Clerk of works	13614		
Contractors' personnel			
Site manager	23 200	31 460	
Resident engineer	15 050	20 408	
Resident surveyor	18 240	24 730	
Junior engineer	13 250	16 175	
Junior surveyor	13 490	16 470	
Planner	19 850	26 920	
Consultants' personnel			
Senior architect	20 950	28 400	
Senior engineer	20 750	28 130	
Senior surveyor	21 830	29 600	
Qualified architect	18 900	25 630	
Qualified engineer	18 400	24 950	
Qualified surveyor	20 400	27 663	

Cost of materials

The figures that follow are the costs of main construction materials, delivered to site in the London area, as incurred by contractors in January 1991. These assume that the materials would be in quantities as required for a medium sized construction project and that the location of the works would be neither constrained nor remote.

All the costs in this section exclude value added tax (VAT - see below).

	Units	*Cost £*
Cement and aggregate		
Ordinary portland cement in 50kg bags	tonne	64.01
Coarse aggregates for concrete	m^3	9.80
Fine aggregates for concrete	m^3	8.40
Ready mixed concrete (11.50N/mm^2)	m^3	39.49
Ready mixed concrete (26.00N/mm^2)	m^3	44.32
Steel		
Mild steel reinforcement (12mm)	tonne	369.45
High tensile steel reinforcement (12mm)	tonne	373.05
Structural steel sections	tonne	388.13
Bricks and blocks		
Common bricks (215 x 102.5 x 65mm)	1000	120.00
Good quality facing bricks (215 x 102.5 x 65mm)	1000	320.00
Hollow concrete blocks (450 x 225 x 140mm)	1000	736.00
Solid concrete blocks (450 x 225 x 140mm)	1000	789.00
Precast concrete cladding units with exposed aggregate finish	m^2	75.00
Timber and insulation		
Softwood sections for carpentry	m^3	154.85
Softwood for joinery	m^3	310.00
Hardwood for joinery	m^3	500.00
Exterior quality plywood (18mm)	m^2	9.80
Plywood for interior joinery (6mm)	m^2	3.11
Softwood strip flooring (22mm)	m^2	45.05
Chipboard sheet flooring (18mm)	m^2	2.70
100mm thick quilt insulation	m^2	2.17
100mm thick rigid slab insulation	m^2	3.50
Softwood internal door complete with frames and ironmongery (826 x 2 040 x 40mm)	nr	40.65
Glass and ceramics		
Float glass (5mm)	m^2	23.40
Sealed double glazing units	m^2	51.88

	Units	Cost £
Plaster and paint		
Good quality ceramic wall tiles (198 x 64.5mm)	m^2	19.95
Plaster in 50 kg bags	tonne	49.82
Plasterboard (6mm thick)	m^2	1.25
Emulsion paint in 5 litre tins	litre	8.69
Gloss oil paint in 5 litre tins	litre	10.62
Tiles and paviors		
Clay floor tiles (150 x 150 x 12.5mm)	m^2	8.89
Vinyl floor tiles (250 x 250 x 2.5mm)	m^2	3.19
Precast concrete paving slabs (200 x 100 x 65mm)	m^2	6.37
Clay roof tiles (plain 265 x 165mm)	1000	262.20
Precast concrete roof tiles (419 x 330mm)	1000	595.65
Drainage		
WC suite complete	nr	135.67
Lavatory basin complete	nr	96.83
100mm diameter clay drain pipes	m	3.17
150mm diameter cast iron drain pipes	m	24.15

Unit rates

The descriptions below are generally shortened versions of standard descriptions listed in five languages (English, French, Italian, German and Spanish) in section 6. Where an item has a two digit reference number (e.g. 05 or 33), this relates to the full description against that number in section 6. Where an item has an alphabetic suffix (e.g. 12A or 34B) this indicates that the standard description has been modified. Where a modification is major the complete modified description is included here and the standard description should be ignored; where a modification is minor (e.g. the insertion of a named hardwood) the shortened description has been modified here but, in general, the full description in section 6 prevails.

The unit rates below are for main work items on a typical construction project in the London area in January 1991. The rates include all necessary labour, materials, equipment and, where appropriate, allowances for contractors overheads and profit, preliminary and general items and contractors profit and attendance on specialist trades. All the rates in this section exclude value added tax (VAT - see below).

		Unit	Rate £
Excavation			
01	Mechanical excavation of foundation trenches	m^3	5.50
02	Hardcore filling making up levels	m^2	1.38
03	Earthwork support	m^2	1.05

		Unit	Rate £
Concrete work			
04	Plain insitu concrete in strip foundations in trenches	m^3	55.00
05	Reinforced insitu concrete in beds	m^3	54.60
06	Reinforced insitu concrete in walls	m^3	64.50
07	Reinforced insitu concrete in suspended floor or roof slabs	m^3	63.30
08	Reinforced insitu concrete in columns	m^3	77.80
09	Reinforced insitu concrete in isolated beams	m^3	72.00
10	Precast concrete slab	m^2	30.90
Formwork			
11	Softwood formwork to concrete walls	m^2	19.10
12	Softwood or metal formwork to concrete columns	m^2	21.10
13	Softwood or metal formwork to horizontal soffits of slabs	m^2	17.20
Reinforcement			
14	Reinforcement in concrete walls	tonne	528.00
15	Reinforcement in suspended concrete slabs	tonne	487.00
16	Fabric reinforcement in concrete beds	m^2	1.93
Steelwork			
17	Fabricate, supply and erect steel framed structure	tonne	1050.00
Brickwork and blockwork			
18	Precast lightweight aggregate hollow concrete block walls	m^2	19.50
19	Solid (perforated) concrete blocks	m^2	20.50
20	Solid (perforated) sand lime bricks	m^2	25.20
21	Facing bricks	m^2	37.40
Roofing			
22	Concrete interlocking roof tiles 430 x 380mm	m^2	13.50
23	Plain clay roof tiles 260 x 160mm	m^2	31.80
24	Fibre cement roof slates 600 x 300mm	m^2	18.40
25	Sawn softwood roof boarding	m^2	9.18
26	Particle board roof coverings	m^2	5.75
27	3 layers glass-fibre based bitumen felt roof covering	m^2	16.30
28	Bitumen based mastic asphalt roof covering	m^2	13.40
29	Glass-fibre mat roof insulation 160mm thick	m^2	6.90
30	Rigid sheet loadbearing roof insulation 75mm thick	m^2	3.49
31	Troughed galvanized steel roof cladding	m^2	18.30
Woodwork and metalwork			
32	Preservative treated sawn softwood 50 x 100mm	m	2.61
33	Preservative treated sawn softwood 50 x 150mm	m	3.42
34A	Single glazed casement window in Philippine Mahogany hardwood, size 650 x 900mm	each	88.90

		Unit	Rate £
35A	Two panel glazed door in W African Mahogany hardwood, size 850 x 2000mm	each	265.00
36	Solid core half hour fire resisting hardwood internal flush doors, size 800 x 2000mm	each	90.30
37A	Aluminium double glazed window, size 1200 x 1200mm	each	398.00
38A	Aluminium double glazed door, size 850 x 2100mm	each	1160.00
39A	Hardwood skirtings (W African Mahogany)	m	4.29
40	Framed structural steelwork in universal joist sections	tonne	1050.00
41	Structural steelwork lattice roof trusses	tonne	1290.00
Plumbing			
42	UPVC half round eaves gutter	m	4.67
43	UPVC rainwater pipes	m	5.63
44	Light gauge copper cold water tubing	m	3.57
45	High pressure plastic for cold water supply	m	2 49
46	Low pressure plastic pipes for cold water distribution	m	3.02
47	UPVC soil and vent pipes	m	8.34
48	White vitreous china WC suite	each	184.00
49	White vitreous china lavatory basin	each	133.00
50	Glazed fireclay shower tray	each	157.00
51	Stainless steel single bowl sink and double drainer	each	117.00
Electrical work			
52	PVC insulated and copper sheathed cable	m	3.07
53	13 amp unswitched socket outlet	each	14.60
54	Flush mounted 20 amp, 1 way light switch	each	40 90
Finishings			
55	2 coats gypsum based plaster on brick walls	m^2	6.17
56	White glazed tiles on plaster walls	m^2	22.60
57	Red clay quarry tiles on concrete floor	m^2	19.10
58	Cement and sand screed to concrete floors	m^2	7.38
59	Thermoplastic floor tiles on screed	m^2	4.84
60	Mineral fibre tiles on concealed suspension system	m^2	28.30
Glazing			
61	Glazing to wood	m^2	19.00
Painting			
62	Emulsion on plaster walls	m^2	1.53
63	Oil paint on timber	m^2	4.11

Approximate estimating

The building costs per unit area given below are averages incurred by building clients for typical buildings in the United Kingdom as at January 1991. They are based upon the total floor area of all storeys, measured between external walls and without deduction for internal walls.

Approximate estimating costs generally include for mechanical and electrical installations but exclude furniture, loose or special equipment, and external works; they also exclude fees for professional services. The costs shown are for specifications and standards appropriate to United Kingdom and this should be borne in mind when attempting comparisons with similarly described building types in other countries. A discussion of this issue is included in section 4. Comparative data for countries covered in this publication, including construction cost data are presented in Part Three.

Approximate estimating costs must be treated with caution; they cannot provide more than a rough guide to the probable cost of building.

All the rates in this section exclude value added tax (VAT - see below).

	Cost m^2 *£*	*Cost* ft^2 *£*
Industrial buildings		
Factories for letting	200	19
Factories for owner occupation (light industrial use)	350	33
Factories for owner occupation (heavy industrial use)	570	53
Factory/office (high-tech) for letting (shell and core only)	360	33
Factory/office (high-tech) for letting (ground floor shell, first floor offices)	570	53
Factory/office (high tech) for owner occupation (controlled environment, fully finished)	750	70
High tech laboratory (air conditioned)	1530	142
Warehouses, low bay (6 to 8m high) for letting (no heating)	190	18
Warehouses, low bay for owner occupation (including heating)	270	25
Warehouses, high bay for owner occupation (including heating)	365	34
Cold stores/refrigerated stores	405	38
Administrative and commercial buildings		
Civic offices, non air conditioned	715	66
Civic offices, fully air conditioned	865	80
Offices for letting, 5 to 10 storeys, non air conditioned	615	57
Offices for letting, 5 to 10 storeys, air conditioned	765	71
Offices for letting, high rise, air conditioned	970	90
Offices for owner occupation 5 to 10 storeys, non air conditioned	715	66
Offices for owner occupation 5 to 10 storeys, air conditioned	935	87
Offices for owner occupation high rise, air conditioned	1500	139

	Cost m² £	Cost ft² £
Prestige/headquarters office, 5 to 10 storeys, air conditioned	1190	111
Prestige/headquarters office, high rise, air conditioned	1550	144
Health and education buildings		
General hospitals	765	71
Teaching hospitals	740	70
Private hospitals	855	79
Health centres	585	54
Nursery schools	705	66
Primary/junior schools	545	51
Secondary/middle schools	495	46
University (arts) buildings	615	57
University (science) buildings	725	67
Management training centres	740	69
Recreation and arts buildings		
Theatres (over 500 seats) including seating and stage equipment	1090	101
Theatres (less than 500 seats) including seating and stage equipment	805	75
Concert halls including seating and stage equipment	1530	142
Sports halls including changing and social facilities	525	49
Swimming pools (international standard) including changing and social facilities	915	85
Swimming pools (schools standard) including changing facilities	650	60
National museums including full air conditioning and standby generator	1885	175
Local museums including air conditioning	765	71
City centre/central libraries	775	72
Branch/local libraries	625	58
Residential buildings		
Social/economic single family housing (multiple units)	315	29
Private/mass market single family housing 2 storey detached/semidetached (multiple units)	355	33
Purpose designed single family housing 2 storey detached (single unit)	545	51
Social/economic apartment housing, low rise (no lifts)	435	40
Social/economic apartment housing, high rise (with lifts)	465	43
Private sector apartment building (standard specification)	425	39
Private sector apartment buildings (luxury)	695	66
Student/nurses halls of residence	505	47
Homes for the elderly (shared accommodation)	555	52

	Cost m^2 £	Cost ft^2 £
Homes for the elderly (self contained with shared communal facilities)	455	42
Hotel, 5 star, city centre	1250	116
Hotel, 3 star, city/provincial	975	91
Motel	715	66

Regional variations

The approximate estimating costs are based on average UK rates. Adjust these costs by the following factors for regional variations:

Greater London	:+12%	North West	: +1%
South East	: +4%	Northern	: -3%
South West	: -1%	Scotland	: +2%
Midlands	: -3%	Wales	: -7%
East Anglia	: -3%	Northern Ireland	: -24%
Yorkshire and Humberside	: -3%		

Value added tax (VAT)

The standard rate of value added tax (VAT) is currently 17.5% (15% in January 1991), chargeable on general building work.

EXCHANGE RATES AND INFLATION

The combined effect of exchange rates and inflation on prices within a country and price comparisons between countries is discussed in section 4.

Exchange rates

The graph below plots the movement of sterling and the US dollar since 1980. The figures used for the graph are quarterly and the method of calculating these, and other related issues are discussed in section 4. The exchange rate at January 1991 was £0.518 to the US dollar.

Sterling and the US Dollar

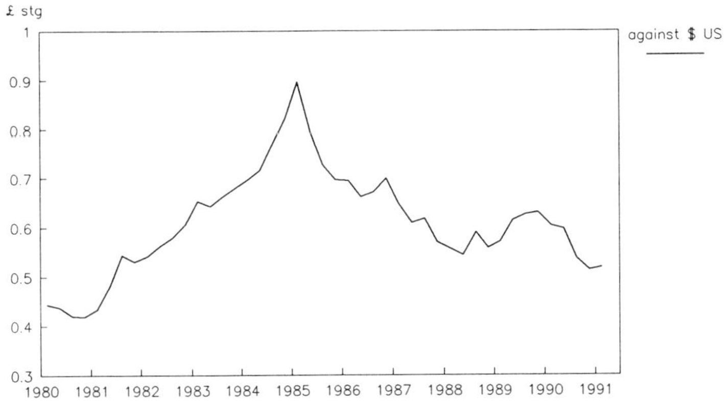

Inflation

The graph below presents a number of indices showing the trends in inflation, prices and costs. All the indices have been rebased to 1980=100. Notes on inflation and the distinction between costs and prices are included in section 4.

Cost and price indices

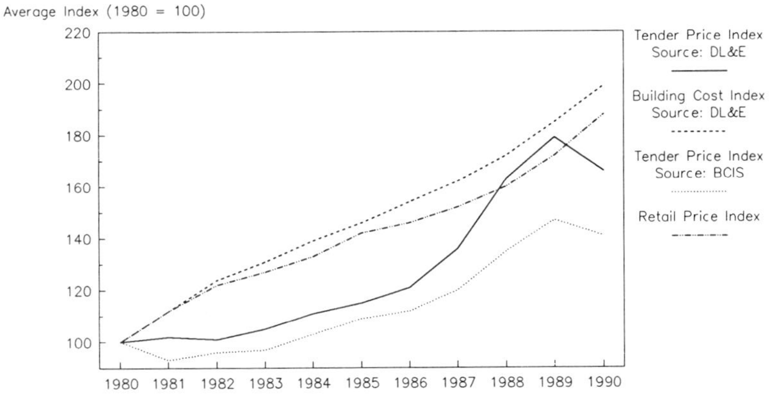

Consumer price inflation

The basis of the figures is the official consumer price index converted to quarterly averages. Inflation peaked in 1981; but since 1986 has been rising steadily to over 9% in 1990, the highest rate since 1981.

The table below presents general inflation in United Kingdom since 1980. The annual change is the percentage change between the last quarters of consecutive years.

Retail price index 1980 = 100

	1981	1982	1983	1984	1985	1986	1987	1988	1989	1990
average index	112	122	127	133	142	146	152	160	172	188
annual change (%)	12.0	8.9	4.1	4.7	6.8	2.8	4.1	5.3	7.5	9.3

Building cost and price inflation

Two of the three tables that follow have been produced by Davis Langdon & Everest: the building cost index provides an index of price movements in general building costs, and the tender price index indicates movements in general building prices in the Greater London area.

Building cost index 1980 = 100

	1981	1982	1983	1984	1985	1986	1987	1988	1989	1990
average index	112	124	131	139	146	154	162	172	185	199
annual change (%)	12.0	10.7	5.6	6.1	5.0	5.5	5.2	6.2	7.6	7.6

Tender price index 1980 = 100

	1981	1982	1983	1984	1985	1986	1987	1988	1989	1990
average index	102	101	105	111	115	121	136	163	179	166
annual change (%)	2.0	-1.0	4.0	5.7	3.6	5.2	12.4	19.9	9.8	-7.3

The table below is prepared by the Building Cost Information Service (BCIS) of the RICS and reflects movements in general building prices across the UK.

BCIS tender price index 1980 = 100

	1981	1982	1983	1984	1985	1986	1987	1988	1989	1990
average index	93	96	97	103	109	112	120	135	147	141
annual change (%)	-7.0	3.2	1.0	6.2	5.8	2.8	7.1	12.5	8.9	-4.1

USEFUL ADDRESSES

National Economic Development Office
Millbank Tower, Millbank
London SW1P 4QX

Department of the Environment
2 Marsham Street
London SW1P 3EB
Tel: (071) 212 3000

Building Research Establishment
Bucknalls Lane, Garston
Watford
Tel: Garston 674040

Construction Industry Research and Information Association
6 Storeys Gate
London SW1P 3AL

Engineering Council
Canberra House
10-16 Maltravers Street
London WC2 3ER

Royal Institute of British Architects, (RIBA)
66 Portland Place
London W1N 4AD
Tel: (071) 580 5533

Institution of Civil Engineers
1-7 Great George Street
London SW1P 3AA

The Association of Consulting Engineers, (ACE)
Alliance House
12 Caxton Street
London SW1H 0QL
Tel: (071) 222 6557

Royal Institution of Chartered Surveyors (RICS)
12 Great George Street
London SW1P 3AD
Tel: (071) 222 7000

British Standards Institution
2 Park Street
London W1A 2BS
Tel: (071) 629 9000

Chartered Institute of Building Services (CIBS)
Delta House, 222 Balham High Road
London SW8 9BS
Tel: (081) 675 5211

Building Employers Confederation
82 New Cavendish Street
London W1M 8AD
Tel: (071) 580 5588

Transport and General Workers Union, Building Crafts Section
Transport House
Smith Square
London SW1P 3AJ
Tel: (071) 828 7788

Union of Construction and Allied Trades and Technicians
UCATT House
Abbeville Road
London SW4 9RL
Tel: (071) 622 2442

United States of America

KEY DATA

Population

Population (1989)	246.33m
Urban population	74%
Population under 15	21.5%
Population over 65	12.6%
Average annual growth rate (1983 to 1988)	1.0%

Geography

Land area	9 373 000 km²
Agricultural area	4 314 000 km²
Capital city	Washington DC (population 617 000)

Economy

Monetary unit	US Dollar ($)
Exchange rate (January 1991) to	
the ECU	$ 1.35
the pound sterling	$ 1.93
Average annual inflation (1984 to 1989)	3.6%
Gross Domestic Product (GDP)	$ 4881bn
GDP per capita (1989)	$ 19 815
GDP growth (real) (1980 to 1988)	3.0%
Private consumption (% of total)	66.4%
Public consumption (% ption (% of total)	19.8%
Investment (% of total)	15.6%
Central government expenditure as percentage of GDP	34.8%

Construction

Total value of construction output (1989)	$ 492bn
Construction output per capita	$ 2000
Construction output as a percentage of GDP	10%

THE CONSTRUCTION INDUSTRY

Construction output

The total value of gross output of the US construction industry in 1989 is estimated at US$492 billion. Of this, the value of new construction in 1989 was US$414 billion and repair and maintenance was US$78 billion. Construction output represents 10% of GDP. A breakdown by type of work is shown in the diagram below and in more detail in the accompanying table.

Construction output 1989

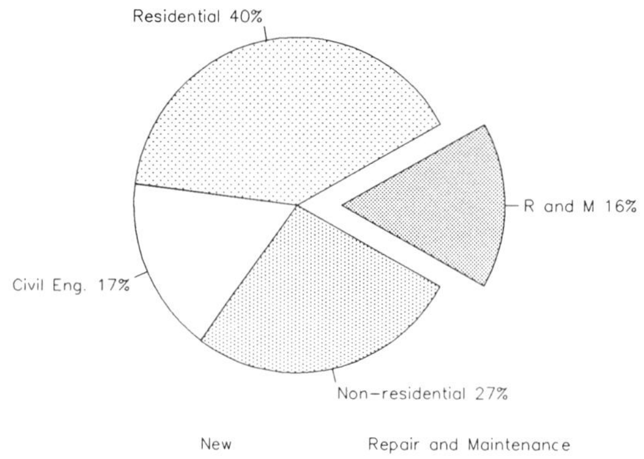

In 1990 the US construction output fell substantially from the high levels of 1989 in all regions of the USA. It can be seen from the table of output that the public sector accounts for only 20% of total construction output.

The US residential market is dominated by single unit housing and the market is drifting more towards this type of housing rather than multiple unit housing. The housing market in 1990 was depressed and is not expected to recover until about 1994.

With the exception of expenditure on educational facilities, public highways and sewerage systems, which has increased, the level of expenditure on public construction has remained fairly static since 1985.

Construction output 1989: detailed breakdown

		Billion US$	%
New work			
Residential			
Private	housing and redevelopment	195.4	39.7
Public	housing and redevelopment	1.8	0.4
Total residential building		197.2	40.0
Non-residential building			
Private	Industrial	16.8	3.4
	Commercial	29.7	6.0
	Offices	27.8	5.6
	Hotels	7.7	1.6
	Education	3.3	0.7
	Health	7.4	1.5
	Misc.	10.8	2.2
	Total private	103.5	21.0
Public	Industrial	1.3	0.3
	Education	13.1	2.7
	Hospitals	2.1	0.4
	Misc.	11.8	2.4
	Total public	28.3	5.7
Total Non-residential building		131.8	26.8
Civil engineering			
Private	Telecommunications	8.6	1.7
	Railways)		
	Electricity)	22.8	4.6
	Gas)		
	Petroleum pipelines)		
	Total private	31.4	6.4
Public	Highways and streets	27.7	5.6
	Sewerage and water supply	12.7	2.6
	Misc.	13.5	2.7
	Total public	53.9	10.9
Total civil engineering		85.3	17.3
Total new work		414.3	84.1
Repair and maintenance		78.0	15.9
TOTAL ALL WORK		492.3	100

Source: US Department of Commerce

The geographical distribution of work other than housing, compared to that of population, is fairly even. The most notable exception is California, which, with a little over 10% of the population, has a 14% share of construction output. New York, for example, with 8% of the population has a 7% share of construction output.

Characteristics and structure of the industry

Measured in terms of the value of contracts won, the USA was the base for six of the world's top ten contractors in 1989, the other four being Japanese. The table below shows the leading US contractors in 1989 from the Engineering News Record's "Top 250 International Contractors". The leading international contractor was Brown and Root.

Leading US contractors 1989 (by contracts won in US$ millions)

Company	*Total US$ millions*	*% of work in USA*
Fluor-Daniel	16 646.9	25.4
Bechtel	12 009.8	55.1
Brown and Root	10 976.7	65.3
Parsons Corporation	9711.0	28.6
M W Kellogg	9394.0	59.2
CRSS Inc.	7984.0	1.3
Rust International	5923.0	0.9
ABB Lummus Crest	5493.0	80.0
Foster Wheeler	4108.0	71.8
Jacobs Engineering Group	3710.0	10.2
Turner Corporation	3386.4	2.0
Ebasco Services	3222.9	3.1
Kiewit Construction Group	1898.2	6.6
The Austin Company	1600.0	25.0
Stone and Webster	1280.0	41.0
United Engineers and Construction International	1260.0	0.8
Perini Corporation	1236.0	10.7
Guy F Atkinson Co. of California	1109.6	34.2
Schal Associates	1037.2	32.1
McDermott Inc.	831.0	40.7

Source: McGraw-Hill Inc.; Engineering News Record

There are a large number of specialist trade contractors and they play an important role. They usually have to provide working drawings, they are used to organising and managing the work on site with little direction from the main contractor and they often supply major items of plant and equipment. Labour-only subcontracting is very little used.

Some states require contractors to be licenced, though this is rarely strictly administered. It is a way of gaining some revenue from the licence fee.

The title of architect is protected in the USA and the regulations for registration vary from state to state. However, the National Council of Architectural Registration Boards grants a certificate to a qualified architect which is usually recognized in all states. There are about 60 000 architects (1987 data, 250 per million inhabitants), mostly in very small practices. The US architect is not much involved in site operations and much detailed design is done by contractors. There are very few quantity surveyors in the USA and the architect is more involved with cost than in the UK. However, there are construction cost consultants who may originally have been architects or engineers but are increasingly being augmented by quantity surveyors. The contractors are often prepared to give cost advice to the architect. There are many more building engineers than architects: 341 000, or 1427 per million inhabitants. Engineers have to be registered, which generally requires a recognized engineering degree and 4 years work experience. There are a number of substantial multi-disciplinary practices in the USA.

The contracting industry employs about 3.5 million persons and there are about 15 million working partners or self-employed proprietors. About 40% of main contractors and trade contractors are union contractors which use union registered employees and negotiate wages with the unions. Open-shop contracting has grown significantly over the last 20 years, especially in housebuilding. This growth has moderated the behaviour and wage demands of unions. One adverse affect, however, is that with the decline in union influence has come a decline in training, as the unions traditionally run good training programmes.

Selection of design consultants

In 1972 Congress established as Federal Law a policy to select architects and engineers on the basis of the highest qualification for each project and at a fair and reasonable price. For large public projects, invitations are published for interested architects and engineers (usually only in the State where the project is located) to indicate their interest and to submit detailed, specific information on their qualifications. A panel of private sector architects and engineers, who are not paid, recommends five firms. These five make presentations, attend interviews, and so on, and three are chosen. The most favoured of these enters into negotiations with the client. If these break down, the second negotiates, and so on. The process is costly to firms entering for a project.

Several states have followed the Federal example and the American Institute of Architects (AIA) recommends the procedure for private clients. It is often followed by the large corporations, though sometimes in a form which gives earlier prominence to estimated construction prices and fees.

Contractual arrangements

The most usual methods of selecting a general contractor are by competitive bidding, by negotiation or by a combination of the two. There are two types of competitive bidding, "open" and "closed". Open is the predominant type, where all contractors use the same proposal form. In the closed type, the competing contractors are required to submit their qualifications along with their bids and are encouraged to suggest cost saving proposals. There are numerous forms of negotiated contract, but most are of the cost-plus-fee type. Negotiated contracts are normally limited to privately financed work since competitive bidding is a legal requirement for most public projects.

Fixed price contracts are the most common. Tenders for buildings are customarily prepared on a lump sum basis, whereas engineering projects are generally bid as a series of unit prices. It is standard practice for contractors to prepare their own quantities which do not form part of the contract. With few exceptions, bids should be accompanied by a bid bond guaranteeing that the contractor will enter into a contract if declared successful.

Standard contract conditions have been developed by various bodies, including the American Institute of Architects, the National Society of Professional Engineers, the Associated General Contractors of America and various Federal, State and Municipal governments. Where a contract provides for arbitration, most stipulate that it shall be conducted under the Construction Industry Arbitration Association.

There has been an increasing use for large projects of management fee or construction management arrangements, but often still retaining a guaranteed maximum price. However the tendency is now less apparent. Design and build projects are also becoming more popular though there are often other provisions for the contractor to offer advice at the design stage.

Specialist trade contractors are usually invited to bid, often from a list selected or approved by the architect. Eight to ten bidders are usual. Nomination is virtually unknown.

Development control and standards

The planning process in the USA is known as planning control and zoning control. It is very fragmented and every town has its own system. There may be 50 separate zoning authorities in one state. There is normally a Zoning Commission Board, a Zoning Board of Appeal and often a Planning Commission or Board in each town. The ease with which development zones of a town can be changed varies according to the attitude of the town or the state.

There is no single national building code for the whole of the USA. Approximately 19 000 municipalities are involved and many have their separate codes. Nevertheless, various national codes have been prepared. The most widely used is the International Conference of Building Officials (ICBO) Uniform Building Code. Others are the Building Officials and Code Administrators International (BOCA), and the Southern Building Code Congress International (SBCC). There are also specialist codes for fire safety, etc. The codes are basically performance codes rather than specifications for the form of construction. Several organizations are working on harmonization of codes, notably the National Institute of Building Sciences (NIBS), a non-governmental institution set up with representation from all parts of the building community.

The specific arrangements for obtaining planning permission and the statutory period for approval varies from State to State. Once the plans have been passed and construction has commenced, field inspection takes place. This is generally regarded as very important and the number of visits are often specified in the codes.

Standards are referred to continually in the building codes. They may be mandatory or discretionary. There are some 150 organizations which develop standards of which perhaps a dozen or so are important. These include the American Society for Testing and Materials (ASTM), the Amercian National Standards Institute (ANSI) and the Amercian Insurance Association (AIA).

Liability and insurance

The contractor is liable for damages caused by his own acts of omission and commission. He must therefore obtain comprehensive liability insurance to protect himself and his subcontractors.

The liability of designers and contractors varies with the contract used and from state to state. In the USA the architect or engineer has a contractual obligation to check the shop drawings of specialist trade contractors and this affects the liability. Normally, professional liability extends 3 to 4 years, but in some circumstances it can extend up to 10 years.

Professional indemnity insurance covers the liability of parties involved in design, except that trade contractors may not be covered or, if they are, then insufficiently so. Professional indemnity insurance is, in any case, very expensive in the USA. A survey of 1986 found that liability insurance was the major concern of the industry. Many companies in 1989 were paying double the insurance premiums of 1984, often with a reduced coverage.

Further reading

Engineering News Record

Statistical publications of US Department of Commerce, for example: US Department of Commerce, Bureau of the census, Statistical Abstract of the United States 1990, Washington DC, 1990.

McGraw Hill and MDIS Constructis, October 1990 and other issues.

CONSTRUCTION COST DATA

Cost of labour

The figures below are typical of labour costs in Washington DC as at January 1991. The "wage rate" is the basis of an employee's income, while the "cost of labour" indicates the cost to a contractor of employing that employee. The difference between the two covers a variety of mandatory and voluntary contributions - a list of items which could be included is given in section 4.

	Wage rate (per hour) US$	*Cost of labour (per hour) US$*	*Number of hours worked per year*
Site operatives			
Mason/bricklayer	22.75	34.20	2000
Carpenter	22.00	33.75	2000
Plumber	25.45	37.95	2000
Electrician	25.15	37.15	2000
Structural steel erector	24.10	42.20	2000
HVAC installer	25.00	38.00	2000
Semi-skilled worker	17.50	26.85	2000
Unskilled labourer	17.10	26.50	2000
Equipment operator	22.50	33.75	2000
Site supervision			
General foreman	23.15	35.55	2000
Trades foreman	24.65	38.30	2000
	(per week)	*(per week)*	
Clerk of works	585	890	2000
Contractors' personnel	215	330	2000
Site manager	1005	1530	2000
Resident engineer	670	1020	2000
Resident surveyor	670	1020	2000
Junior engineer	505	765	2000
Junior surveyor	505	765	2000

Cost of materials

The figures that follow are the US national average costs for main construction materials, delivered to site in the Washington DC area, as incurred by contractors in January 1991. These assume that the materials would be in quantities as required for a medium sized construction project and that the location of the works would be neither constrained nor remote.

	Units	*Cost US$*
Cement and aggregate		
Ordinary portland cement in 50kg bags	bag	6.55
Coarse aggregates for concrete	ton	9.70
Fine aggregates for concrete	ton	10.88
Ready mixed concrete (mix 2500 psi)	yd^3	49.45
Ready mixed concrete (mix 3000 psi)	yd^3	51.05
Steel		
Mild steel reinforcement	ton	555.00
High tensile steel reinforcement	ton	605.00
Bricks and blocks		
Common bricks (8" x 2.67" x 4")	1000	250.00
Good quality facing bricks (8" x 2.67" x 4")	1000	385.00
Hollow concrete blocks (8" x 8" x 16") 4000 psi	each	1.12
Solid concrete blocks (4" x 8" x 16") 4000 psi	each	0.98
Precast concrete cladding units with exposed aggregate finish	ft^2	10.65
Timber and insulation		
Softwood sections for carpentry	m^3	937.00
Exterior quality plywood	m^2	360.00
Plywood for interior joinery	ft^2	1.50
Softwood strip flooring	ft^2	1.85
Chipboard sheet flooring	ft^2	0.33
100mm thick quilt insulation	ft^2	0.21
100mm thick rigid slab insulation	ft^2	0.88
Softwood internal door complete with frames and ironmongery	nr	175.00
Glass and ceramics		
Float glass (0.25" thick)	ft^2	3.05
Sealed double glazing units (0.625")	ft^2	7.15

	Units	Cost US$
Plaster and paint		
Good quality ceramic wall tiles (4.5" x 4.5")	ft^2	1.85
Plaster in 50 kg bags	bag	16.05
Plasterboard (0.5" thick)	yd^2	3.54
Gloss oil paint in one gallon tins	gallon	17.25
Tiles and paviors		
Clay floor tiles (4" x 4" x 0.5")	ft^2	2.90
Vinyl floor tiles (12" x 12" x 0.125")	ft^2	1.75
Clay roof tiles (158 pcs per 100 sq ft)	$100ft^2$	275.00
Precast concrete roof tiles (158 pcs per 100 sq ft)	$100ft^2$	54.00
Drainage		
WC suite complete	nr	339.00
Lavatory basin complete	nr	287.00
100mm diameter clay drain pipes	ft	2.95
150mm diameter cast iron drain pipes	ft	6.12

Unit rates

The descriptions below are generally shortened versions of standard descriptions listed in five languages (English, French, Italian, German and Spanish) in section 6. Where an item has a two digit reference number (eg 05 or 33), this relates to the full description against that number in section 6. Where an item has an alphabetic suffix (eg 12A or 34B) this indicates that the standard description has been modified. Where a modification is major the complete modified description is included here and the standard description should be ignored; where a modification is minor (eg the insertion of a named hardwood) the shortened description has been modified here but, in general, the full description in section 6 prevails.

The unit rates below are US national average rates for main work items on a typical construction project as at January 1991. The rates include all necessary labour, materials, equipment and, where appropriate, allowances for contractors overheads and profit, preliminary and general items and contractors profit and attendance on specialist trades.

		Unit	Rate US$
Excavation			
01	Mechanical excavation of foundation trenches	yd^3	6.97
02	Hardcore filling making up levels	yd^2	3.36
03	Earthwork support	ft^2	1.39

		Unit	Rate US$
Concrete work			
04	Plain insitu concrete in strip foundations in trenches	yd^3	11.85
05	Reinforced insitu concrete in beds	yd^3	26.00
06	Reinforced insitu concrete in walls	yd^3	15.80
07	Reinforced insitu concrete in suspended floor or roof slabs	yd^3	22.00
08	Reinforced insitu concrete in columns	yd^3	41.00
09	Reinforced insitu concrete in isolated beams	yd^3	41.00
10	Precast concrete slab	ft^2	4.39
Formwork			
11	Softwood formwork to concrete walls	ft^2	2.70
12	Softwood or metal formwork to concrete columns	ft	21.00
13	Softwood or metal formwork to horizontal soffits of slabs	ft^2	3.19
Reinforcement			
14	Reinforcement in concrete walls	ton	1025.00
15	Reinforcement in suspended concrete slabs	ton	1050.00
16	Fabric reinforcement in concrete beds	$100ft^2$	44.00
Brickwork and blockwork			
18	Precast lightweight aggregate hollow concrete block walls	ft^2	3.96
19	Solid (perforated) concrete blocks US$275 per m^2 delivered	1000	1175.00
20	Sand lime bricks US$265 per m^2 delivered	1000	1150.00
21	Facing bricks US$280 per m^2 delivered	1000	1200.00
Roofing			
22	Concrete interlocking roof tiles 430 x 380mm	100 ft^2	260.00
23	Plain clay roof tiles 260 x 160mm	100 ft^2	465.00
27	3 layers glass-fibre based bitumen felt roof covering	100 ft^2	130.00
28	Bitumen based mastic asphalt roof covering	100 ft^2	120.00
29	Glass-fibre mat roof insulation 160mm thick	ft^2	0.63
30	Rigid sheet loadbearing roof insulation 75mm thick	ft^2	1.39
31	Troughed galvanized steel roof cladding	ft^2	1.68
Woodwork and metalwork			
32A	Preservative treated sawn softwood 50mm x 100mm (7' high)	ft	8.05
33A	Preservative treated sawn softwood 50 x 150mm (1000 board feet)	mbf	955.00
34	Single glazed casement window in hardwood, size 650 x 900mm	each	170.00

		Unit	Rate US$
36	Solid core half hour fire resisting hardwood internal flush doors, size 800 x 2000mm	each	306.48
41	Structural steelwork lattice roof trusses	ton	1200.00

Plumbing

42	UPVC half round eaves gutter	ft	3.15
43	UPVC rainwater pipes	ft	2.29
44	Light gauge copper cold water tubing	ft	5.05
45	High pressure plastic for cold water supply	ft	6.30
46	Low pressure plastic pipes for cold water distribution	ft	6.70
47	UPVC soil and vent pipes	ft	15.00
48	White vitreous china WC suite	each	565.00
49	White vitreous china lavatory basin	each	313.00
50	Glazed fireclay shower tray	each	575.00
51	Stainless steel single bowl sink and double drainer	each	379.00

Electrical work

52	PVC insulated and copper sheathed cable	100 ft	345.00
53	13 amp unswitched socket outlet	each	66.41
54	Flush mounted 20 amp, 1 way light switch	each	63.00

Finishings

55	2 coats gypsum based plaster on brick walls	yd^2	15.25
56	White glazed tiles on plaster walls	ft^2	4.46
57	Red clay quarry tiles on concrete floor	ft^2	7.00
60	Mineral fibre tiles on concealed suspension system	ft^2	1.88

Glazing

61	Glazing to wood	ft^2	7.50

Painting

62	Emulsion on plaster walls	ft^2	1.75
63	Oil paint on timber	ft^2	0.95

Approximate estimating

The building costs per unit area given below are US national averages incurred by building clients for typical buildings as at January 1991. They are based upon the total floor area of all storeys, measured between external walls and without deduction for internal walls.

Approximate estimating costs generally include for mechanical and electrical installations but exclude furniture, loose or special equipment, and external works; they also exclude fees for professional services. The costs shown are for specifications and standards appropriate to the United States and this should be borne in mind when attempting comparisons with similarly described building types in other countries. A discussion of this issue is included in section 4. Comparative data for countries covered in this publication, including construction cost data are presented in Part Three.

Approximate estimating costs must be treated with reserve; they cannot provide more than a rough guide to the probable cost of building.

	Cost *m² US$*	*Cost* *ft² US$*
Industrial buildings		
Factories for letting	280	26.05
Factories for owner occupation (light industrial use)	390	36.35
Factories for owner occupation (heavy industrial use)	670	62.30
Warehouses, low bay (6 to 8m high) for letting (no heating)	225	20.70
Warehouses, low bay for owner occupation (including heating)	305	28.55
Warehouses, high bay for owner occupation (including heating)	475	44.05
Administrative and commercial buildings		
Offices for letting, 5 to 10 storeys, non air conditioned	570	52.95
Offices for letting, 5 to 10 storeys, air conditioned	700	65.05
Offices for letting, high rise, air conditioned	670	62.05
Offices for owner occupation, high rise, air conditioned	855	79.25
Prestige/headquarters office, 5 to 10 storeys, air conditioned	1060	98.30
Health and education buildings		
General hospitals	1110	103
Teaching hospitals	1320	123
Private hospitals	1850	172
Health centres	825	76.50
Primary/junior schools	705	65.40
Secondary/middle schools	710	65.95
Recreation and arts buildings		
Theatres (over 500 seats) including seating and stage equipment	660	61.35
Theatres (less than 500 seats) including seating and stage equipment	1000	92.85
Sports halls including changing and social facilities	555	92.85
Swimming pools (international standard) including changing and social facilities	1130	105
Swimming pools (schools standard) including changing facilities	800	74.15
City centre/central libraries	1060	98.75
Branch/local libraries	695	64.45

	Cost m² US$	Cost ft² US$
Residential buildings		
Private/mass market single family housing 2 storey detached/semidetached (multiple units)	450	41.65
Purpose designed single family housing 2 storey detached (single unit)	605	56.40
Social/economic apartment housing, low rise (no lifts)	380	35.50
Social/economic apartment housing, high rise (with lifts)	570	53.10
Private sector apartment building (standard specification)	595	55.40
Private sector apartment buildings (luxury)	735	68.20
Student/nurses halls of residence	930	86.35
Hotel, 5 star, city centre	935	67.00
Hotel, 3 star, city/provincial	790	73.20
Motel	635	58.85

Regional variations

The approximate estimating costs are based on US national average costs. Adjust these costs by the following factors for regional variations:

Los Angeles, CA.	:+12%
Hartford, CONN.	: -2%
Miami, FLA.	: -11%
New York, N.Y.	:+27%
Philadelphia, PENN.	: +9%
Providence, R.I.	: 0%
Fort Worth, Texas	: -12%
Columbia, S Carolina	: -20%
Seattle, Washington	: +3%
Dallas, Texas	: -13%

EXCHANGE RATES AND INFLATION

The combined effect of exchange rates and inflation on prices within a country and price comparisons between countries is discussed in section 4.

Exchange rates

The graph below plots the movement of the American dollar against sterling since 1980. The figures used for the graph are quarterly and the method of calculating these, and other related issues are discussed in section 4. The exchange rate at January 1991 was US$1.93 to the pound sterling.

The £ Sterling against the US Dollar

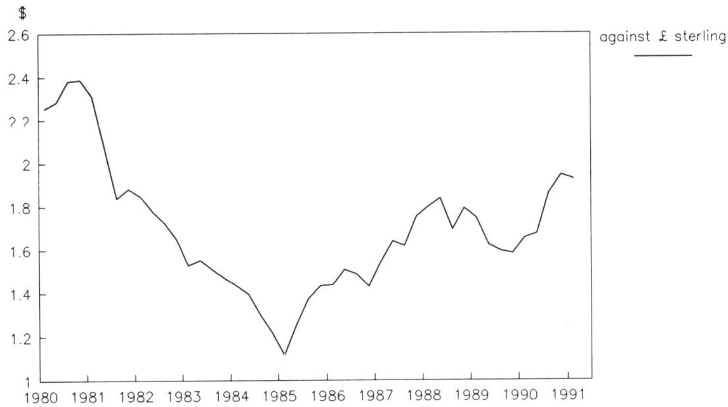

Inflation

The graph and the table below present the consumer price index showing the trend in inflation since 1982. The index has been rebased to 1982=100. Notes on inflation and the distinction between costs and prices are included in section 4.

Consumer price inflation

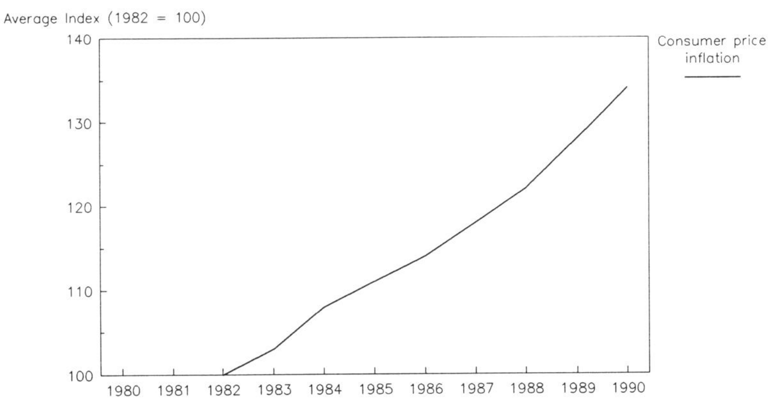

Consumer price inflation

The basis of these figures is the official consumer price index converted to quarterly averages. Inflation peaked in 1984; since then inflation has fallen to a low of 1.9% in 1986 but has increased steadily since then almost to its 1984 level.

Consumer price inflation 1982 = 100

	1983	1984	1985	1986	1987	1988	1989	1990
average index	103	108	112	114	118	122	128	134
annual change (%)	3.2	4.9	3.6	1.9	3.6	4.1	4.8	4.5

USEFUL ADDRESSES

National Bureau of Standards
US Department of Commerce
Centre for Building Technology
Gaithersburg MD 20899
Tel: (301) 921 3377

American National Standards Institute
1430 Broadway
New York
N.Y. 10018

USSR

KEY DATA

Population

Population (1989)	286m
Urban population	67%
Population under 15	n.a.
Population over 65	n.a.
Average annual growth rate (1980 to 1985)	1.0%

Geography

Land area	22 402 000 km²
Agricultural area	n.a.
Capital city	Moscow (population 8 703 000)

Economy

Monetary unit	Rouble (Rb)
Exchange rate (January 1991) to	
the ECU	Rb 2.23
the pound sterling	Rb 3.19
the US dollar	Rb 1.65
Average annual inflation (1984 to 1989)	1.4%
Gross Domestic Product (GDP)	Rb 940bn
GDP per capita (1989)	Rb 3287
GDP growth (real) (1980 to 1988)	n.a.
Private consumption (% of total)	n.a.
Public consumption (% of total)	n.a.
Investment (% of total)	n.a.
Central government expenditure as percentage of GDP	n.a.

Construction

Total value of construction output (1989)	Rb 156bn
Construction output per capita	Rb 545
Construction output as a percentage of GDP	16.6%

NATIONAL PLANNING IN THE ECONOMY (USSR)

Until 1985 and the introduction of *perestroika* (restructuring) national plans were the main instrument of economic policy. Now the planning mechanism has been largely disbanded and with the political restructuring control is in the hands of the governments of republics rather than of the USSR. However, there is no effective replacement of the planning system and there is little guidance in the projected move to a market economy. The economy is in a state of hiatus.

The overall shape of the economy was determined by a long term plan stretching 20 years ahead. This covered the main objectives, the percentages of consumption, accumulation, or investment; the shares of the main sectors of the economy; and geographical distribution. In more detail there was the 5 Year Plan and, in even more detail, the 1 Year Plan. These were all the responsibility of the State Planning Organization - *Gosplan.*

During the current changes, the proportion of investment taking place outside the state budget has already risen. In 1986, 96% was from the state budget; in 1990 it was to be about 57%; and in 1991 about 30%. The remaining investment is to be funded by funds generated by enterprises themselves. However, there is no capital market. Just because an enterprise is large, and hence generates a surplus it does not mean that the reinvestment of this surplus in the same enterprise is particularly efficient.

Historically, the planning system has not worked either at the macro or the micro level. At the macro level, the rate of growth of national income began to slow from the late 1960s. In the five year plan 1966 to 1970, growth in national income was 41%; in 1981 to 1985 growth was only 16%. The basic level of growth - excluding exports of oil at very high prices and the very high sales of alcohol - actually declined in the early 1980s. Technological progress faltered and, at the beginning of the 1980s, real per capita income started to fall. The country has vast natural resources but shortages were experienced of virtually everything from steel and machinery to consumer goods and food. Grain is being imported from abroad. The quality of home produced goods has fallen.

As a result of poor economic performance and political change, the policy under *perestroika* is to move towards a market-type economy. The problems are enormous. Not only does this involve a very difficult transition period at a time when the standard of living is already very low, but it also requires a new understanding of the meaning of "market" and a change in culture to one which encourages and accepts individual initiative and the rewards for success as well as the penalties for failure. There is still, even at senior levels in the government organizations, no clear understanding of competition or profit in a capitalist country. In 1987 the value of GDP was reckoned to be 825 billion roubles and by 1989 it was 924 billion, both at current prices. However, in real terms it is now decreasing. Comparison of the GDP with that of other countries is not really possible because official exchange rates have little meaning. Soviet internal prices bear no relation to world market prices.

THE CONSTRUCTION INDUSTRY

Construction output

Total investment is high and, in 1989, was nearly 25% of GDP. This was divided as follows: 56% going to construction, building and civil engineering, 36% to machinery and equipment, and 8% to design, research and development etc. Nearly 45% of this investment was in the form of modernization and rehabilitation of assets. Statistics for investment include plant, machinery and vehicles, as well as construction of buildings and civil engineering works and correspond approximately to the definition of gross domestic fixed capital formation in the UK. However all investment is often referred to as "capital construction" and it is often difficult to know whether the output of the construction industry or total investment is referred to.

In 1987 the total value of construction output, excluding repair and maintenance was about 14% of GDP. This implies a total for 1989 of about 130 billion roubles equivalent to US$78.7 billion. If a 20% allowance for repair and maintenance is added, the total comes to about 156 billion roubles. From indicators described in the introductory section, however, it seems that the output in real terms in the USSR is rather over a third of that in Western Europe, or about 200 billion ECUs. It is quite possible that the level of construction output and, possibly, of GDP in the Soviet Union is substantially undercounted compared to the UK and many countries of Western Europe. It is not known, for example, whether the work in progress (of which there is a large amount) is included in output figures; whether the cost at which work is included in the statistics is the official estimated cost or the actual cost, which may be two or three times higher; or whether the costs include interest on capital utilized. Indeed, the official exchange rate gives a total construction output of about 230 billion ECUs.

There are very few data on the breakdown of construction by type of work. However, the table below gives some indication of the situation for 1986 and 1987. These data include rehabilitation but not repair and maintenance, which is estimated at about 20% of total work. "Other" is a residual figure and seems very low. The figures may exclude much civil engineering work. The data should be treated with caution.

Breakdown of output by type of work

	1986 (billion roubles)		1987 (billion roubles)		1989 (billion roubles)	
Industrial projects						
· technical improvement and reconstruction of existing firms	10	)	12	)	15	)
· expansion of existing firms	13	) 48	13	) 50	12	) 51
· new construction projects	24	)	24	)	22	)
· single projects of existing firms	1	)	1	)	2	)
Non industrial work						
· housing	31		34		38	
· other	26		21		22	
Total value of output	99		105		111	

Source: *Capital Construction in the USSR, 1988 pp 16, 26, 153*
Narodnoje Hozajstvo SSSR v 1989, pp 280, 533.

Output on construction, including repair and maintenance work nearly doubled between 1970 and 1987, though housing increased by only about 25%. However, these increases have not been maintained. In the first half of 1990, for example, the volume of housing construction was 6% below the level of the previous year.

The table below shows the distribution of construction output in the Republics in 1987 compared with the distribution of area and population. The dominance of the Russian Federation, followed by the Ukraine, is evident. However, whereas Russia has a higher percentage of output than of population - reflecting the importance of Moscow and Leningrad (now St Petersburg) - the Ukraine has a lower share of construction than population.

Geographical distribution of construction output and population, 1987

Republic	*Construction output (%)*	*Area (%)*	*Population (%)*
Russian Federation	61.2	76.2	51.4
Ukraine	13.5	2.7	18.0
Belorussia	3.6	0.9	3.5
Uzbekistan	4.1	2.0	6.9
Kazakhstan	6.4	12.1	5.7
Georgia	1.6	0.3	1.9
Azerbajan	1.7	0.4	2.5
Lithuania	1.5	0.3	1.2
Moldavia	1.1	0.2	1.5
Latvia	0.8	0.3	0.9
Kirghizia	0.9	0.9	1.5
Tadjikistan	1.0	0.6	1.8
Armenia	1.0	0.1	1.1
Turkomenistan	1.0	2.2	1.2
Estonia	0.5	0.2	0.5
	100.0	100.0	100.0

Note: Totals do not sum exactly due to rounding
Source: Capital Construction in the USSR, 1988, p 16.

Finance for construction

The sources of finance for construction work were 50% from the central budget funds, 10% credit provided to state enterprises and 40% from the funds of the enterprises themselves. The share of central budget financing is to be reduced substantially. In the next five years it was supposed to fall to 20% to 30% of total investment. The construction of military facilities, conservation projects and low income housing were to be financed at the central level. The construction of health care and educational structures should be financed through funds from the republics, with supervision of the projects by central or local authorities. All other construction projects were to be financed by private business including foreign capital. However, with the change in the political system involving the autonomy of the republics, the economic arrangements will be determined within each republic.

The rate of interest is 2% to 3% at present but the Department of Finance is planning to raise the long-term interest rate to 8% to 10% and the Department expects that the share of finance provided on credit will rise to 20% to 25% of the total.

The structure of the industry

The construction industry in the USSR consisted of a few very large organizations - in 1989 there were 3650 organizations known as "trusts". They each had a number of enterprises or firms which in total amounted to 33 000; a fraction of the firms in the much smaller United Kingdom. These trusts not only undetook construction work but also produced and installed plant and machinery. Of the 33 000 construction organizations, half were concerned with specialist work, including contractors and subcontractors. Over 20 000 of the firms were in the Russian republic.

Over 2000 of the trusts (57%) had turnovers between 15 and 40 million roubles. At the lower level, below 15 million roubles, there were less than 500 organizations and at the top there were well over 1000 giant organizations with turnovers greater than 40 million roubles - many of them well in excess of this figure.

Twenty two ministries controlled the 3650 trusts. Of these, four ministries, each with their trusts, operated all over the Soviet Union. One dealt with transport works and had a turnover of 6 billion roubles per annum; it built railways, roads, ports, underground railways, bridges and tunnels. Another dealt with specialized buildings with a turnover of 11 billion. An oil and gas trust constructed 20 thousand kilometres of pipelines per annum as well as gas field installations with a total turnover of 7 billion roubles per annum. An energy trust constructed nuclear and thermal plants and electric power lines; its turnover was 7 billion a year. Lastly, the coal trust had a construction turnover of 3.5 billion a year. These five together had a turnover of 34.5 billion roubles or around 18% of all construction output. The remainder of the trusts operated mainly in their home republic. Each of the 14 republics - other than Russia - had one construction ministry. The Russian republic had four such ministries.

There were no very small firms in the construction industry, though there were a few co-operative type subcontractors. However, a recent co-operative Development Law has enabled the formation of a number of construction co-operatives and they are now growing rapidly. In 1990 it was expected that co-operatives would carry out around 7% of construction output. This structure will change but, in summer 1991, there had been, so far as is known, little dismantling of the trusts.

Design was normally undertaken by large design offices. Integration of design and construction was found only in housebuilding enterprises. Architects do not play a leading role in construction.

The resources of the construction industry

In 1987, the proportion of expenditure on construction work on various inputs was 55% on materials, 13% on wages and salaries, 14% on use of equipment and 18% on overheads.

The production and distribution of building materials has been inadequate to deal with requirements and work on site is frequently held up by shortages. Furthermore, quality is poor. One reason is that the building materials factories work to capacity so that maintenance is not carried out. For example, there are modern brick kilns where the bricks go through a controlled kiln on a moving belt. However, because maintenance is neglected, the temperature is not even and the bricks are either burnt or not fired sufficiently.

Construction technology in the USSR involved a large amount of heavy concrete construction. Consequently there is a very high usage of heavy materials and low usage of lighter, more sophisticated products. If the consumption of selected materials in the USA and USSR per million dollars of construction output is expressed as a ratio of US consumption to USSR consumption, the following ratios are found:

	Materials usage ratios USA/USSR
Cement	0.51 :1
Precast concrete	0.17 :1
Bricks	0.08 :1
Steel structures	1.84 :1
Aluminium structures	13.3 :1
Plastics	2.8 :1

Even allowing for the artificial exchange rate and the differences in mix of output, these show, unmistakably, the heavy construction technology of the USSR compared to the relatively light technology of the USA.

The number of persons employed in construction assembly organizations is 12.1 million or 9% to 10% of the employed population. Of these, 9.8 million are operatives both skilled and unskilled and the remainder managers and service workers. Co-operatives employed about 1 million but the figure will now be greater. Wages are generally low - in 1987 the average wage was 259 roubles a month. A young graduate engineer in construction will receive a similar amount while senior managers receive about 500 roubles a month. The official rate of exchange of 1 rouble = 1£ is irrelevant. However the commercial rate of 3.2 roubles to the £ sterling or 2.2 to the ECU would mean that the monthly average wage was £81 or 118 ECUs.

The construction process and change in the construction industry

In the past, the arrangements were that the client for building work - state agency, local community or council - would chose a design organization, usually in the Ministry concerned; effectively there was little choice. Following the completion of the design,

Gosplan included the project in the Plan, but even then it might be years before construction began.

Once projects were agreed, the building material requirements were calculated by the organizations designated as responsible for the construction and by the material producers. The calculation of material requirements was based on quantity of materials used per million roubles in past "representative" projects of particular types, e.g. iron and steel factories. These norms were revised every 5 to 7 years. Costing of projects was by means of norms laid down in large, detailed documents revised only every 5 to 10 years.

The planned norm for duration of contracts was typically exceeded by a factor of 2 to 3. It is reported that costs similarly exceeded the norms. The average construction time of a project costing $37 million was 9.5 years in the USSR compared with 2.4 years in the USA. There is a large stock of uncompleted buildings so that, in the present environment, it is likely that these buildings will be sold to Western firms to complete and manage when finished.

Relations between members of the construction team are complicated and ineffective, partly because the State is both the client and the contractor: the lender of money and the borrower. There is, for example, in theory a penalty for late delivery, but in practice there are always extenuating circumstances, e.g. non-availability of materials. In any case, the 3% penalty, if paid at all, is paid to central government, not to the owner of the project.

Partly in response to poor performance, there has been a slow move towards decentralization for some years now. At one time the whole industry was completely under inflexible centralized control.

Fairly recently, the building materials ministry was abolished and production of cement, glass and some other building materials is now under the control of concerns organized in association on a voluntary basis.

In 1988 the construction trusts were given more freedom to manage their own affairs. A recent development (1st July 1990) is that they have become enterprises which are responsible for their own business, with opportunities to develop as they wish and powers to reinvest their own surpluses or profits. It is expected that there will be more decentralization. However, these changes have had little practical effect and the difficulties in moving towards greater competition and a market orientation are enormous. Enterprises have very little idea of their true costs as they have had no need to calculate them because they are laid down in books of norms. At present if the cost overrun is small, "arrangements are made"; if it is large, a new contract is drawn up. Construction firms are now allowed to fix prices on a contractual basis but penalties for non-observance of contract conditions, credit and tax regulations are still insignificant. Essentially all expenses, including losses, are covered by the State budget.

Great changes were planned in the years 1986 to 1988, including:

- an increase in the share of contracts obtained by competition
- a limit on financing from state budget, financing from own funds of enterprises and bank credits
- competition between state collective and co-operative sectors
- increasing sanctions for non-performance, and bonuses for good performance
- integration of design and construction phases and wide application of turnkey and project management contracts.

Laws have been promulgated to enable greater opportunities for business development notably through different forms of ownership, including private ownership. These include the Ownership Law, the Socialist Enterprise Law, the Tax Law, the Individual Business Activity Law, the Co-operative Development Law, the Small Enterprises Development Decree and the Joint-stock Companies Decree.

The new tax law has been greatly simplified and is now less bureaucratic with a universal 22% tax on capital income (that is, profit) and 23% local tax rate so that the total tax will not exceed 45%. For state and commercial banks the tax rate will be 55% and for joint ventures (with a foreign capital share of 30% or more) 30%. There are also a number of tax incentives:

- 30% investment tax credit for ecological and R&D investment as well as exemptions for small enterprises and joint ventures doing business in basic industries for their first two years
- the Small Enterprises Development Law for the construction industry to apply to all firms with over 200 employees
- the President's August 1990 decree to fund the Department of State Property to value State-owned assets and to offer them for sale in the near future.

Development control and standards

Building norms and regulations (SNiP) were approved by *Gosstroy*. They cover materials, details, construction and equipment, norms for building - cost and duration - and the organization of, and techniques for building construction.

Throughout the USSR, permission to carry out building work in implementing a particular project design was given by the State architectural and building control bodies of local Councils of Workers' deputies. Permission was given when an approved design, a capital construction plan and a guarantee of the financing of the work were submitted. Quality of design was controlled at all stages by design organizations, by technical departments, in approval bodies, by technical inspection departments and by other specialist organizations.

There were national standards (GOST) approved by *Gosstroy* as well as special standards for certain sectors, republic standards and special standards for certain enterprises. The quality of building components, structures, materials and equipment was controlled. For a "quality mark" additional controls are necessary.

These norms and standards were published annually by *Gosstroy*. It seems likely that this system will remain substantially in place.

However, all these changes were planned at the level of the Soviet Union and now they will be in the remit of the republics. It is not known how far the changes will follow those envisaged in the USSR. Change is taking place but the impression currently in the construction industry is of uncertainty and indecision. The state and state enterprises are still almost the only source of capital. They will continue to be the clients of the industry for some time to come.

Research and development

Scientific research work in the building industry in the USSR was planned as part of the overall State plan. A network of scientific research institutes had been built up according to the following divisions:

- general theoretical research
- industrial construction
- urban development
- housing construction
- civil construction
- building materials.

Research was often linked to design work. The institutes were under various organizations according to subject and areas: *Gosstroy, Gosstroy* in the Republics, Ministries of the USSR and in the Republics. Some continuation of the work of these institutions may be expected in future but in a different direction than previously.

Foreign participation and future opportunities

There is scope for participation by western countries in all aspects of construction projects and the construction industry. This includes:

- property development linked with construction, for example, possible purchase of unfinished hotels by Western contractors in conjunction with Western hotel firms. The contractor would complete the hotel and the hotel firm would own and manage it
- provision of contractor management in various forms
- provision of management training
- material and product manufacture.

The Soviet Union is well aware of its deficiencies in design, the lack of CAD hardware and software and of the need for better equipment and more sophisticated use of materials in lighter structures. It also sees prospects for the participation of foreign contractors in remote territories.

The problem is to find the right mechanism for this co-operation, given the problems of a non-convertible currency and the present poor prospects for the economy. On 1st November 1990 a new commercial rate of 1 US dollar for 1.8 roubles was introduced. This rate was to be used for international trade transactions, foreign investments in the USSR and Soviet investments abroad. Progress was also being made in the area of foreign ownership of Soviet companies. Though land cannot be purchased, outright long leases are possible. It is still difficult for foreigners to undertake development as a commercial proposition.

There are several forms of participation:

- joint ventures
- multilateral partnerships. West German firms have participated in consortia for large scale developments
- joint entries to third world countries
- purchase of unfinished construction products. A special consulting office is open to promote these purchases by foreign firms at Stroyserviceinform, Moscow 103001, Boljshaja Sadovajalo, Russia.

The Association for International Economic Co-operation of Investors and Constructors (AMECC) has recently been established to promote construction projects. The address is shown below.

Costs and prices

Some published figures of "prime costs" of construction show a fall from 1971 to 1975, a small rise during 1976 to 1980, and further falls between 1981 to 1985 and in 1986 and 1987. These statistics are not in keeping with experience in the Soviet Union and, in the last three or four years at least, price rises have been significant. The 1989/90 level of inflation is officially put at 5% but observers believe it was at least 15% to 18% per annum. Construction prices have also risen and cost overruns are usual, sometimes by two or three times the original figures.

It is possible that the statistics of "prime" costs in construction refer to the norms for prices which are quite out of keeping with reality.

Further reading

1. State Committee on Statistics, Capital Construction in the USSR Moscow: Finance and Statistics, 1988.
2. J A Rekitar, The Economic Restructuring in the USSR: Dead Ends and New Departures, unpublished paper, 1989.
3. J A Rekitar, Certain Macro-Economic Problems of the Construction Complex in the Economy of Industrialised Countries, unpublished paper, 1989.
4. V B Kondratiev, Current Economic Reforms and Construction Industry in the USSR, unpublished paper, 1990.

Acknowledgements

The authors are grateful to Professor Rekitar, Dr Kondratiev, Mr Lebediv and Mr Mlovidov for their patience and diligence in explaining the operations of the Soviet construction industry.

EXCHANGE RATES AND INFLATION

The combined effect of exchange rates and inflation on prices within a country and price comparisons between countries is discussed in section 4.

The graph below plots the movement of the Russian rouble against sterling and the US dollar since 1980. The figures used for the graph are quarterly and the method of calculating these, and other related issues are discussed in section 4. The exchange rate at January 1991 was Rouble 3.19 to the pound sterling and Rouble 1.65 to the US dollar.

The Russian Rouble against Sterling and the US Dollar

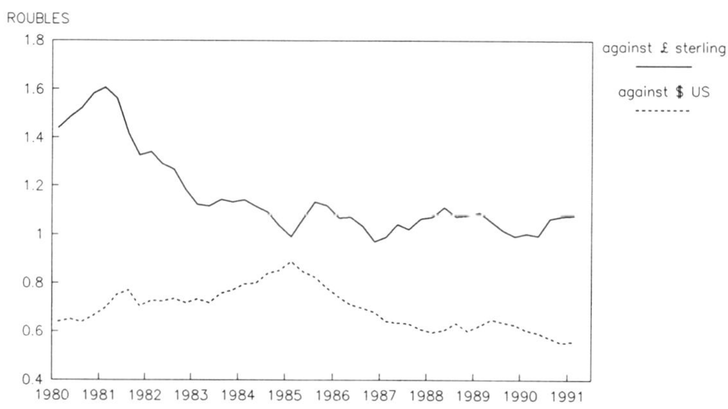

Inflation

The graph and the table below present the consumer price index showing the trend in inflation since 1980. The index has been rebased to 1980=100. Notes on inflation and the distinction between costs and prices are included in section 4.

Consumer price inflation

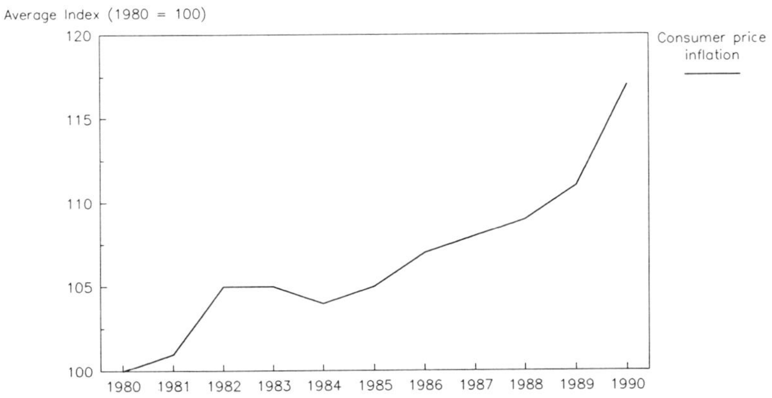

Consumer price inflation

The basis of the figures is the official consumer price index converted to quarterly averages. Inflation peaked in 1990; however, there is considerable doubt about the reliability of official figures on which the inflation rate is based.

Notes on inflation are included on section 4.

Consumer price inflation 1980 = 100

	1981	1982	1983	1984	1985	1986	1987	1988	1989	1990
average index	101	105	105	104	105	107	108	109	111	117
annual change (%)	1.0	4.0	nil	-1.0	1.0	2.0	1.0	1.0	2.0	5.3

USEFUL ADDRESSES

AMECC
23, Profsoyuzhaya Str. GSP-7
Moscow 117859
Tel: (095) 12-1790 or 120 9166
Telex: 411687 IMEMO SU
Fax: 310-70-27

Ministry for the Building Materials Industry
Ploshchad Nogina 2/5
Moscow 103713
Tel: 928 4482

Ministry for Construction, Roads and Municipal Machine Building
Prospekt Kalinina 23/2
Moscow 121019
Tel: 202 8081

Ministry for Industrial Construction
Petrovka 14
Moscow 103031
Tel: 209 1147

Yugoslavia

KEY DATA

Population

Population (1989)	23.6m
Urban population	49%
Population under 15	22.9%
Population over 65	9.1%
Average annual growth rate (1983 to 1988)	0.7%

Geography

Land area	256 000 km^2
Agricultural area	n.a.
Capital city	Belgrade (population 1 455 000)

Economy

Monetary unit	Yugoslav Dinar (YuD)
Exchange rate (January 1991) to	
the ECU	YuD 18.4
the pound sterling	YuD 26.2
the US Dollar	YuD 13.6
Average annual inflation (1984 to 1989)	345.2%
Gross Domestic Product (GDP)	n.a.
GDP per capita (1989)	n.a.
GDP growth (real) (1980 to 1988)	n.a.
Private consumption (% of total)	n.a.
Public consumption (% of total)	n.a.
Investment (% of total)	n.a.
Central government expenditure as percentage of GDP	n.a.

Construction

Total value of construction output (1989)	n.a.
Construction output per capita	n.a.
Construction output as a percentage of GDP	n.a.

NATIONAL CONTEXT

Introduction

At the time of writing (mid-1991) Yugoslavia was in the throes of a civil war between Serbia and Croatia. It is not yet possible to know what consequences this will have, either for the construction industry or the wider economy. What follows is a brief description of the industry as at the end of 1990.

THE CONSTRUCTION INDUSTRY

Construction output

The output of the Yugoslav construction industry for 1989 is not available. However, indirect measures suggest that it may be equivalent to about 15 billion ECUs. The share of the construction industry in GDP and of gross investment has been declining during the 1980s and from 1980 to 1987 the rate of growth of construction was negative in real terms (-6.3% annually).

The decline in the construction industry was believed to be having severe consequences for the rest of the economy.

Characteristics and structure of the industry

The Yugoslav construction industry was dominated by large firms in the state sector. In 1987, 87 firms employing more than 1000 persons did 25% of the work, while those employing over 500 did 48% of the work.

Falling construction output in the 1990s changed the attitudes of clients and companies towards costs and performance. At the beginning of 1989, the state planning of this sector was largely abandoned in favour of the market. The market chose between the successful firms and unsuccessful ones and the successful firms sought opportunities abroad. The search for work abroad was backed by regional governments and the banks including JUBMES - the Associated Yugoslav Bank for International Economic Co-operation. Measures adopted included the creation of enterprise zones, the development of infrastructure programmes, the stimulation of foreign investment, the reduction in rent subsidies, sales of communal housing. At the same time, the construction companies were improving their management and diversifying into other businesses, reducing involvement in less profitable activities and shedding unnecessary resources. Those companies which did not respond were likely to be permitted to go out of business.

There are no Yugoslav companies in the Building magazine's "Top 300 European Contractors", but there are eight in Engineering News Record's "Top 250 International Contractors" as shown below.

Major Yugoslav constractors

Major contractors	*Head office*	*Place in ENR's "Top 250 International Contractors"*	*Exports (% of turnover)*	*Main work types*
Energoprojekt Holding Co	Belgrade	72	80	Building, general civil engineering
CC Pelagonija	Skopje	90	70	Building, general civil engineering
KMG Trudenik Co	Belgrade	133	15	Building and design
Ingra	Zagreb	152	n.a.	Water works
EM Hidromontaza	Maribov	180	60	Power, industrial, petroleum, design
Petrolinvest	Sarajevo	182	90	Industrial, petroleum
Granit Construction Co	Skopje	204	15	Building, transport, design
Slovenija Cesta Tehnika (SCT)	Ljubliana	218	n.a.	Transport

Source: Engineering News Record

Development control and standards

Physical planning was under the control of the State, the Republic, the Region and the Commune in increasing order of detail. In theory, every spatial change was legally regulated. In practice this was not wholly achieved, especially not in villages, under-developed areas and those cities which were exposed to high immigration pressure. More particularly, there was no spatial plan for the whole country, none for Serbia and many urban plans were outdated. Where plans were adopted there were irregularities which resulted in large uncontrolled developments, especially in Serbia and Montenegro.

The basic technical aspects of building quality and technology were regulated by Federal laws and regulations especially through standards, technical norms and quality norms. These technical regulations were mandatory and were uniformly applied to all structures regardless of whether they were dwellings, civil works or other buildings.

The regulations of republics and autonomous provinces covered matters of building design, performance of work, procedures for applying for and issuing permits, relations between participants in projects and so on. They were concerned with application of the federal laws and the broader economic and social implications of a project.

The norms and standards were provided for in the Standardization Act. The responsible institution was the Yugoslav Institution for Standardization (JZS). Product testing could be carried out by the producer or by a registered organization.

Building permits were issued on the basis of full drawings. Designs were checked either by the design organizations, if they were approved for this purpose, or by construction inspectors, who also controlled the building process. After completion, the building or structure was inspected and changes may have been required if the terms of the building permit had not been followed. A guarantee period was fixed for latent defects.

Further reading

Juraj Marusic and Mirko Oreskovic, "Circumstances and Strategies in Building Economics in Yugoslavia", in CIB 90, Building Eocnomics and Construction Management, Vol 1: The Building Market; Building Economics and Urban Development.

Barbara Medanic, "Long-term Development Planning in Construction and Related Industries in Yugoslavia", in CIB 90, Building Eocnomics and Construction Management, Vol 1: The Building Market; Building Economics and Urban Development.

EXCHANGE RATES AND INFLATION

The combined effect of exchange rates and inflation on prices within a country and price comparisons between countries is discussed in section 4.

Exchange rates

The graph below plots the movement of the Yugoslav Dinar against sterling and the US dollar since 1980. The figures used for the graph are quarterly and the method of calculating these, and other related issues are discussed in section 4. The exchange rate at January 1991 was YuD 26.2 to the pound sterling and YuD 13.6 to the US dollar.

The Yugoslav Dinar against Sterling and the US Dollar

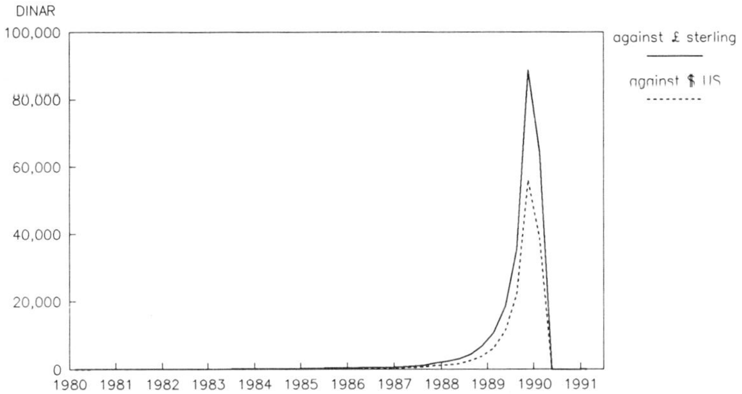

6. Multi-lingual descriptions of construction items

Excavation (Assume excavation in firm soil)	**Terrassement** (Terrassement en terre ferme)	**Scavi** (In terreno consistente)
1. Mechanical excavation of foundation trenches. Starting from ground level (including removal of excavation material from site). Over 0.30m wide, not exceeding 2.00m deep.	1. Fouilles par engins mécaniques en tranchées pour fondations commençant du niveau existant du sol (compris nivellement du fond, chargement des déblais par camion et enlèvement aux décharges). Dimensions des tranchées ou rigoles de plus de 0,30m de largeur et ne dépassant pas 2,00m de profondeur.	1. Scavo di fondazioni in trincea con mezzi meccanici, di oltre 0,30m di larghezza, e non superiore a 2,00m di profondità dal livello terra. E'compreso il trasporto del materiale di risulta fuori dal cantiere.
2. Hardcore filling in making up levels. Hard brick, broken stone (or sand where appropriate). Crushed to pass a 100mm ring 150mm deep.	2. Remblais pour élever les niveaux (comprenant des briques, des caillous, des pierres cassées et du sable), broyés pour pouvoir passer à travers un anneau de 10cm de diamètre et de 15 cm de longueur.	2. Riempimento in sottofondo con vecchi scarti di muratura, pietre spaccate o sabbia, frantumato per passare un anello di diametro 100mm, in uno strato di 150mm spessore.
3. Earthwork support. Sides of trench excavation. Distance between opposing faces not exceeding 2.00m. Maximum depth 2.00m.	3. Alignement des parois de fouilles de tranchées ou rigoles. Distance entre les parois n'excéde pas 2,00m. Profondeur maximum de 2,00m.	3. Sbadacchiature ai lati di trincee, distanza tra le faccie opposte non superiore a 2,00m, fino a una profondità massima 2.00m.
Concrete work (Formwork and reinforcement measured separately)	**Travaux de béton** (Le coffrage et l'acier d'armatures sont mesurés séparément)	**Opere in calcestruzzo** (Casseforme e ferri di armatura misurati a parte)
4. Plain insitu concrete in strip foundations in trenches 20N/mm². Ordinary Portland Cement, 20mm coarse aggregate. Size 500mm wide x 300mm thick.	4. Béton non armé de remplissage pour tranchées de fondations 20N/mm². Ciment de Portland ordinaire, 2cm dimension maximum des granulats. Dimensions: 0,5m de largeur x 0,30m de profondeur.	4. Calcestruzzo semplice in trincea di fondazione 20N/mm²: cemento tipo Portland normale, inerti di 20mm. Dimensioni 500mm di larghezza x 300mm spessore.

Aushub	**Excavacion**	**Excavation**
(Anzunehmen ist, daβ der Aushub in festem Baugrund erfolgt)	(Excavaciones en terreno compacto)	(Assume excavation in firm soil)
1. Maschineller Aushub der Fundamentgräben, ausgehend vom bestehenden Planum (mit Abtransport des Aushubmaterials vom Grundstück). Breite über 0.30m, Tiefe bis maximal 2.00m.	1. Excavación mecánica de zanjas de cimentación. Iniciandose a nivel de tierra e incluyendo el transporte de tierras. De más de 30cm de ancho, y una profundidad máxima de 2m.	1. Mechanical excavation of foundation trenches. Starting from ground level (including removal of excavation material from site). Over 0.30m wide, not exceeding 2.00m deep.
2. Grobkörnige Unterpackung zur Erhöhung der Baufläche. Harteziegel, Steinschlag (wenn angemessen auch Sand), zerschlagen, um durch einen Ring, Durchmesser 100mm, zu kommen. Tiefe 150mm.	2. Sub-base granular de cantera (o arena en su caso). Granulometría de 100 a 150mm.	2. Hardcore filling in making up levels. Hard brick, broken stone (or sand where appropriate). Crushed to pass a 100mm ring 150mm deep.
3. Grabenspriessungen. Grabenbreite maximal 2.00m, Grabentiefe maximal 2.00m.	3. Entibación de zanjas de cimentación de menos de 2m de ancho y 2m de profundidad.	3. Earthwork support. Sides of trench excavation. Distance between opposing faces not exceeding 2.00m. Maximum depth 2.00m.
Betonarbeiten	**Trabajos en hormigón**	**Concrete work**
(Schalungen und Bewehrung seperat gemessen)	(Encofrados y armaduras, medidas aparte)	(Formwork and reinforcement measured separately)
4. Unbewehrter Ortbeton für Streifenfundamente in Gräben, 20N/mm^2. Normaler Portlandzement, Korngröße 20mm. Breite 500mm, Dicke 300mm.	4. Hormigón 20N/mm^2 de cemento portland y áridos de 20mm en zanjas de cimentación de 500 x 300mm.	4. Plain insitu concrete in strip foundations in trenches 20N/mm^2. Ordinary Portland Cement, 20mm coarse aggregate. Size 500mm wide x 300mm thick.

English	French	Italian
5. Reinforced insitu concrete in beds 20N/mm². Ordinary Portland Cement, 20mm coarse aggregate. 200mm thick.	5. Béton armé pour planchers en dalle pleine 20N/mm². Ciment de Portland ordinaire, 2cm dimension maximum des granulats. Dimensions: 0,20m d'épaisseur.	5. Calcestruzzo armato gettato in opera in solette 20N/mm²: cemento tipo Portland normale, inerti di 20mm. Dimensione 200mm spessore.
6. Reinforced insitu concrete in walls 20N/mm². Ordinary Portland Cement, 20mm coarse aggregate. 200mm thick.	6. Béton armé pour murs 20N/mm². Ciment de Portland ordinaire, 2cm dimension maximum des granulats. Dimensions: 0,20m d'épaisseur.	6. Calcestruzzo armato gettato in opera in muri 20N/mm²: cemento tipo Portland normale, inerti di 20mm. Dimensione 200mm spessore.
7. Reinforced insitu concrete in suspended floor or roof slabs 20N/mm². Ordinary Portland Cement, 20mm coarse aggregate. 150mm thick.	7. Béton armé pour planchers portés ou dalles de toiture 20N/mm². Ciment de Portland ordinaire, 2cm dimension maximum des granulats. Dimensions: 0,15m d'épaisseur.	7. Calcestruzzo armato gettato in opera in solai di pavimenti o tetti 20N/mm²: cemento di tipo Portland normale, inerti di 20mm. Dimensione 150mm spessore.
8. Reinforced insitu concrete in columns 20N/mm². Ordinary Portland Cement, 20mm coarse aggregate. Size 400 x 400mm.	8. Béton armé pour poteaux 20N/mm². Ciment de Portland ordinaire, 2cm dimension maximum des granulats. Dimensions: 0,40m x 0,40m.	8. Calcestruzzo armato gettato in opera in pilastri 20N/mm²: cemento tipo Portland normale, inerti di 20mm. Dimensioni 400 x 400mm.
9. Reinforced insitu concrete in isolated beams 20N/mm². Ordinary Portland Cement, 20mm coarse aggregate. Size 400 x 600mm deep.	9. Béton armé pour poutres isolées 20N/mm². Ciment de Portland ordinaire, 2cm dimension maximum des granulats. Dimension: 0,40m de largeur x 0,60m d'épaisseur.	9. Calcestruzzo armato gettato in opera in travi isolate 20N/mm²: cemento tipo Portland normale, inerti di 20mm. Dimensioni 400 x 600mm di profondità.
10. Precast concrete slab (including reinforcement as necessary). Contractor designed for total loading of 3N/mm² 5.00m span.	10. Planchers préfabriqués en béton (compris l'armature). Portée maximum de 5m entre appui acceptant un poids maximum de 3 N/mm².	10. Calcestruzzo in soletta prefabbricata (compreso il ferro di armatura necessario), progettata dall'impresa per un carico totale di 3N/mm², luce 5,00m.

5. Ort-Stahlbeton für Bodenplatten, 20N/mm^2. Normaler Portlandzement, Korngröße 20mm. Dicke 200mm.	5. Hormigón para armar de 20N/mm^2 de cemento portland y áridos de 20mm en pavimento de 200mm de espesor.	5. Reinforced insitu concrete in beds 20N/mm^2. Ordinary Portland Cement, 20mm coarse aggregate. 200mm thick.
6. Ort-Stahlbeton für Wände, 20N/mm^2. Normaler Portlandzement, Korngröße 20mm. Dicke 200mm.	6. Hormigón para armar de cemento portland y áridos de 20mm en muros de 200mm de espesor.	6. Reinforced insitu concrete in walls 20N/mm^2. Ordinary Portland Cement, 20mm coarse aggregate. 200mm thick.
7. Ort-Stahlbeton für Deckenplatten oder Dächer, 20N/mm^2. Normaler Portlandzement, Korngröße 20mm. Dicke 150mm.	7. Hormigón para armar de cemento portland y áridos de 20mm en losas de 150mm de espesor.	7. Reinforced insitu concrete in suspended floor or roof slabs 20N/mm^2. Ordinary Portland Cement, 20mm coarse aggregate. 150mm thick.
8. Ort-Stahlbeton für Stützen, 20N/mm^2. Normaler Portlandzement, Korngröße 20mm. Dimensionen 400 x 400mm.	8. Hormigón para armar de cemento portland y áridos de 20mm en pilares de 400 x 400mm.	8. Reinforced insitu concrete in columns 20N/mm^2. Ordinary Portland Cement, 20mm coarse aggregate. Size 400 x 400mm.
9. Ort-Stahlbeton für isolierte Balkenträger, 20N/mm^2. Normaler Portlandzement, Korngröße 20mm. Breite 400mm, Tiefe 600mm.	9. Hormigón para armar de 20N/mm^2 de cemento portland y áridos de 20mm en vigas de 400 x 600mm.	9. Reinforced insitu concrete in isolated beams 20N/mm^2. Ordinary Portland Cement, 20mm coarse aggregate. Size 400 x 600mm deep.
10. Fertigbetonplatte (einschließlich erferderliches Bewehrung). Für eine Gesamtbelastung von 3N/mm^2 vom Unternehmer hergestellt. Spannweite 5.00m.	10. Losas de hormigón pretensada (incluyendo la armadura necesaria), para una carga total de 3N/mm^2 de 5,00m de luz.	10. Precast concrete slab (including reinforcement as necessary). Contractor designed for total loading of 3N/mm^2 5.00m span.

Formwork	**Coffrage**	**Casseforme**
(Assume a simple repetitive design which allows 3 uses of formwork)	(Assumez simple coffrage permettant 3 usages)	(Considerare casseratura semplice ripetitiva che permetta il riutilizzo delle casseforme tre volte)
11. Softwood or metal formwork to concrete walls. Basic finish. (one side only).	11. Coffrage en bois ou métal pour murs en béton. Face lisse (Une seule face)	11. Casseforme in legno o acciaio per muri in calcestruzzo, faccia liscia. (Un lato soltanto).
12. Softwood or metal formwork to concrete columns. Basic finish. Columns 1600m girth.	12. Coffrage en bois ou métal (prière de rayer ce qui n'est pas utilisé) pour poteaux. Face lisse. Périmètre de 1,60m pour le poteau.	12. Casseforme in legno o acciaio per pilastri in calcestruzzo, facce lisce. Pilastri 1,60m di contorno.
13. Softwood or metal formwork to horizontal soffits of slabs. Basic finish. Slab 150mm thick, not exceeding 3.50m high.	13. Coffrage horizontal en bois ou métal (prière de rayer ce qui n'est pas utilisé) pour dalles en porté. Face lisse. Dalle de 0,15m d'épaisseur avec distance maximum de 3,50m entre niveaux portés.	13. Casseforme in legno o acciaio per solai orizzontali, faccia liscia, solai 150mm di spessore, non oltre 3,50m dal piano di appoggio.
Reinforcement	**Acier d'armature**	**Armature**
14. Reinforcement in concrete walls. Hot rolled high tensile bars cut, bent and laid, 16mm diameter.	14. Armature pour murs en béton. Acier à haute adhérence roulés au chaud, coupées, pliés et posées, 1,60cm de diamètre.	14. Armatura in muri di calcestruzzo: barre di acciaio ad aderenza migliorata, tagliate, piegate e posate, 16mm di diametro.
15. Reinforcement in suspended concrete slabs. Hot rolled high tensile bars cut, bent and laid, 25mm diameter.	15. Armatures pour dalles en porté en béton. Acier à haute adhérence roulées ou chaud, coupées, pliés et posées, 2,50cm de diamètre.	15. Armatura in solai di calcestruzzo: barre di acciaio ad aderenza migliorata, tagliate, piegate e posate, 25mm di diametro.

Schalungen	Encofrados	Formwork
(Anzunehmen ist, daβ eine einfache Konstruktion eine dreimalige Nutzung der Schalungen ermöglicht)	(Se supone una forma simple y repetitiva que permite 3 usos del encofrado)	(Assume a simple repetitive design which allows 3 uses of formwork)
11. Schalungen aus Weichholz oder Metall für Betonwände. Glatte Fläche (nur eine Seite).	11. Encofrado metálico a una cara, en muros, acabado para revestir.	11. Softwood or metal formwork to concrete walls. Basic finish. (one side only).
12. Schalungen aus Weichholz oder Metall für Betonstützen. Glatte Fläche. Umfang der Stützen 1600mm.	12. Encofrado metálico en pilares de 1600mm de perímetro, acabado para revestir.	12. Softwood or metal formwork to concrete columns. Basic finish. Columns 1600m girth.
13. Schalungen aus Weichholz oder Metall für horizontale Unterseiten von Deckenplatten. Glatte Fläche. Dicke der Deckenplatten 150mm, Höhe der Decke maximal 3.50m.	13. Encofrado de madera en losas de 150mm, acabado para revestir a menos de 3,50m de altura.	13. Softwood or metal formwork to horizontal soffits of slabs. Basic finish. Slab 150mm thick, not exceeding 3.50m high.
Bewehrung	**Armaduras**	**Reinforcement**
14. Bewehrung für Betonwände. Warmgewalzte hochzugfeste Stäbe, liefen schneiden biegen und verlegen, Durchmesser 16mm.	14. Armadura en muros de hormigón con acero de alta resistencia de 16mm de diámetro, colocada.	14. Reinforcement in concrete walls. Hot rolled high tensile bars cut, bent and laid, 16mm diameter.
15. Bewehrung für Deckenplatten. Warmgewalzte hochzugfeste Stäbe, liefen schneiden biegen und verlegen, Durchmesser 25mm.	15. Armadura en losas de hormigón con acero de alta resistencia de 25mm de diámetro, colocada.	15. Reinforcement in suspended concrete slabs. Hot rolled high tensile bars cut, bent and laid, 25mm diameter.

16. Fabric (mat) reinforcement in concrete beds (measured separately). Weight approximately 3.0 kg/m². Laid in position with 150mm side and end laps.	16. Acier d'armature treillis soudé pour planchers en dalle pleine. Poids approximatif: 3kg/m² posé en place avec un recouvrement de 0,15m.	16. Rete in acciaio elettrosaldata in solette di calcestruzzo, peso approssimato 3,0 Kg/m², posata in opera con risvolti ai lati di 150mm.
Steelwork	**Charpente métallique**	**Carpenteria metallica**
17. Fabricate, supply and erect steel framed structure. Including painting all steel with one coat primer.	17. Fabrication, fourniture et installation d'une structure métallique y compris l'application d'une couche primaire autirouille.	17. Fabbricazione, fornitura e montaggio di struttura metallica compresa l'applicazione di una mano di minio.
Brickwork and blockwork (Assume a notional thickness of 100mm for bricks and blocks. Rates should be for the nearest standard size to 100mm)	**Maçonnerie** (Les prix doivent être pour les briques et blocs de dimensions standards locales les plus proches de 10cm d'épaisseur)	**Opera di muratura** (Prezzi per mattoni e blocchi di dimensioni di normale produzione, il più vicino possibile a 100mm di spessore)
18. Precast lightweight aggregate hollow concrete block walls. Gauged mortar. 100mm thick.	18. Murs en élévation en blocs creux en béton de granulats légers avec joint de 10mm de mortier. 0,10m épaisseur du mur brut.	18. Blocchi cavi di calcestruzzo alleggeriti, posati con l'impiego di malta bastarda, in muri spessore 100mm.
19. Solid (perforated) clay or concrete common bricks (priced at .... per m² delivered to site). Gauged mortar. 100mm thick walls.	19. Murs en élévation en briques pleines (perforées) en terre cuite ou béton de fabrication courante (prix des briques par m² .... livrées au chantier) avec joint de 10mm de mortier. 0,10m épaisseur du mur brut.	19. Mattoni pieni di laterizio o di cemento (al prezzo di ...... lire/m² franco cantiere), posati con l'impiego di malta bastarda, in muri spessore 100mm.

16. Baustahlmatten für Bodenplatten (separat gemessen). Gewicht ungefähr 3.0 kg/m². Mit 150mm Überlappung an Seiten und Enden verlegt.	16. Mallazo en pavimentos de hormigón (medido aparte) con un peso aproximado de 3.0 kg/m².	16. Fabric (mat) reinforcement in concrete beds (measured separately). Weight approximately 3.0 kg/m². Laid in position with 150mm side and end laps.
Stahlbauarbeiten	**Estructura metálica**	**Steelwork**
17. Fertigung, Lieferung und Montage eines konstruktion Stahlrahmens. Mit Grundanstrich für alle Stahlflächen.	17. Acero en estructura metálica, colocada, incluyendo una capa de pintura de imprimación.	17. Fabricate, supply and erect steel framed structure. Including painting all steel with one coat primer.
Mauererarbeiten (Die Kostenkennwerte sollen für Mauersteine, Dicke 100mm, oder nächstliegendes örtliches Standardformat gelten)	**Paredes de ladrillo y bloques de hormigón** (Se supone un grueso de 100mm para ladrillos y bloques. Los datos tendrian que refereirse a medidas standard lo mas proximas posibles a 100mm)	**Brickwork and blockwork** (Assume a notional thickness of 100mm for bricks and blocks. Rates should be for the nearest standard size to 100mm)
18. Wände aus Betonhohlblock steinen. Zement-kalkmörtel. Wanddicke 100mm.	18. Pared de bloque de hormigón hueco, tomado con mortero, de 100mm de espesor.	18. Precast lightweight aggregate hollow concrete block walls. Gauged mortar. 100mm thick.
19. Normale Vollziegel (bzw. Hochlochziegel) oder Betonsteine (Preis ... pro m² mit Lieferung). Zementkalkmörtel. Wanddicke 100mm.	19. Pared de ladrillo perforado (PVP de 650 pesetas por m², puesto en pie de obra) tomado con mortero, de 100mm de espesor.	19. Solid (perforated) clay or concrete common bricks (priced at .... per m² delivered to site). Gauged mortar. 100mm thick walls.

20. Solid (perforated) sand lime bricks (priced at .... per m^2 delivered to site). Gauged mortar. 100mm thick walls.	20. Murs en élévation en briques pleines (perforées) en chaux et sable (prix des briques par m^2 .... livrées au chantier), avec joint de 10mm de mortier gros. 0,10m épaisseur du mur brut.	20. Mattoni pieni di calce e sabbia (al prezzo di ...... lire/m^2 franco cantiere), posati con l'impiego di malta bastarda, in muri spessore 100mm.
21. Facing bricks (priced at .... per m^2 delivered to site). Gauged mortar, flush pointed as work proceeds. Half brick thick walls.	21. Murs en briques pleines apparentes (prix des briques par m^2 ..... livrées au chantier), avec joint de 10mm de mortier. Faces alignées, parementées et rejointoyées au mortier gros. 0,10m épaisseur du mur.	21. Mattoni "a cortina" (al prezzo di ...... lire/m^2 franco cantiere), posati con l'impiego di malta bastarda con giunti stilati, in muratura ad una testa.
Roofing	**Toiture**	**Tetti**
22. Concrete interlocking roof tiles 430 x 380mm (or nearest equivalent). On and including battens and underfelt. Laid to 355mm gauge with 75mm laps (excluding eaves fittings or ridge tiles).	22. Tuiles à emboîtement en béton. Dimensions 0,43 x 0,38m (ou les plus proches dimensions) y compris les liteaux et carton bituminé. 8 tuiles au m^2. (Non-inclus sont les avances et la faîtière du toit).	22. Tegole di cemento ad incastro 430 x 380mm (o di simili dimensioni), posate a 8 tegole/m^2, su listelli di legno e cartonfeltro bituminato compresi nel prezzo. Sono esclusi colmi ed altri pezzi speciali.
23. Plain clay roof tiles 260 x 160mm (or nearest equivalent). On and including battens and underfelt. Laid to 100mm lap (excluding eaves fittings or ridge tiles).	23. Tuiles plates en terre cuite. Dimensions 0,26m x 0,16m (ou les plus proches dimensions) y compris les liteaux et le carton bituminé. 60 tuiles au m^2. (Non-inclus sont les avances et la faîtière du toit).	23. Tegole piane in laterizio 260 x 160mm (o di simili dimensioni), posate a 60 tegole/m^2, su listelli di legno e cartonfeltro bituminato compresi nel prezzo. Sono esclusi colmi ed altri pezzi speciali.

20. Kalksandvollstein (bzw. -lochstein) (Preis ... pro m² mit Lieferung). Zementkalkmörtel. Wanddicke 100mm.	20. Pared de ladrillo calcáreo (PVP de 960 pesetas por m², puesto en pie de obra) tomado con mortero de 100mm de espesor.	20. Solid (perforated) sand lime bricks (priced at .... per m² delivered to site). Gauged mortar. 100mm thick walls.
21. Verblendsteine (Preis ... pro m² mit Lieferung). Zementkalkmörtel, Vollfugen auszuführen bei der Hochmauerung. Halbsteindicke Wände.	21. Pared de ladrillo en cara vista (PVP de 1100 pesetas por m², puesto a pie de obra), con junta enrasada, de medio pie de espesor.	21. Facing bricks (priced at .... per m² delivered to site). Gauged mortar, flush pointed as work proceeds. Half brick thick walls.
Bedachungsarbeiten	**Cubiertas**	**Roofing**
22. Betonflachdachsteine, 430 x 380mm (oder nächstliegendes Standardformat). Mit Lattung und Dachpappenunterlage. Lattenweite 355mm, Überdeckung 75mm. (Ohne Traufenabschluß und Firststeine).	22. M² de cubierta con tejas de hormigón de 430 x 380mm incluyendo piezas especiales con longitud vista de 355m y un solape de 75mm (no se incluyen uniones con los canalones o las piezas de remate).	22. Concrete interlocking roof tiles 430 x 380mm (or nearest equivalent). On and including battens and underfelt. Laid to 355mm gauge with 75mm laps (excluding eaves fittings or ridge tiles).
23. Biberschwannzziegel 260 x 160mm (oder nächstliegendes Standardformat). Mit Lattung und Dachpappenunterlage. Überdeckung 100mm. (Ohne Traufenabschluß und Firststeine).	23. M² de cubiertas de tejas planas cerámicas de 260 x 160mm incluyendo piezas especiales con un solape de 100mm y excluyendo las entregas a canalones o piezas de remate.	23. Plain clay roof tiles 260 x 160mm (or nearest equivalent). On and including battens and underfelt. Laid to 100mm lap (excluding eaves fittings or ridge tiles).

24. Fibre cement roof slates 600 x 300mm (or nearest equivalent). On and including battens and underfelt. Laid flat or to fall as coverings for roofs.

24. Plaques type ardoise en amiante - ciment. Dimensions 0,60m x 0,30m (ou les plus proches dimensions) y compris les liteaux et le carton bituminé. Posées plat ou en pente en tant que couverture pour toits.

24. Lastre di fibro-cemento 600 x 300mm (o di simili dimensioni), posate in piano o in pendenza su listelli di legno e cartonfeltro bituminato compresi nel prezzo.

25. Sawn softwood roof boarding, preservative treated 25mm thick. Laid flat or to fall.

25. Planches en sapin de pays traité pour toits. 2,50cm d'épaisseur. Posées plat ou en pente.

25. Tavolame in abete impregnato con protettivo, 25mm di spessore, posato in piano o in pendenza.

26. Particle board roof coverings with tongued and grooved joints 25mm thick. Laid flat or to fall.

26. Planches en particules de bois avec languettes et rainures pour toits. 2,50cm d'épaisseur. Posées plat ou en pente.

26. Lastre di truciolare 25mm di spessore, con giunti ad incastro a linguetta, posati in piano o in pendenza.

27. 3 layers glass-fibre based bitumen felt roof covering. Finished with limestone chippings in hot bitumen. To flat roofs.

27. 3 couches de feutre bituminé à base de laine minérale pour toits. Finis avec émulsion bitumineuse et granulés minéreaux. Pour toits plats.

27. Manto impermeabile prefabbricato, composto da 3 strati di feltro bituminoso a base di fibre di vetro, coperto con emulsione bituminosa ed inerti reflettenti, su coperture piane.

28. Bitumen based mastic asphalt roof covering in 2 layers. On and including sheathing felt underlay, with white chippings finish. To flat roofs.

28. 2 couches d'asphalte coulé neuf à base de bitume pour toits, avec feutre bituminé et finis avec granulés minéreaux. Pour toits plats.

28. Due strati di asfalto con manto di feltro sottostante e finitura di inerti reflettenti, per coperture piane.

29. Glass-fibre mat roof insulation 160mm thick. Laid flat between ceiling joists.

29. Isolation de toitures par laine minérale en rouleaux, posé entre chevrons sans fixation. 160mm d'épaisseur.

29. Coibentazione del tetto con materassino di lana di vetro, spessore 160mm, posato piatto tra morali.

24. Faserzementplatten für Dach 600 x 300mm (oder nächstliegendes Standardformat). Mit Lattung und Dachpappenunterlage. Flach oder im Gefälle als Dachhaut verlegt.	24. M^2 cubierta con fibrocemento en piezas de 600 x 300mm incluyendo las juntas y un fieltro inferior.	24. Fibre cement roof slates 600 x 300mm (or nearest equivalent). On and including battens and underfelt. Laid flat or to fall as coverings for roofs.
25. Weichholzdachschalung mit Schutzbehandlung. Dicke 25mm. Flach oder im Gefälle verlegt.	25. Remate de madera tratada previamente de 25mm de espesor, colocado plano o inclinado.	25. Sawn softwood roof boarding, preservative treated 25mm thick. Laid flat or to fall.
26. Spanplattendachdeckung, Stöße mit Nut und Feiler. Dicke 25mm. Flach oder im Gefälle verlegt.	26. Cubierta con tablero de particulas, incluidas las juntas, de 25mm de espesor, colocado plano o inclinado.	26. Particle board roof coverings with tongued and grooved joints 25mm thick. Laid flat or to fall.
27. Dreilagige Dachhaut aus Glasfaser-Bitumendachdichtungsbahnen. Überdeckt mit Kalksteinsplitt, in Heißbitumen eingebettet. Für Flachdächer.	27. Impermeabilización con tela bituminosa de tres capas de fibra de vidrio, acabada con árido visto sobre emulsión asfáltica en caliente en cubiertas.	27. 3 layers glass-fibre based bitumen felt roof covering. Finished with limestone chippings in hot bitumen. To flat roofs.
28. Zweilagiger Bitumenmastixdachbelag. Mit Dachpappenunterlage, überdeckt mit Weißsplitt. Für Flachdächer.	28. Impermeabilización de cubiertas con dos capas de tela asfáltica incluyendo el fieltro inferior, acabado con árido visto.	28. Bitumen based mastic asphalt roof covering in 2 layers. On and including sheathing felt underlay, with white chippings finish. To flat roofs.
29. Wärmedämmung aus Glasfasermatten, Dicke 160mm. Zwischen Unterzügen verlegt.	29. Aislamiento con manto de fibra de vidrio de 160mm de espesor colocado entre las viguetas del techo.	29. Glass-fibre mat roof insulation 160mm thick. Laid flat between ceiling joists.

30. Rigid sheet resin-bonded loadbearing glass-fibre roof insulation 75mm thick. Laid on flat roofs.	30. Isolation de toits plats par laine minérale en panneaux rigide. 75mm d'épaisseur.	30. Coibentazione di coperture piane composta di pannelli rigidi di lana di vetro, spessore 75mm.
31. 0.8mm troughed galvanized steel roof cladding in single spans of 3.00m with loading of 0.75 KN/m². Fixed to steel roof trusses with bolts. To pitched roofs.	31. Couverture de toits par tôles ondulées en acier galvanisé. 0,80 mm d'épaisseur. Portée maximum de 3m acceptant un poids maximum de 0,75KN/m². Compris fixation à des chevrons d'acier. Pour toits en pente.	31. Copertura di tetti a pendenza con lamiera grecata zincata spessore 0,8mm, luce di 3,00m, con sovraccarico di 0,75 KN/m², imbullonata a capriate metalliche.
Woodwork and metalwork	**Menuiserie et métallerie**	**Opere in legno e in ferro**
(Hardwood should be assumed to be of reasonable exterior quality)	(Bois éxotique ou chêne doit être considéré d'une qualité extérieure raisonable)	(Il legno duro sarà di buona qualità, adatto per l'uso esterno)
32. Preservative treated sawn softwood. Size 50 x 100mm. Framed in partitions.	32. Sapin de pays traité. Dimensions 5cm x 10cm. Pour ossature de cloisons.	32. Legno segato impregnato con protettivo, dimensioni 50 x 100mm, per ossatura di tramezzi.
33. Preservative treated sawn softwood. Size 50 x 150mm. Pitched roof members.	33. Sapin de pays traité. Dimensions 5cm x 15cm. Pour toiture en pente.	33. Legno segato impregnato con protettivo, dimensioni 50 x 150mm, per struttura di tetto a pendenza.
34. Single glazed casement window in (.............) hardwood including hardwood frame and sill. Including steel butts and anodised aluminium espagnolette bolt. Size approx. 650 x 900mm with 38 x 100mm frame and 75 x 125mm sill.	34. Fenêtre à un vantail en chêne ou bois éxotique avec un paneau de verre simple de ...mm d'épaisseur y compris châssis en chêne, rebord en chêne, charnières en acier et verrous type espagnolette en aluminium anodisé. Dimensions approximatives 65cm x 90cm avec châssis de 3,80cm x 10cm et rebord de 7,50cm x 12,50cm.	34. Finestra ad anta semplice in legno duro del tipo ...., compreso telaio (38 x 100mm), soglia (75 x 125mm), cerniere e chiusura in alluminio. Dimensioni approssimate 650 x 900mm.

30. Wärmedämmung aus tragenden harzverleimten Glasfaserplatten, Dicke 75mm. Auf Flachedächern verlegt.

30. Placa autoportante de resina rígida con aislamiento de fibra de vidrio de 75mm de espesor, colocada en techos planos.

30. Rigid sheet resin-bonded loadbearing glass-fibre roof insulation 75mm thick. Laid on flat roofs.

31. Trapezprofildachdeckung aus verzinktem Stahl, Dicke 0.8mm, Spannweiten maximal 3.00m, Belastung 0.75KN/m². Mit Bolzen an Stahlbinder befestigt. Für Schrägdächer.

31. Cubierta de acero galvanizado de 0,8mm de espesor colocada en vanos de 3m, con carga de 0,75KN/m². Fijado a soportes metálicos de techo con tornillos.

31. 0.8mm troughed galvanised steel roof cladding in single spans of 3.00m with loading of 0.75 KN/m². Fixed to steel roof trusses with bolts. To pitched roofs.

Holz und Metallarbeiten

(Anzunehmen ist, daß das Hartholz von orderntlicher äußerer Qualität ist)

Carpinteria y cerrajeria

(Se considera la madera de una calidad exterior razonable)

Woodwork and metalwork

(Hardwood should be assumed to be of reasonable exterior quality)

32. Schnittweichholz, 50mm x 100mm, mit Schutzbehandlung. Für Trennwandgerippe.

32. M de montante de madera de 50 x 100mm, acabado natural con tratamiento prévio.

32. Preservative treated sawn softwood. Size 50 x 100mm. Framed in partitions.

33. Schnittweichholz, 50mm x 150mm, mit Schutzbehandlung. Bauteile für Schrägdach.

33. Travesaño de 50 x 150mm con tratamiento previo, acabado natural colocado en cubiertas.

33. Preservative treated sawn softwood. Size 50 x 150mm. Pitched roof members.

34. Flügelfenster, einfach verglast, aus (.....) Hartholz, mit Rahm und Fenstersbank aus Hartholz. Mit Stahleinstemmbändern und Espagnolettenverschluß aus adonisiertem Aluminium. Ungefähre Dimensionen 650 x 900mm, Rahmen 38 x 100mm, Fensterbank 75 x 125mm.

34. Ventana de una hoja de pino melis incluyendo marco y hoja, incluso bisagras de acero y españoletas de aluminio anodizado. Medidas aproximadas de 650 x 900mm con huecohojas de 38 x 100mm, marcos de 75 x 125mm.

34. Single glazed casement window in (.............) hardwood including hardwood frame and sill. Including steel butts and anodised aluminium espagnolette bolt. Size approx. 650 x 900mm with 38 x 100mm frame and 75 x 125mm sill.

35. Two panel door with panels open for glass in (............) hardwood including hardwood frame and sill. Including glazing with 6mm wired polished plate security glass fixed with hardwood beads and including steel butts, anodised handles and push plates and security locks. Size approximately 850 x 2000mm with 38 x 100mm frame and 38 x 150mm sill.

36. Solid core half hour fire resisting hardwood internal flush door lipped on all edges. Unpainted, including steel butts, anodised handles and push plates and mortice lock. Size approximately 800 x 2000mm.

37. Aluminium double glazed window and hardwood sub-frame. Standard anodised horizontally sliding double glazed in (..............) hardwood sub-frame and sill. Including double glazing with 4mm glass, including all ironmongery. Size approximately 1200 x 1200mm with 38 x 100mm sub-frame and 75 x 125mm sill.

35. Porte à un vantail en chêne ou bois éxotique à deux carreaux pour paneaux de verre armé de 6mm d'épaisseur y compris le châssis en chêne, le seuil en chêne, les charnières en acier, les poignées, serrures de sûreté et plaques à pousser toutes anodisées. Dimensions approximatives 85cm x 200cm avec un châssis de 3,80cm x 10cm et un seuil de 3,80 cm x 15cm.

36. Porte plane pleine intérieure à un vantail en chêne ou bois éxotique avec une demi heure de coupe-feu. Non-painte y compris charnières en acier, poignées, serrures à mortaiser et plaques à pousser anodisées. Dimension approximatives 80cm x 200cm.

37. Fenêtre coulissante en aluminium anodisé avec double vitrage 4/6/4, sous-châssis et rebord en chêne y compris toute quincaillerie et ferrage. Dimensions approximatives 120cm x 120cm avec un sous-châssis de 3,80cm x 10cm et un rebord 7,50cm x 12,50cm.

35. Porta a vetri a due ante in legno duro tipo ...., compreso telaio (38 x 100mm), soglia (38 x 100mm), vetri retinati spessore 6mm fissati con bacchette in legno duro, cerniere in acciaio, maniglie e piastre in alluminio e serratura di sicurezza. Dimensioni 850 x 2000mm.

36. Porta interna del tipo antincendio REI 30' in legno duro non verniciato, compreso cerniere in acciaio, maniglie e piastre in alluminio e serratura ad incasso. Dimensioni approssimate 800 x 2000mm.

37. Finestra in alluminio di normale produzione con ante scorrevoli e controtelaio in legno duro del tipo ...., vetro camera con lastre di vetro di 4mm, compresa tutta la ferramenteria. Dimensioni approssimate 1200 x 1200mm, contro telaio di 38 x 100mm e soglia 75 x 125mm.

35. Zweifüllungstür mit Sicherheitsverglasung, aus (....) Hartholz, mit Rahmen und Fensterbank aus Hartholz. Zwei Füllungen aus Spiegel-drahtglas, Dicke 6mm, Hartholzglasleisten befestigt. Mit Stahleinstemmbändern, adonisierten Griffen und Stoßschildern, und Sicherheitsschlössern. Ungefähre Dimensionen 800 x 2000mm, Rahmen 38 x 100mm, Schwelle 38 x 150mm.

35. Puerta vidriera de dos hojas en pino de melis incluyendo marco y hojas, incluso vidrio de seguridad armado pulido de 6mm, con bisagras de acero, y herrajes de aluminio anodizado, así como cerradura de seguridad. Medidas aproximadas de 850 x 2000mm con armazón de 38 x 100mm y marco de 38 x 150mm.

35. Two panel door with panels open for glass in (............) hardwood including hardwood frame and sill. Including glazing with 6mm wired polished plate security glass fixed with hardwood beads and including steel butts, anodised handles and push plates and security locks. Size approximately 850 x 2000mm with 38 x 100mm frame and 38 x 150mm sill.

36. Innen-Flächenvolltür aus Hartholz, mit halb-stündiger Feuerwider-standszeit, alle Ränder mit Umleimern. Unbemalt, mit Stahleinstemmbändern, adonisierten Griffen und Stoßschildern, und Einsteckscholß. Ungefähre Dimensionen 800 x 2000mm.

36. Puerta anti-incendios RF-30 de madera, anclada en todo su perímetro, sin pintar, incluyendo bisagras de acero, herrajes de aluminio anodizado y cerradura de muesca. Medidas aproximadas de 800 x 2000mm.

36. Solid core half hour fire resisting hardwood internal flush door lipped on all edges. Unpainted, including steel butts, anodised handles and push plates and mortice lock. Size approximately 800 x 2000mm.

37. Horizontalschiebefenster aus anodisiertem Aluminium mit Doppel-verglasung, Glasdicke 4mm. Innenprofil und Fensterbank aus (....) Hartholz. Mit allen Beschlägen. Ungefähre Dimensionen 1200 x 1200mm, Fensterprofil 38 x 100mm, Fensterbank, 75 x 125mm.

37. Ventana corredera de aluminio anodizado standard con doble vidrio y premarco de flandes, incluyendo doble vidrio de 4mm y toda la cerrajería. Medidas aproximadas de 1200 x 1200mm, con sección de premarco de 38 x 100mm y 75 x 125mm de sección de alfeizar.

37. Aluminium double glazed window and hardwood sub-frame. Standard anodised horizontally sliding double glazed in (.............) hardwood sub-frame and sill. Including double glazing with 4mm glass, including all ironmongery. Size approximately 1200 x 1200mm with 38 x 100mm sub-frame and 75 x 125mm sill.

38. Aluminium double glazed door set and hardwood sub-frame. Standard anodised aluminium, double glazed in (..............) hardwood sub-frame and sill. Including double glazing with 4mm glass, including all ironmongery. Size approximately 850 x 2100mm with 38 x 100mm sub-frame and 75 x 125mm sill.

38. Porte à un vantail en aluminium anodisé avec double vitrage 4/6/4, sous-châssis et seuil en chène, compris toute quincaillerie et ferrage. Dimensions approximatives 85cm x 210cm avec un sous-châssis de 3,80cm x 10cm et un seuil de 7,50cm x 12,50cm.

38. Porta finestra in alluminio di normale produzione con contro telaio in legno duro del tipo ...., vetro camera con lastre di vetro di 4mm, compresa tutta la ferramenteria. Dimensioni approssimate 850 x 2100mm, controtelaio 38 x 100mm e soglia 75 x 125mm.

39. Hardwood skirtings. Wrought (...............) hardwood. Fixed on softwood grounds. Size 20 x 100mm.

39. Plinthes en chêne ou bois éxotique fixées à des liteaux en sapin de pays. Dimensions 2cm x 10cm.

39. Zoccolino battiscopa in legno duro piallato del tipo ...., fissato con tasselli di legno. Dimensioni 20 x 100mm.

40. Framed structural steelwork in universal joist sections. Bolted or welded connections, including erecting on site and painting one coat at works.

40. Ouvrages assemblés de charpente métallique en fer pour poitrails, filets, poutres, poteaux ... etc en âme pleine, y compris les percages, boulons, rivets ou soudures nécessaires à l'exécution des ouvrages et une couche de painture aux ateliers.

40. Profilati di ferro a doppio T intelaiati, compresi i giunti saldati o bullonati, il montaggio in cantiere e una mano di minio protettivo.

41. Structural steelwork lattice roof trusses. Bolted or welded connections, including erecting on site and painting one coat at works.

41. Ouvrages assemblés de charpente métallique en fer pour pannes, faîtages, et sablières ... etc en ferme triangulée, y compris les percages, boulons, rivets ou soudures nécessaires à l'exécution des ouvrages et une couche de painture aux ateliers.

41. Capriate in ferro, compresi i giunti saldati o bullonati, il montaggio in cantiere e uno mano di minio protettivo.

38. Türanlage aus anodisiertem Aluminium mit Doppelverglasung, Glasdicke 4mm. Innenprofil und Schwelle aus (....) Hartholz. Mit allen Beschlägen. Ungefähre 850 x 2100mm, Innenprofil 38 x 100mm, Schwelle 75 x 125mm.

38. Balconera de aluminio con anodizado standard, con doble vidrio y premarcos de madera en flandes, incluyendo doble vidrio de 4mm de espesor, y toda la cerrajería. Medidas aproximadas de 850 x 100mm y 75 x 125mm de sección de alfeizar.

38. Aluminium double glazed door set and hardwood sub-frame. Standard anodised aluminium, double glazed in (..............) hardwood sub-frame and sill. Including double glazing with 4mm glass, including all ironmongery. Size approximately 850 x 2100mm with 38 x 100mm sub-frame and 75 x 125mm sill.

39. Sockel aus gehobeltem (.....) Hartholz. An Weichholzstreifen befestigt. Dimensionen 20 x 100mm.

39. Zócalo de madera de melis colocado sobre pavimentos de madera de medidas de 20 x 100mm.

39. Hardwood skirtings. Wrought (...............) hardwood. Fixed on softwood grounds. Size 20 x 100mm.

40. Rahmentragwerk aus Doppel-T-Stahl. Bolzen- oder Schweißverbindungen, mit Montage vor Ort und einem Anstrich vor Lieferung.

40. Acero estructural en elementos principales, en perfiles universales, atornillado o soldado, colocado, incluso una capa de pintura de imprimación.

40. Framed structural steelwork in universal joist sections. Bolted or welded connections, including erecting on site and painting one coat at works.

41. Sägedachbinder aus Stahl. Bolzen- oder Schweißverbindungen, mit Montage vor Ort und einem Anstrich vor Lieferung.

41. Acero estructural para estructuras espaciales, atornillado o soldado, colocado, incluso una capa de pintura de imprimación.

41. Structural steelwork lattice roof trusses. Bolted or welded connections, including erecting on site and painting one coat at works.

Plumbing	**Plomberie**	**Impianto idrico sanitario**
(Sizes of sanitary installations and pipes are indicative)	(Les dimensions des installations sanitaires et des tuyaux sont indicatives)	(Le dimensioni degli impianti e delle tubazioni sono puramente indicative)
42. UPVC half round eaves gutter. Screwed to softwood at 1.00m centres. 110mm external diameter (excluding bends, outlets etc.).	42. Goutières à demi cercle en PVC vissées à du bois du sapin de pays à 1m d'intervalle. 11cm de diamètre extérieur (non inclus sont les coudes, les moignons ..etc)	42. Canale di gronda di sezione ad U in UPVC, di diametro esterno 110mm, con un supporto ad ogni metro di lunghezza, fissato con viti. Sono esclusi angoli, uscite ecc.
43. UPVC rainwater pipes with pushfit joints. Screwed to brickwork at 1.50m centres. 100mm external diameter (excluding bends, outlets etc.).	43. Tuyaux en PVC pour canalisation d'eau pluviale avec joints, vissés à un mur de briques à 1,50m d'intervalles. 10cm de diamètre extérieur (non inclus sont les coudes, les culottes ...etc)	43. Pluviali in UPVC di diametro esterno 100mm, con giunti ad incastro, fissati al muro ogni 1,5m. Sono esclusi gommiti, uscite ecc.
44. Light gauge copper cold water tubing with compression or capillary fittings. Screwed to brickwork horizontally at 1.00m centres. 15mm external diameter.	44. Tuyaux en cuivre pour distribution d'eau froide avec tous raccords, joints ...etc. Vissés à un mur en briques à 1m d'intervalle. 1,50cm de diamètre extérieur.	44. Tubazione leggera di rame per acqua fredda, 15mm diametro, con giunti saldati o a compressione, fissata orizzontalmente ogni metro di lunghezza.
45. High pressure polypropylene, polythene or UPVC (as appropriate) pipes for cold water supply. Fixed horizontally to brick walls at 1.00m centres. 15mm external diameter, complete with fittings.	45. Tuyaux à haute pression en polypropylène, polyéthylène ou PVC pour provision d'eau froide. Fixés extérieurement à un mur en briques à 1m d'intervalle. 1,50cm de diamètre extérieur, compris toutes fournitures.	45. Tubazione di UPVC, polipropilene o polietilene ad alta pressione per distribuzione di acqua fredda, fissata orizzontalmente alla muratura ogni metro di lunghezza, completa di gomiti, raccordi ecc. Dimensione 15mm diametro esterno.

Installationen	**Instalaciones sanitarias**	**Plumbing**
(Die Dimensionen der Sanitäranlagen und leitungen gelten als Richtlinien)	(Las medidas de las instalaciones sanitarias y cañerias son indicativas)	(Sizes of sanitary installations and pipes are indicative)
42. Halbrunde Dachrinne aus Hart-PVC. In Entfernungen von 1.00m an Weichholz geschraubt. Außendurchmesser 110mm. (Ohne Bögen, Auslauf usw.).	42. Canalón semicircular de UPVC de atornillado a madera cada 1m, diámetro exterior de 110mm (excluyendo curvas salidas).	42. UPVC half round eaves gutter. Screwed to softwood at 1.00m centres. 110mm external diameter (excluding bends, outlets etc.).
43. Regenfallrohr aus Hart-PVC, mit Einsteckverbindung. In Entfernungen von 1.50m an Mauerwerk geschraubt. Außendurchmesser 100mm. (Ohne Bögen, Auslauf usw.).	43. Canalones de UPVC de recogida de lluvias con juntas. Atornillado a muro de ladrillo cada 1,50m de diámetro exterior de 100mm (excluyendo curvas y salidas).	43. UPVC rainwater pipes with pushfit joints. Screwed to brickwork at 1.50m centres. 100mm external diameter (excluding bends, outlets etc.).
44. Leichtkupferrohre für Kaltwasser, mit Verbindungsstücken. In Entfernungen von 1.00m horizontal an Mauerwerk geschraubt. Außendurchmesser 15mm.	44. Tubos de cobre para agua fría con juntas de unión a compresión. Atornillados a muro de ladrillo horizontalmente cada 1m, diámetro exterior de 15mm.	44. Light gauge copper cold water tubing with compression or capillary fittings. Screwed to brickwork horizontally at 1.00m centres. 15mm external diameter.
45. Hochdruckrohre aus Polypropylene, Polyäthylen oder Hart-PVC (wie angemessen) für Kaltwasserversorgung. In Entfernungen von 1.00m horizontal an Mauerwerk befestigt. Außendurchmesser 15mm. Mit Formstücken.	45. Conducto de polipropileno de alta presión, polietileno o PVC, para suministro de agua fría, fijado horizontalmente en paredes de ladrillo cada 1m, de 15mm de diámetro exterior, acabado incluso complementos.	45. High pressure polypropylene, polythene or UPVC (as appropriate) pipes for cold water supply. Fixed horizontally to brick walls at 1.00m centres. 15mm external diameter, complete with fittings.

46. Low pressure polypropylene, polythene or UPVC (as appropriate) pipes for cold water distribution. With plastic compression fittings 20mm external diameter, laid in trenches.

46. Tuyaux à pression basse en polypropylène, polyéthylène ou PVC pour distribution d'eau froide avec fournitures posés en tranchée. 2cm de diamètre extérieur.

46. Tubazione di UPVC, polipropilene o polietilene a bassa pressione per distribuzione di acqua fredda, posata in trincea, completa di giunti in plastica a compressione, manicotti, raccordi ecc. Dimensione 20mm diametro esterno.

47. UPVC soil and vent pipes with solvent welded or ring seal joints. Fixed vertically to brickwork with brackets at 1.50m centres. 100mm external diameter.

47. Tuyaux en PVC pour canalisation d'eau usée avec joints, fixés verticallement à une base de brique avec anneaux à 1,50m d'intervalle. 10cm de diamètre extérieur.

47. Tubi per fognatura in UPVC, con giunti saldati o con anelli di gomma, fissati verticalmente alla muratura ogni metro di lunghezza. Dimensione 100mm diametro esterno.

48. White vitreous china WC suite with black plastic seat and cover and plastic low level cistern, 9 litre capacity. Complete with ball valve and float and flush pipe to WC suite. Fixed to concrete.

48. WC en procelaine vitrifiée de couleur blanche avec abattant et couvercle de couleur noir et citerne en plastic de 9 litres de capacité complet avec robinets de chasse et d'arrêt, tube de chasse ...etc. Fixé à une base de béton.

48. Vaso igienico bianco con sedile in plastica nera con coperchio, casetta di 9 litri capacità, valvola, galleggiante, tubo di collegamento con rosone ecc, fissato al pavimento di cemento.

49. White vitreous china lavatory basin with 2 No. chrome plated taps (or medium quality chrome plated mixer taps). Including plug, overflow and waste connections (excluding trap). Size approximately 560 x 400mm, fixed to brickwork with concealed brackets.

49. Lavabos en porcelaine vitrifiée de couleur blanche avec 2 robinets chromé (ou un robinet mélangeur de qualité moyenne) avec bouchon, autres fournitures necessaires et colliers de canalisation (non-inclus sont les trappes). Dimensions approximatives 56cm x 40cm, fixés à une base de briques.

49. Lavabo in porcellana dura vetrochina da 560 x 400mm, compresi due rubinetti cromati, (o miscellatore), piletta, tappo, catenella ed allacciamenti (escluso sifone), fissato alla muratura con mensole.

46. Niederdruckrohre aus Polypropylen, Polyäthylen, oder Hart-PVC (wie benötigt) für Kaltwasserverteilung. Mit Kunststoff-verbindungsstücken. Außendurchmesser 20mm. In Gräben verlegt.

46. Conducto de polipropileno de baja presión, polietileno o PVC, para distribución de agua fría de 20mm de diámetro exterior con juntas de presión plástica colocado por capas.

46. Low pressure polypropylene, polythene or UPVC (as appropriate) pipes for cold water distribution. With plastic compression fittings 20mm external diameter, laid in trenches.

47. Lüftungs- und Abwasserrohre aus Hart-PVC, mit Quellschweiß- oder Dichtungsringvergindungen. In Entfernungen von 1.50m mit Rohrhaltern vertikal an Mauerwerk befestigt. Außendurchmesser 100mm.

47. Conducto de ventilación con soldadura disuelta o juntas de cierre de anilla. Fijados verticalmente a muro de ladrillos con soportes cada 1,5m, de diámetro exterior de 100mm.

47. UPVC soil and vent pipes with solvent welded or ring seal joints. Fixed vertically to brickwork with brackets at 1.50m centres. 100mm external diameter.

48. Klosettanlage aus weißem Kristallporzellan mit Brille und Deckel aus schwarzem Kunststoff und Tiefspülkasten mit neun Litern Kapazität. Mit Schwimmerventil und Spülrohr ausgestattet. An Betonunterlage befestigt.

48. WC de porcelana vitrificada blanca con asiento y tapa de plástico negro, cisterna de plástico de nivel bajo, (capacidad 9 litros) completado con válvula esférica, flotador y tubo para el suministro de agua al WC. Fijado a hormigón.

48. White vitreous china WC suite with black plastic seat and cover and plastic low level cistern, 9 litre capacity. Complete with ball valve and float and flush pipe to WC suite. Fixed to concrete.

49. Waschbecken für WC aus weißen Kristallporzellan mit zwei verchromten Wasserhähnen (bzw. einer verschromten Misch-batterie mittlerer Qualität). Mit Stöpsel, Überlaufloch und Anschluß an Abwasserleitung (ohne Siphon). Ungefähre Dimensionen 560 x 400mm, mit verdeckten Konsolen an Mauerwerk befestigt.

49. Lavabo de porcelana vitrificada blanca con 2 grifos cromados, calidad media. Incluyendo tapón, rebosadero y desagüe (excluyendo colector), de medidas aproximadas de 560 x 400mm fijado al ladrillo con soporte oculto.

49. White vitreous china lavatory basin with 2 No. chrome plated taps (or medium quality chrome plated mixer taps). Including plug, overflow and waste connections (excluding trap). Size approximately 560 x 400mm, fixed to brickwork with concealed brackets.

50. Glazed fireclay shower tray. Including overflow and waste (excluding trap). Size approximately 750 x 750 x 175mm, fixed to concrete.

50. Bac à douche en terre cuite vitrifiée de couleur blanche y compris fournitures (non-inclu est la trappe). Dimensions approximatives 75cm x 75cm x 17,50cm fixé à un base de béton.

50. Piatto doccia in maiolica vetrificata bianca da 750 x 750 x 175mm, compreso piletta e griglia (escluso sifone), fissato al pavimento di cemento.

51. Stainless steel single bowl sink and double drainer (excluding taps). Including plug, overflow and connections (excluding trap). Size approximately 1500 x 600mm, fixed to softwood sink unit (excluding sink base).

51. Eviers en acier inoxydable avec un bac et deux égoutierès (non-inclus sont les robinets et la trappe), y compris bouchon et autres fournitures. Dimensions approximatives 150cm x 60cm fixés à une base de sapin de pays.

51. Lavello in acciaio inox con vasca singola e doppia scolatura, da 1500 x 600mm, da montare su base di legno. Compreso piletta, tappo ed allacciamenti, escluso rubinetti, sifone e base.

Electrical work

Electricité

Impianto ellettrico

52. PVC insulated and copper sheathed cable, 450/750 volt grade, twin core and ECC 6mm^2 cross section area. Fixed to timber with clips.

52. Cable flexible isolé avec PVC et blindé avec cuivre, grade de 450/750 volte, à double âme et conducteur continuel à terre. Section nominale de 6mm^2. Fixé à une base de bois par attaches plastic.

52. Cavo bipolare in rame isolato con PVC, di tipo 450/750 volt con conduttore di terra, 6mm^2 sezione, fissato su legno con graffette.

53. 13 amp, 2 gang flush mounted white, unswitched socket outlet. Including 6.0m of 2.5mm^2 concealed PVC insulated copper cable (excluding conduit). Flush mounted to brickwork including all fittings and fixing as necessary.

53. Double prise de courant sans interrupteurs de 13 ampères, installations en encastré y compris 6m de cable en cuivre de 2,50mm^2 isolé par PVC et fournitures. Fixé au ras d'un mur en briques (non-inclu est le conduit de canalisation).

53. Doppio presa di corrente ad incasso 13 amp, montata al muro. Sono compresi 6m cavo di rame 2,5mm^2 di sezione isolato in PVC, scatole, plache, piastre ecc. Esclusa canalizazzione.

50. Keramik Duschwanne. Mit Überlaufloch und Anschluß an Abwasserleitung (ohne Siphon). Ungefähre Dimensionen 750 x 750 x 175mm, an Betonunterlage befestigt.	50. Plato de ducha de cerámica vitrificada. Incluyendo rebosadero de agua y desagüe (excluyendo colector). Medidas aproximadas de 750 x 750 x 175mm fijado a hormigón.	50. Glazed fireclay shower tray. Including overflow and waste (excluding trap). Size approximately 750 x 750 x 175mm, fixed to concrete.
51. Spültisch aus Edelstahl mit einem Becken und zwei Ablaufplatten (ohne Wasserhähne). Mit Stöpsel, Überlaufloch und Anschlüssen (ohne Siphon). Ungefähre Dimensionen 1500 x 600mm, an Weicholtzunterteil befestigt (Unterteil nicht enthalten).	51. Fregadero de acero inoxidable de un seno y dos escurridores. Incluyendo rebosadero de agua, tapón y desagüe (excluyendo colector). Medidas aproximadas de 1500 x 600mm, fijado a soporte de madera (excluyendo el soporte).	51. Stainless steel single bowl sink and double drainer (excluding taps). Including plug, overflow and connections (excluding trap). Size approximately 1500 x 600mm, fixed to softwood sink unit (excluding sink base).
Elektroanlagen	**Instalaciones electricas**	**Electrical work**
52. PVC-isoliertes Kupferkabel, Klasse 450/750V, zweiadrig mit Erdschutzleiter, Querschnitt 6mm². Mit Klemmen an Holz befestigt.	52. Cable de 450/750v., con dos conductos y 6mm² de sección y tubo de PVC fijado a la pared con grapas.	52. PVC insulated and copper sheathed cable, 450/750 volt grade, twin core and ECC 6mm² cross section area. Fixed to timber with clips.
53. Zweifache weiße Steckdose ohne Schalter, 13A. Mit 6.0m verdecktem PVC-Isoliertem Kupferkabel, 2.5mm² (ohne Kabelrohr). Putzbündig an Mauerwerk versetzt, mit allen anfallenden Nebenarbeiten.	53. Bateria de dos cajas de derivación de 13 amp con toma de corriente desconectada, de color blanco, incluyendo 6m de cable de cobre de 2,5mm, protegido con PVC (excluyendo el tubo protector). Colocado sobre pared, incluidos los accesorios y elementos de fijación necesarios.	53. 13 amp, 2 gang flush mounted white, unswitched socket outlet. Including 6.0m of 2.5mm² concealed PVC insulated copper cable (excluding conduit). Flush mounted to brickwork including all fittings and fixing as necessary.

54. Flush mounted 20 amp, 2 gang, 1 way white light switch. Including 6.0m of 1.5mm^2 concealed mineral insulated copper cable (excluding conduit). Flush mounted to brickwork including all fittings and fixings as necessary.

54. Foyer lumineux simple à double interrupteur de 20 ampère (installation en encastré) y compris 6m de cable en cuivre de 1,5mm^2 isolé, et fournitures fixé au ras d'un mur en briques (non-inclu est le conduit de canalisation).

54. Doppio interrutore ad incasso ad azione singola, montata al muro. Sono compresi 6m di cavo di rame 1,5mm^2 sezione con isolamento minerale, scatola e tutti gli accessori necessari. Esclusa canalizazzione.

Finishings

Revêtements

Finiture

55. 2 coats gypsum based plaster on brick walls 13mm thick. Floated finish.

55. Deux couches d'enduits de plâtre d'épaisseur de 13mm sur mur de briques. Surface lisse.

55. Due strati di intonaco a base di gesso, applicato sul muro di mattoni, 13mm di spessore, finito a fratazzo.

56. White glazed tiles on plaster walls size 100 x 100 x 4mm. Fixed with adhesive and grouted between tiles.

56. Carreaux en porcelaine vitrifiée de couleur blanche fixés avec colle à un mur en plâtre avec joints remplis. Dimensions 10cm x 10cm x 0,40cm.

56. Piastrelle di maiolica bianche, 100 x 100 x 4mm, fissate con colla alle superfici di muri intonacati, compresa la sigillatura dei giunti con cemento bianco.

57. Red clay quarry tiles on concrete floors size 150 x 150 x 16mm. Bedded and jointed in mortar.

57. Carreaux en terre cuite rouge posés sur mortier sur sol en béton, avec joints remplis. Dimensions 15cm x 15cm x 1,60cm.

57. Piastrelle di gres rosso, 150 x 150 x 16mm, fissate e sigillate con malta di cemento al pavimento.

58. Floor screed. Cement and sand screed to concrete floors 1:3 mix. 50mm thick. Floated finish.

58. Forme composé d'une partie de ciment et trois parties de sable, coulée sur sol. 5cm d'épaisseur, surface lisse.

58. Massetto di cemento e sabbia (1:3), 50mm di spessore, gettato sul pavimento di calcestruzzo e finito liscio.

59. Thermoplastic floor tiles on screed 2.5mm thick. Fixed with adhesive.

59. Dalles thermoplastiques de 0.25cm d'épaisseur posées sur forme et fixées avec colle.

59. Pavimentazione in piastrelle termoplastiche di spessore 2,5mm, fissate con colla.

54. Weißer Lichtschalter mit zwei serienschaltern, 20A. Mit 6.0m verdecktem mineral- isoliertem Kuperkabel, 1.5mm^2, (ohne Kabelrohr). Putzbündig an Mauerwerk versetzt, mit allen anfallenden Nebenarbeiten.

54. Bateria de dos interruptores empotrados de 20 amperios de color blanco de un solo encendido. Incluyendo 6m de cable de cobre de 1,5mm protegido con tubo (se excluye este). Colocado sobre pared incluidos todos los accesorios y elementos de fijación necesarios.

54. Flush mounted 20 amp, 2 gang, 1 way white light switch. Including 6.0m of 1.5mm^2 concealed mineral insulated copper cable (excluding conduit). Flush mounted to brickwork including all fittings and fixings as necessary.

Verkleidungen

Acabados

Finishings

55. Zwei lagen Gipsputz auf Ziegelmauerwerk, Dicke 13mm, abgeglättet.

55. Revoque y enlucido con yeso en paredes de ladrillo, acabado a "buena vista".

55. 2 coats gypsum based plaster on brick walls 13mm thick. Floated finish.

56. Weiße Glasurplatten auf Gipswänden, Dimensionen 100 x 100 x 4mm, mit Kleber befestigt. Mit Fugenausfüllung.

56. Alicatado en paredes con cerámica vidriada de 100 x 100 x 4mm. Tomado con cemento cola, incluso la borada de las juntas.

56. White glazed tiles on plaster walls size 100 x 100 x 4mm. Fixed with adhesive and grouted between tiles.

57. Rote Keramikplatten auf Betonböden, Dimensionen 150 x 150 x 16mm. In Mörtel eingebettet, mit Fugenausfüllung.

57. Solado con tobas de 150 x 150 x 16mm sobre suelo de hormigón, tomado con mortero de c.p., incluso borada.

57. Red clay quarry tiles on concrete floors size 150 x 150 x 16mm. Bedded and jointed in mortar.

58. Bodenestrich aus Zement und Sand (1:3) für Betonböden. Dicke 50mm, abgeglättet.

58. Recrecido de suelos con 50mm de mortero de c.p., de dosificación 1:3, acabado enlucido.

58. Floor screed. Cement and sand screed to concrete floors 1:3 mix. 50mm thick. Floated finish.

59. Thermoplastische Bodenplatten auf Estrich verklebt. Dicke 2.5mm.

59. Pavimento termoplástico en piezas de 2,5mm de espesor, fijado con adhesivo sobre embaldosado o capa de hormigón.

59. Thermoplastic floor tiles on screed 2.5mm thick. Fixed with adhesive.

60. Suspended ceiling system. Fissured mineral fibre tiles size 300 x 300 x 15mm. On galvanized steel concealed suspension system. Fixed to concrete soffits with 500mm drop (excluding lamp fittings).

60. Faux plafonds en plaques de fibres minéreaux fissurées de dimensions de 30cm x 30cm x 1,5cm suspendus d'une hauteur de 50cm d'un plafond en béton à l'aide d'un système de suspension dissimulé en acier galvanisé, (non-inclus sont les fournitures pour lampes).

60. Controsoffitto composto da un telaio in acciaio galvanizzato, sospeso 500mm dal soffitto di cemento, e piastrelle di fibra minerale ad incasso, 300 x 300 x 15mm, (escluse le plafoniere).

Glazing

Vitrerie

Opere in vetro

61. Glazing to wood. Ordinary quality 4mm glass. Softwood beads.

61. Verre simple de 4mm d'épaisseur fixé à une base en bois avec parcloses en sapin de pays.

61. Lastre di vetro semplice 4mm di spessore, fissate a finestre di legno con listelli.

Painting

Peinture

Opere da pittore

62. Emulsion on plaster walls. One coat diluted sealer coat and 2 coats full vinyl emulsion paint.

62. Peinture - émulsion pour mur en plâtre composée d'une couche intermédiare d'émulsion dilué et deux couches d'émulsion à plein pour finition.

62. Tinteggiatura lavabile (idropittura) con emulsione vinilica, la prima mano diluita con acqua e due mani successive non diluite.

63. Oil paint on timber. One coat primer and 2 coats oil based paint.

63. Peinture à l'huile sur bois composée d'une couche primaire et deux couches à base d'huile pour finition.

63. Una mano di primer e due mani di smalto a base d'olio su superfici in legno.

60. Abgehängte aus Decken Mineralfaserplatten, Dimensionen 300 x 300 x 15mm. 500mm Abhänghöhe, verdeckte Aufhängevorrichtung aus verzinktem Stahl, an Betondecke befestigt. (Ohne Leuchten).

60. Falso techo formado por piezas de fibra mineral de 300 x 300 x 15mm sobre estructura de acero galvanizado fijado al techo con varillas de 500mm (no se incluyen los accesorios de iluminación).

60. Suspended ceiling system. Fissured mineral fibre tiles size 300 x 300 x 15mm. On galvanised steel concealed suspension system. Fixed to concrete soffits with 500mm drop (excluding lamp fittings).

Verglasung

61. Verglasung für Holzelemente. Glas normaler Qualität, Dicke 4mm. Weichholzglasleisten.

Vidrio cristanina

61. Vidrio cristanina de 4mm colocado sobre madera.

Glazing

61. Glazing to wood. Ordinary quality 4mm glass. Softwood beads.

Malerarbeiten

62. Emulsionsfarbe auf Putzwänden. Ein verdünnter Grundanstrich und zwei Anstriche Voll.

63. Ölfarbe auf Holz. Ein Grundanstrich und zwei Anstriche Ölfarbe.

Pintura

62. Pintura de paredes enyesadas, con una capa de imprimación y dos de pintura vinílica.

63. Pintura de aceite sobre madera con una capa de imprimación y dos de aceite.

Painting

62. Emulsion on plaster walls. One coat diluted sealer coat and 2 coats full vinyl emulsion paint.

63. Oil paint on timber. One coat primer and 2 coats oil based paint.

PART THREE

COMPARATIVE DATA

7. Introductory notes

Part Three brings together data from a variety of sources but mainly Part Two, and presents them in the form of tables to allow rapid comparison among the countries included in the book. This also helps place countries, their main statistical indicators and their construction costs in an international context.

There are ten tables derived from Part Two arranged in three sections:

8 Key Macroeconomic indicators

- Population
- Geography
- Finance and economics

9 Construction output indicators

- Construction output
- Construction output per capita

10 Construction cost data

- Skilled and unskilled labour costs
- Material costs - Cement and aggregates for concrete
- Material costs - Reinforcing steel and softwood
- Approximate estimating - Warehouses and offices
- Approximate estimating - Housing

The first five tables are based on the Key data sheets at the beginning of each country section, the remainder are drawn from Construction cost data in each country section. Each table is prefaced by explanatory notes. There are inherent dangers in attempting to compare international data, particularly where two sets of data are used (e.g.: construction output and population) and, even more so, when exchange rates are used. While these tables can provide useful initial comparisons between countries they should, nevertheless, be used with caution.

The summary tables are followed by a section on the European Construction Price Study. This describes an annual exercise undertaken by the European Statistical Office and presents a summary of results based on the exercise.

8. Key macroeconomic indicators

POPULATION

The table below summarises population statistics from the country Key data sheets.

The table highlights not only the differences in total population between the countries included in this book but also the variations in population growth rates, the proportions of the population of working age and the proportions of the population living in urban areas. Most European countries have low and declining population growth rates. Developed countries generally have a higher proportion of population of working age than developing countries. The proportion of urban population is generally high (over 50%) in developed countries though it can also be affected by geography.

Country	*Population*				
	Total (m)	*Urban %*	*Under 15 %*	*Over 65 %*	*Growth % pa*
Austria	7.6	57.0	17.0	15.0	0.1
Belgium	9.9	97.0	18.1	14.7	0.1
Cyprus	0.7	n.a.	n.a.	10.3	1.2
Czechoslovakia	15.6	67.0	23.3	11.6	0.3
Denmark	5.1	86.0	17.0	15.5	0.1
Finland	5.0	60.0	19.3	13.2	0.4
France	55.9	74.0	20.2	15.0	0.4
Germany FRG	61.2	86.0	14.9	15.4	(0.1)
German GDR	16.7	77.0	19.8	13.1	n.a.
Greece	10.0	62.0	13.7	18.4	0.3
Hungary	10.6	60.0	19.9	13.4	(0.2)
Ireland	3.5	58.0	27.7	10.3	0.2
Italy	57.4	68.0	17.1	14.2	0.2
Japan	122.6	77.0	18.5	11.7	0.5
Luxembourg	0.4	2.0	17.1	13.4	0.0
Malta	0.3	n.a.	n.a.	10.2	n.a.
Netherlands	14.8	88.0	17.8	12.9	0.7
Norway	4.2	74.0	18.8	16.4	0.3
Poland	37.9	61.0	25.2	9.9	0.8
Portugal	10.4	32.0	21.2	12.9	0.8
Spain	39.1	77.0	20.4	13.0	0.5
Sweden	8.4	84.0	16.5	18.3	0.3
Switzerland	6.6	61.0	16.4	15.3	0.4
Turkey	53.8	47.0	35.4	4.3	2.3
United Kingdom	57.1	92.0	18.9	15.5	0.3
United States	246.3	74.0	21.5	12.6	1.0
USSR	286.0	67.0	n.a.	n.a.	1.0
Yugoslavia	23.6	49.0	22.9	9.1	0.7

GEOGRAPHY

The table below summarises geographic statistics from the country Key data sheets.

As with population, the table highlights the differences between countries. The figures for population density and the percentage of national population in the capital city are however perhaps more useful indicators of land use than the total area of a country. The percentage of national population in the capital city gives an indication of the relative importance of the capital. Belgium, the old FRG (West Germany), Japan, Malta, the Netherlands and the United Kingdom all have population densities of 300 persons or more per km^2.

Country	*Area km2*		*Population*		
	Total	*Agricultural*	*per km2*	*Capital city*	
	000's	*000's*		*000's*	*% of total*
Austria	84	32	90	1531	20.1
Belgium	33	15	300	977	9.9
Cyprus	9	n.a.	78	167	23.9
Czechoslovakia	128	68	122	1191	7.6
Denmark	43	28	119	468	9.2
Finland	338	25	15	486	9.7
France	547	313	102	2176	3.9
Germany FRG	249	120	246	277	0.5
German GDR	108	n.a.	155	1285	7.7
Greece	132	92	76	3027	30.3
Hungary	93	n.a.	114	2072	19.5
Ireland	69	57	51	503	14.4
Italy	301	171	191	2821	4.9
Japan	378	53	324	11719	9.6
Luxembourg	3	n.a.	133	76.0	19.0
Malta	1	n.a.	300	14.0	4.7
Netherlands	37	20	400	1031	7.0
Norway	324	10	13	451	10.7
Poland	313	189	121	1649	4.4
Portugal	92	33	113	2128	20.5
Spain	505	308	77	4731	12.1
Sweden	450	35	19	1435	17.1
Switzerland	41	20	161	135	2.0
Turkey	779	366	69	3196	5.9
United Kingdom	244	186	234	6735	11.8
United States	9373	4314	26	617	0.3
USSR	22402	n.a.	13	8703	3.0
Yugoslavia	256	n.a.	92	1455	6.2

FINANCE AND ECONOMICS

This table sumarises financial and economic data from the country Key data sheets. Gross Domestic Product (GDP) and GDP per capita for each country are given in ECU and US$. The average real rate of growth in GDP and the average inflation rate are given for the years 1984 to 1989. Denmark, Finland, the old FRG (West Germany), Japan, Norway, Sweden, Switzerland and the United States all have relatively high GDP per capita (US$18 000 or more in 1989).

Country	*1989*				*1984 – 1989*	
	GDP		*GDP per capita*		*GDP growth*	*Average*
	ECU bn	*US$ bn*	*ECU 000's*	*US$ 000's*	*(real) % pa*	*inflation % pa*
Austria	107	115	14.1	15.2	1.8	2.2
Belgium	129	139	13.0	14.0	1.5	2.4
Cyprus	4	4	n.a.	n.a.	n.a.	3.7
Czechoslovakia	37	40	2.4	2.6	2.1	1.4
Denmark	90	96	17.5	18.8	1.9	4.3
Finland	94	101	18.9	20.3	3.2	6.6
France	733	788	13.1	14.1	1.9	3.5
Germany FRG	1025	1099	16.7	18.0	1.7	1.3
German GDR	129	139	7.8	8.3	4.0	n.a.
Greece	42	45	4.2	4.5	1.4	13.8
Hungary	22	23	2.0	2.2	1.8	8.9
Ireland	27	29	7.7	8.2	1.7	3.7
Italy	718	772	12.5	13.4	2.2	6.2
Japan	2415	2596	19.7	21.2	4.1	1.1
Luxembourg	6	6	15.9	17.1	n.a.	4.2
Malta	2	2	4.9	5.3	n.a.	1.9
Netherlands	193	208	13.0	14.0	1.4	0.7
Norway	78	84	18.6	20.0	2.7	6.6
Poland	n.a.	n.a.	n.a.	n.a.	n.a.	75.6
Portugal	35	37	3.3	3.6	2.2	12.5
Spain	300	323	7.7	8.3	2.6	6.9
Sweden	156	167	18.4	19.8	1.8	5.6
Switzerland	152	164	22.9	24.7	1.9	2.1
Turkey	73	79	1.4	1.5	5.3	n.a.
United Kingdom	682	725	12.0	12.7	2.8	5.3
United States	4519	4881	18.3	19.8	3.0	3.6
USSR	1403	1492	4.9	5.2	n.a.	n.a.
Yugoslavia	n.a.	n.a.	n.a.	n.a.	n.a.	345.2

9. Construction output indicators

CONSTRUCTION OUTPUT

The table below summarises construction output statistics from the country key data sheets.

In this summary table, figures in national currency for 1989 are listed but, in addition, in order to facilitate (crude) comparisons, US dollar, £ sterling and ECU equivalents are presented for each figure. The currency conversions have been carried out using the average exchange rates for 1989. As noted earlier, construction statistics, including those for construction output, are notoriously unreliable and, in addition, national definitions of construction output vary widely. It would, therefore, be unwise to draw too many conclusions from the table.

Country	*National unit of currency*	*Construction output (1989)*			*billions*
		National currency	*UK £*	*US $*	*ECU*
Austria	ASch	264	12.32	19.44	18.09
Belgium	Bfr	772	12.10	19.09	17.76
Cyprus	Cy£	0.4	0.51	0.80	0.74
Czechoslovakia	Kcs	136	5.67	8.94	8.32
Denmark	Dkr	100.8	8.51	13.42	12.50
Finland	Fmk	85	12.39	19.54	18.19
France	Fr	557.9	54.01	85.18	79.27
Germany FRG	DM	260	85.53	134.72	125.54
Germany GDR	EM	51	16.78	26.42	24.62
Greece	Dr	923	3.52	5.56	5.17
Hungary	Ft	260	2.71	4.27	3.98
Ireland	I£	2.59	2.26	3.55	3.32
Italy	L	115000	52.17	82.29	76.57
Japan	Y	70100	314.70	496.46	461.92
Luxembourg	LFr	30	0.47	0.74	0.69
Malta	ML	0.06	0.11	0.17	0.16
Netherlands	G	53.7	15.66	24.75	22.98
Norway	Nkr	93.5	8.39	13.22	12.31
Poland	Zl	14668	11.21	17.67	16.45
Portugal	Esc	1000	3.92	6.19	5.76
Spain	Pta	5537	28.74	45.34	42.19
Sweden	SKr	187	18.07	28.51	26.52
Switzerland	Swfr	50.8	19.46	30.79	28.57
Turkey	TL	20000	5.98	9.44	8.78
UK	£	46.2	46.20	72.19	67.81
USA	$	492	311.39	492.00	457.06
USSR	Rb	156	157.58	247.62	231.29
Yugoslavia	YuD	22	n.a.	n.a.	n.a.

CONSTRUCTION OUTPUT PER CAPITA

This table is based on the previous one, but has each figure for construction output divided by the population in that country. Bearing in mind the uncertainty of both construction and population data and the limitations of exchange rates, the table reveals some useful indicators of construction activity. Only nine of the 27 countries covered in the book have construction output per capita figures for 1989 less than 1000 ECUs, and four of these are former Eastern Block countries.

Country	*National unit of currency*	*Construction output per capita (1989)*			*billions*
		National currency	*UK £*	*US $*	*ECU*
Austria	ASch	34690	1619	2554	2377
Belgium	Bfr	77979	1222	1928	1794
Cyprus	Cy£	571	723	1143	1062
Czechoslovakia	Kcs	8718	363	573	533
Denmark	Dkr	19649	1660	2616	2436
Finland	Fmk	17171	2503	3947	3674
France	Fr	9971	965	1522	1417
Germany FRG	DM	4250	1398	2202	2052
Germany GDR	EM	3050	1003	1580	1473
Greece	Dr	92300	352	556	517
Hungary	Ft	24528	256	403	375
Ireland	I£	740	646	1014	949
Italy	L	2000000	907	1431	1332
Japan	Y	571778	2567	4049	3768
Luxembourg	LFr	80000	1254	1978	1841
Malta	ML	159	285	455	418
Netherlands	G	3628	1058	1672	1553
Norway	Nkr	22260	1996	3149	2930
Poland	Zl	387000	296	466	434
Portugal	Esc	96154	377	595	553
Spain	Pta	141970	737	1163	1082
Sweden	SKr	22150	2140	3377	3141
Switzerland	Swfr	7670	2939	4648	4313
Turkey	TL	371747	111	175	163
UK	£	810	810	1266	1189
USA	$	2000	1266	2000	1858
USSR	Rb	545	551	865	808
Yugoslavia	YuD	n.a.	n.a.	n.a.	n.a.

10. Construction cost data

SKILLED AND UNSKILLED LABOUR COSTS

This table summarises labour costs for each country. The figures in national currency are taken from each country's construction cost data and have been converted into pound sterling, US dollar and ECU equivalents using January 1991 exchange rates.

It is probable that the definitions of skilled and unskilled, and what is included in labour costs varies between countries, thus these figures should not be taken as strictly comparable. The ranking and relative level of labour costs are, however, broadly similar to the GDP per capita figures.

Country	*Skilled labour costs*				*Unskilled labour costs*			*hour*
	National currency	*UK £*	*US $*	*ECU*	*National currency*	*UK £*	*US $*	*ECU*
Austria	270	13.13	25.36	18.76	220	10.70	20.67	15.28
Belgium	n.a.	n.a.	n.a.	n.a.	n.a.	n.a.	n.a.	n.a.
Cyprus	4.48	5.38	10.39	7.69	3.64	4.37	8.45	6.25
Czechoslovakia	40.29	0.78	1.50	1.11	30.26	0.58	1.13	0.83
Denmark	163	14.49	27.99	20.70	163	14.49	27.99	20.70
Finland	111.5	15.82	30.56	22.60	76.7	10.88	21.02	15.55
France	95.87	9.65	18.63	13.79	70.98	7.14	13.79	10.21
Germany FRG	22.82	7.80	15.07	11.14	n.a.	n.a.	n.a.	n.a.
Germany GDR	n.a.	n.a.	n.a.	n.a.	n.a.	n.a.	n.a.	n.a.
Greece	2391	7.74	14.96	11.06	1733	5.61	10.84	8.02
Hungary	n.a.	n.a.	n.a.	n.a.	n.a.	n.a.	n.a.	n.a.
Ireland	6.93	6.33	12.22	9.04	5.79	5.28	10.21	7.55
Italy	28800	13.10	25.32	18.72	26000	11.83	22.85	16.90
Japan	n.a.	n.a.	n.a.	n.a.	n.a.	n.a.	n.a.	n.a.
Luxembourg	692	11.49	22.19	16.41	540	8.96	17.32	12.81
Malta	1.925	3.34	6.45	4.77	1.575	2.73	5.28	3.90
Netherlands	n.a.	n.a.	n.a.	n.a.	n.a.	n.a.	n.a.	n.a.
Norway	185.6	16.23	31.36	23.20	160	14.00	27.04	20.00
Poland	8186	0.44	0.86	0.63	n.a.	n.a.	n.a.	n.a.
Portugal	645	2.47	4.78	3.54	515	1.98	3.82	2.82
Spain	1821	9.90	19.13	14.15	1428	7.76	15.00	11.09
Sweden	190	17.41	33.64	24.88	185	16.95	32.76	24.23
Switzerland	n.a.	n.a.	n.a.	n.a.	n.a.	n.a.	n.a.	n.a.
Turkey	1352	0.22	0.43	0.32	825	0.14	0.26	0.20
UK	5.89	5.89	11.38	8.42	4.63	4.63	8.95	6.62
USA	37.21	19.26	37.21	27.51	26.5	13.71	26.50	19.60
USSR	n.a.	n.a.	n.a.	n.a.	n.a.	n.a.	n.a.	n.a.
Yugoslavia	n.a.	n.a.	n.a.	n.a.	n.a.	n.a.	n.a.	n.a.

MATERIALS COSTS - CEMENT AND AGGREGATES FOR CONCRETE

The table below summarises costs per tonne for cement and costs per m^3 for aggregates for concrete. The figures in national currency are taken from each country's Construction cost data and here converted into US dollar, pound sterling and ECU equivalents using January 1991 exchange rates. Converted figures have been rounded to the nearest whole number.

Despite the fact that there are internationally recognized standards of quality for cement and that it is one of the few internationally traded construction materials, the variation in cost between countries is remarkably large. This may well be a result of controlled prices or of import protection. The variation in aggregate costs is less surprising; quality and availability can be expected to vary widely within, let alone between, countries and a large proportion of aggregate costs can be in their transportation.

Country	*Cement*			*tonne*	*Aggregates for concrete*			*m3*
	National currency	*UK £*	*US $*	*ECU*	*National currency*	*UK £*	*US $*	*ECU*
Austria	1250	61	117	87	200	9.7	18.8	13.9
Belgium	4000	66	128	95	650	10.8	20.8	15.4
Cyprus	27	32	63	46	4.75	5.7	11.0	8.1
Czechoslovakia	650	13	24	18	100	1.9	3.7	2.8
Denmark	1030	92	177	131	65	5.8	11.2	8.3
Finland	475	67	130	96	35	5.0	9.6	7.1
France	704	71	137	101	163	16.4	31.7	23.4
Germany FRG	237	81	156	116	70.15	24.0	46.3	34.3
Germany GDR	n.a.	n.a.	n.a.	n.a.	n.a.	n.a.	n.a.	n.a.
Greece	10960	35	69	51	910	2.9	5.7	4.2
Hungary	n.a.	n.a.	n.a.	n.a.	n.a.	n.a.	n.a.	0.0
Ireland	73	67	129	95	10.65	9.7	18.8	13.9
Italy	110000	50	97	72	20350	9.3	17.9	13.2
Japan	17500	68	131	97	3200	12.4	23.9	17.7
Luxembourg	5700	95	183	135	875	14.5	28.1	20.8
Malta	32	55	107	79	3.83	6.6	12.8	9.5
Netherlands	180	55	105	78	50	15.2	29.3	21.7
Norway	1200	105	203	150	50	4.4	8.4	6.2
Poland	343200	19	36	27	50100	2.7	5.2	3.9
Portugal	12480	48	93	68	2240	8.6	16.6	12.3
Spain	13000	71	137	101	2400	13.0	25.2	18.6
Sweden	n.a.	n.a.	n.a.	n.a.	n.a.	n.a.	n.a.	n.a.
Switzerland	275	112	216	160	50	20.4	39.3	29.1
Turkey	220000	37	70	52	27000	4.5	8.6	6.4
UK	64	64	124	91	9.1	9.1	17.6	13.0
USA	131	68	131	97	13.46	7.0	13.5	10.0
USSR	n.a.	n.a.	n.a.	n.a.	n.a.	n.a.	n.a.	n.a.
Yugoslavia	n.a.	n.a.	n.a.	n.a.	n.a.	n.a.	n.a.	n.a.

MATERIALS COSTS - REINFORCING STEEL AND SOFTWOOD

The table below summarises costs per tonne for reinforcing steel and costs per m^3 for softwood. The figures in national currency are taken from each country's Construction cost data and here converted into US dollar, pound sterling and ECU equivalents using January 1991 exchange rates.

Similar to cement, the range of costs for reinforcing steel is surprisingly large from a low of 120 ECUs per tonne in Czechoslovakia to a high of 879 ECUs per tonne in Germany. The lowest costs for timber are in Czechoslovakia and Poland.

Country	*Reinforcing steel*			*tonne*	*Softwood*			*m3*
	National currency	*UK £*	*US $*	*ECU*	*National currency*	*UK £*	*US $*	*ECU*
Austria	10000	486	939	695	3160	154	297	220
Belgium	33300	553	1068	790	15000	249	481	356
Cyprus	185	222	429	317	180	216	418	309
Czechoslovakia	4350	84	162	120	1600	31	60	44
Denmark	5750	511	987	730	2380	212	409	302
Finland	3500	497	959	709	1800	255	493	365
France	5480	551	1065	788	1872	188	364	269
Germany FRG	1800	615	1188	879	n.a.	n.a.	n.a.	n.a.
Germany GDR	n.a.	n.a.	n.a.	n.a.	n.a.	n.a.	n.a.	n.a.
Greece	74700	242	467	346	53630	174	336	248
Hungary	n.a.	n.a.	n.a.	n.a.	n.a.	n.a.	n.a.	n.a.
Ireland	310	283	547	404	265	242	467	346
Italy	610000	278	536	397	230000	105	202	150
Japan	64500	249	481	356	59000	228	440	326
Luxembourg	22000	365	705	522	11500	191	369	273
Malta	185	321	620	458	106	184	355	263
Netherlands	1525	462	893	661	750	227	439	325
Norway	5750	503	972	719	2100	184	355	262
Poland	3890000	210	406	301	1182000	64	123	91
Portugal	85000	326	630	466	81390	312	603	446
Spain	82000	446	861	637	72000	391	756	559
Sweden	6370	584	1128	834	n.a.	n.a.	n.a.	n.a.
Switzerland	1500	611	1180	872	675	275	531	393
Turkey	1300000	216	416	308	1500000	249	480	356
UK	371	371	717	530	155	155	299	221
USA	580	300	580	429	937	485	937	693
USSR	n.a.	n.a.	n.a.	n.a.	n.a.	n.a.	n.a.	n.a.
Yugoslavia	n.a.	n.a.	n.a.	n.a.	n.a.	n.a.	n.a.	n.a.

APPROXIMATE ESTIMATING - WAREHOUSES AND OFFICES

This table summarises approximate estimating costs per square metre for two non-residential building types: warehouses and offices. The aim has been to take typical costs, where available, for one of the most basic building types, the speculative warehouse, and one of the more common, the speculative air- conditioned, medium rise office. In countries where building types did not exactly match these descriptions, the nearest equivalent has been taken. Where a range of costs has been given, the mid point is shown.

It must be borne in mind that even where costs are given under the same description in one or more countries, this is not to say that they are identical, or even physically similar. Approximate estimating costs for a particular country are for the normal standards prevailing in that country. Quality and technical standards vary widely and there are differences between countries in what is, and is not, included. The table, therefore, should be used with care.

Country	*Warehouses*				*Medium rise offices*			*per m2*
	National currency	*UK £*	*US $*	*ECU*	*National currency*	*UK £*	*US $*	*ECU*
Austria	6500	316	611	452	18000	875	1691	1251
Belgium	20000	332	641	474	52000	863	1667	1233
Cyprus	75	90	174	129	260	312	603	446
Czechoslovakia	n.a.	n.a.	n.a.	n.a.	n.a.	n.a.	n.a.	n.a.
Denmark	n.a.	n.a.	n.a.	n.a.	n.a.	n.a.	n.a.	n.a.
Finland	n.a.	n.a.	n.a.	n.a.	n.a.	n.a.	n.a.	n.a.
France	2550	257	496	367	7500	755	1457	1078
Germany FRG	1770	605	1169	864	2685	918	1773	1311
Germany GDR	n.a.	n.a.	n.a.	n.a.	n.a.	n.a.	n.a.	n.a.
Greece	75000	243	469	347	200000	648	1251	925
Hungary	n.a.	n.a.	n.a.	n.a.	n.a.	n.a.	n.a.	n.a.
Ireland	300	274	529	391	700	639	1234	913
Italy	550000	250	483	358	1750000	796	1538	1138
Japan	200000	773	1493	1104	380000	1468	2836	2097
Luxembourg	n.a.	n.a.	n.a.	n.a.	n.a.	n.a.	n.a.	n.a.
Malta	n.a.	n.a.	n.a.	n.a.	283	491	948	701
Netherlands	1000	303	586	433	2400	728	1406	1040
Norway	1850	162	313	231	5750	503	972	719
Poland	n.a.	n.a.	n.a.	n.a.	n.a.	n.a.	n.a.	n.a.
Portugal	44300	170	328	243	87600	336	649	480
Spain	35000	190	368	272	65000	353	683	505
Sweden	4910	450	869	643	8350	765	1478	1093
Switzerland	1150	468	904	669	2000	814	1573	1163
Turkey	662000	110	212	157	725000	120	232	172
UK	270	270	522	386	765	765	1478	1093
USA	305	158	305	226	700	362	700	518
USSR	n.a.	n.a.	n.a.	n.a.	n.a.	n.a.	n.a.	n.a.
Yugoslavia	n.a.	n.a.	n.a.	n.a.	n.a.	n.a.	n.a.	n.a.

APPROXIMATE ESTIMATING - HOUSING

This table summarises approximate estimating costs per square metre for housing. Two types of housing have been taken: mass market, single family houses and medium quality, medium rise apartments. In countries where housing types did not match exactly these descriptions, the nearest equivalent has been taken. Where a range of costs has been given, the mid point is shown.

Approximate estimating costs for a particular country are for the normal standards prevailing in that country. It should be noted that costs per square metre for housing are particularly sensitive to levels of specification, total area of dwelling unit and the rules of measurement adopted; these all vary significantly among countries and this table should, therefore, be used with care.

Country	*Single family housing*				*Appartments/ flats*			*per m2*
	National currency	*UK £*	*US $*	*ECU*	*National currency*	*UK £*	*US $*	*ECU*
Austria	n.a.	n.a.	n.a.	n.a.	13200	642	1240	917
Belgium	33000	548	1058	783	32000	531	1026	759
Cyprus	170	204	394	292	150	180	348	257
Czechoslovakia	n.a.	n.a.	n.a.	n.a.	n.a.	n.a.	n.a.	n.a.
Denmark	6000	533	1030	762	10000	889	1717	1270
Finland	n.a.	n.a.	n.a.	n.a.	5150	731	1412	1044
France	5090	512	989	732	4010	404	779	577
Germany FRG	1240	424	819	606	1900	649	1254	928
Germany GDR	n.a.	n.a.	n.a.	n.a.	n.a.	n.a.	n.a.	n.a.
Greece	160000	518	1001	740	132000	427	826	611
Hungary	n.a.	n.a.	n.a.	n.a.	n.a.	n.a.	n.a.	n.a.
Ireland	300	274	529	391	480	438	846	626
Italy	1200000	546	1055	780	1200000	546	1055	780
Japan	n.a.	n.a.	n.a.	n.a.	350000	1352	2612	1932
Luxembourg	35000	581	1122	830	50000	830	1603	1186
Malta	n.a.	n.a.	n.a.	n.a.	n.a.	n.a.	n.a.	n.a.
Netherlands	n.a.	n.a.	n.a.	n.a.	n.a.	n.a.	n.a.	n.a.
Norway	5000	437	845	625	7000	612	1183	875
Poland	4250000	230	444	328	2567000	139	268	198
Portugal	68000	261	504	373	85000	326	630	466
Spain	55000	299	578	427	65000	353	683	505
Sweden	8450	774	1496	1107	6600	605	1169	864
Switzerland	1375	560	1081	800	1800	733	1416	1047
Turkey	344000	57	110	82	828000	138	265	196
UK	355	355	686	507	425	425	821	607
USA	450	233	450	333	595	308	595	440
USSR	n.a.	n.a.	n.a.	n.a.	n.a.	n.a.	n.a.	n.a.
Yugoslavia	n.a.	n.a.	n.a.	n.a.	n.a.	n.a.	n.a.	n.a.

11. The European Community Construction Price Study

SUMMARY OF RESULTS 1985 TO 1990

For a number of years the Statistical Office of the European Communities (Eurostat) has collected and analysed construction price data from its twelve member countries. The Eurostat Construction Price Comparison Study contributes to a wider study of prices and national accounts in Member Countries which aims to determine purchasing price parities as an alternative to exchange rates for inter country comparisons.

The exercise is carried out on the basis of simplified bills of quantities for "typical" housing, non-residential and civil engineering projects which are priced by experts from each country in national currencies in July of each year. The same bills are priced in all countries thus the exercise is a comparison using "identical buildings" as discussed in section 4. The pricing is intended to reflect national average prices charged to clients by contractors in competition. National prices (in £ sterling for the UK, Drachma for Greece, etc) are converted by Eurostat into European Currency Units (ECU's). Explanatory notes on the composition and value of the ECU are given in section 2.

With the permission of Eurostat a summary of results are presented here. Changes in the content of the Bills of Quantities over recent years has meant that the results can no longer be used to produce time series of indices for individual countries. The following table presents the relative prices of construction for the different types of construction projects for each year since 1985. In this table UK prices provide the base of 100 against which the other countries can be compared.

Thus, in 1985, Denmark had the highest relative price for housing work (126) and UK, the highest price for other building work and civil engineering work (100). The table also demonstrates how relative construction price levels vary from year to year, thus Denmark had the highest price level for all construction work in 1985, 1986 and 1987, Luxembourg in 1988 and the Netherlands in 1989 and 1990.

The table does not permit comparisons of price levels between countries in different years. Due to a number of factors, including the simplified nature of the bills of quantities used, the Eurostat figures should only be used as indicators of the relative prices of construction among countries in any year. It must be emphasized that these results are of work done by Eurostat using their own methodology and they will not necessarily be consistent with price indicators given elsewhere in this publication.

Inter country comparisions of construction prices by years

Year	*Country*	*Housing*	*Other buildings*	*Civil engineering*	*All construction*
1985	Belgium	90	73	75	79
	Denmark	126	93	83	101
	France	103	97	92	97
	Greece	75	70	66	70
	Ireland	79	73	76	76
	Italy	76	73	86	78
	Luxembourg	96	93	73	87
	Netherlands	118	85	85	96
	Portugal	51	44	74	56
	Spain	66	57	76	66
	Germany	118	82	76	92
1986	Belgium	102	81	82	88
	Denmark	157	100	100	119
	France	117	100	92	103
	Greece	73	55	53	61
	Ireland	82	99	75	85
	Italy	101	75	83	87
	Luxembourg	112	97	91	100
	Netherlands	141	101	95	112
	Portugal	56	101	95	84
	Spain	78	69	92	80
	Germany	142	113	76	110
1987	Belgium	113	87	74	86
	Denmark	158	124	79	121
	France	n.a.	n.a.	n.a.	n.a.
	Greece	68	69	42	60
	Ireland	90	62	78	77
	Italy	82	89	82	85
	Luxembourg	135	113	103	117
	Netherlands	132	100	78	103
	Portugal	48	49	54	50
	Spain	70	71	72	71
	Germany	119	100	76	98

Inter country comparisions of construction prices by years

Year	*Country*	*Housing*	*Other buildings*	*Civil engineering*	*All construction*
1988	Belgium	66	83	78	76
	Denmark	98	113	94	102
	France	60	77	68	68
	Greece	56	64	76	65
	Ireland	43	62	62	55
	Italy	59	83	98	80
	Luxembourg	78	136	146	120
	Netherlands	124	110	108	114
	Portugal	30	53	95	59
	Spain	39	72	88	66
	Germany	77	94	81	84
1989	Belgium	95	71	80	82
	Denmark	141	96	93	110
	France	n.a.	n.a.	n.a.	n.a.
	Greece	102	95	68	88
	Ireland	79	83	88	83
	Italy	79	84	97	87
	Luxembourg	114	98	136	116
	Netherlands	201	83	85	123
	Portugal	50	45	62	52
	Spain	52	49	59	53
	Germany	119	85	82	95
1990	Belgium	101	87	72	86
	Denmark	148	115	76	113
	France	n.a.	n.a.	43	43
	Greece	127	113	81	107
	Ireland	100	91	67	86
	Italy	82	85	86	84
	Luxembourg	107	118	92	106
	Netherlands	145	114	82	114
	Portugal	48	54	62	55
	Spain	48	50	62	53
	Germany	126	96	77	99

Index

COME USARE QUESTO LIBRO

Questo libro è in tre parti - *Parte Prima: Costruzioni Edilizia in Europa; Parte Seconda: Paesi Individuali; Parte Terza: Dati Comparativi.* Ventisette paesi sono e lencati in questo libro. Due de questi - L'America è Giappone che sono fuori dall'Europa, ma sono stati inclusi per ragioni comparativi.

Parte Prima: Costruzioni edilizia in Europa

Tre soggetti sono presentati in Parte Prima: *L'industria Edilizia in Europa* descrive i cambi che stanno effettuandosi attraverso L'Europa del'est è dell'ovest. *La Legislazione della Comunita Europea* presenta una breve quida di cambi nella legistratura la quale costringerà a l'industria edilizia a dei cambiamenti nelle costruzioni. *Il Programma Finanziario di Sopporto della Comunità Europea*, sottolinea i maggiori fondi che sono a disposizione per l'aiuto della ricostruzione.

Parte Seconda: Paesi individuali

Nella Parte Seconda ventisette paesi sono elencati in ordine aflabetico, ed ognlino sará presentato nello stesso formato sottolineando i maggiori paragrafi.

· *I Dati importanti* che provvedono l'indicatore economico e delle costruzioni
· *L'industria edilizia* sottolinea la structtura dell'industria per gli appalti, è i contratti di procedura, le regole, è la qualitá
· *Costi di costruzioni* che includono i dati sulla mano d'opera e costi di materiali misuramenti di rati per articoli per l'avoro di costruzioni, costi di nucleo (sezioni) (estimazione approssimata) per differenti tipi di fabbricati
· *Rate di cambio è prente inflazione* i dati sul cambio del valore della sterlina e sul dollaro USA. Incluso i dati sull'indice dei movimenti de prezzi al minuto, è costruzioni
· *Indirizzi utili* i quali forniscono, no mi e indirizzi di autorità. istituzioni professional, associazioni di professioni edilizia ecc.

Le note (4) provvedono delle note generali spiegazioni e definizioni. Di tutti gli articoli (o materiali) cha appargano nel prezzo i dati sono completamente descritti nella sezione 6 in cinque lingue (Inglese, Francese, Tedesco, Italiano, Spagnolo).

Parte Terza: Dati comparativi

Per fare una rapida differenza trá paesi elencati nel libro la terza parte porta insieme i dati della parte seconda e sotto pone gli stessi sotto tre paragrafi importanti.

· *Chiave indicatore macroeconomico* incluso i dati domografici e finanziari
· *Indicatore di produzione di costruzioni* incluso produzione per capitale
· *Dati di costi per costruzione edilizia* incluso costi di capitale, costi di nucleo (o sezione) è misuramento di rate.